Physical Kinetics

Serguei N. Burmistrov

Physical Kinetics

Classical and Quantum Problems and Solutions

Springer

Serguei N. Burmistrov
Kurchatov Institute
Moscow, Russia

ISBN 978-981-19-1651-9 ISBN 978-981-19-1649-6 (eBook)
https://doi.org/10.1007/978-981-19-1649-6

This Springer imprint is published by the registered company Springer Nature Singapore Pte Ltd.
The registered company address is: 152 Beach Road, #21-01/04 Gateway East, Singapore 189721, Singapore

For students studying the physical kinetics

Preface to the English Edition

The Russian edition of this book came to light 5 years ago. The present edition is augmented with new problems, increasing their number from 40 to 60. All the novel problems refer to new sections of physical kinetics. They are the following: the Lindblad master equation, ballistic transport in quantum microcontacts, weak localization effects, kinetics of first-order phase transitions (the Stefan problem), kinetic phenomena at the superfluid-solid helium, and rarefied Bose and Fermi gases. The misprints revealed in the Russian edition are corrected as well.

Using this opportunity, the author would like to express his gratitude to all colleagues at the Department of Theoretical Physics in Moscow Institute of Physics and Technology. In addition, I am also grateful to Prof. N. M. Shchelkachev for the fruitful discussion of questions concerning the selection of problems for the practical classes on the physical kinetics at the MIPT General and Applied Physics Faculty.

Moscow, Russia
January 2021

Serguei N. Burmistrov

Preface

To study physical kinetics and, most importantly, to master its methods of studying the thermodynamically non-equilibrium systems, it is not sufficient to be restricted with listening to the lecture course and considering the theoretical assertions. Thus, the physical kinetics course includes the practical classes aimed at facilitating the students to master in solving the problems. Solving the problems gives the students an opportunity to test their real knowledge which, in ideal, should not be a set of memorized information alone. For the teacher, every problem solved independently by the student is the most effective indicator of comprehending the subject. The problem self-solving develops and educates the analytical and creative thinking.

It is much more difficult to master the physical kinetics and learn how to apply its methods in practice compared as to master in such fields of theoretical physics as statistical physics, quantum mechanics, or field theory. The fact is that the physical kinetics consists of many completely versatile fields, for example, kinetic theory of gases, plasma theory, transport phenomena in metals, semiconductors, dielectrics, and disordered media, kinetic phenomena in quantum crystals and liquids, and phase transition kinetics. In general, in order to describe the kinetic phenomena in each such field of physics, we have to use and apply specific models and methods. As a rule, the models and methods are very complicated and require the cumbersome calculations which can simply be unattainable without special mathematical training. On the other hand, the study of physical kinetics undoubtedly requires the mastering of practical methods for solving various problems. Solving the specific tasks is a perfect addition to the lecture course. The analysis of the solution helps the student to understand better the lecture course as a whole.

The book is based on the practical classes for the students studying the physical kinetics in the general course of theoretical physics at the General and Applied Physics Faculty of Moscow Institute of Physics and Technology. The problems have been included in the students' homeworks. In order to present the basics of physical kinetics in a form accessible to physics students who do not necessarily specialize as future theoretical physicists, we tried to avoid using complex diagrammatic, renorm-group, and other special mathematical tools of theoretical physics. As a rule, the presentation of the material is based on simple representations of the Boltzmann transport equation or Lindblad master equation. All the problems are supplied with the detailed solutions and comments explaining the motivation of the problems and their connection with various problems of modern physical kinetics. The section with the problem conditions is preceded by the brief introduction that contains the necessary information and notions useful for solving and understanding the problems proposed.

The book can be used as a supplement to the textbooks published on the physical kinetics. The purpose of the book is to help students in training the practical skills and mastering the basic elements of physical kinetics. To understand the subject matter, it is sufficient to know the traditional courses of theoretical physics at the physical universities. The physical universality of the presentation as well as the analysis of a large number of specific physical problems allows this book to serve as a textbook for all students studying the physical kinetics.

In conclusion, the author expresses his sincere gratitude to the entire team of teachers at the Theoretical Physics Department of Moscow Institute of Physics and Technology.

Moscow, Russia Serguei N. Burmistrov

About This Book

The book contains 60 problems of various degrees of complexity. The problems are based on the materials for the practical classes in the course of physical kinetics with the students of General and Applied Physics Faculty at Moscow Institute of Physics and Technology. All the problems are provided with the detailed solutions. The comments to the problems reflect the connection with the problems and methods of modern physical kinetics. A brief introduction gives the necessary information for solving and understanding the problems.

The book is proposed for students and postgraduates studying the theoretical physics.

Contents

Chapter 1
Brief Reminders

1.1 The Boltzmann Transport Equation

The subject of *statistical physics* is a study of thermodynamically equilibrium physical systems, i.e. systems in which the probability to find a system in one or another state is governed by the Gibbs distribution. In its turn, the subject of physical kinetics in the broad sense is a study of thermodynamically non-equilibrium systems and kinetics of physical processes in them. The kinetic processes, occurring in the thermodynamically non-equilibrium systems, are irreversible and, therefore, energy dissipative.

The main task of physical kinetics is to infer and establish the *transport equation*, the solution of which would allow us to find and calculate the physical characteristics of a non-equilibrium system. The complexity of this problem is associated with a huge variety of physical systems that assumes the division of physical kinetics into a large number of independent sections, such as the kinetic theory of gases, theory of non-equilibrium processes in plasma, kinetic phenomena in dielectrics, metals, semiconductors, disordered media, quantum liquids and crystals, and the kinetics of first- and second-order phase transitions. In each of these sections or fields of physics, one has to use its own specific physical models and mathematical methods for describing the kinetic phenomena. So, it is not possible to specify a single universal model or general method.

The range of issues that are usually the focus of physical kinetics can be divided into several groups. The first group includes the issues with a small deviation of the system from equilibrium as a result of small external effects. In this case, the response of the system is usually linear and proportional to the magnitude of the impact. The linear response can be characterized with the generalized susceptibilities which, depending on the field of physical phenomena, are referred to as kinetic or transport coefficients as well. The typical examples of phenomena are the appearance of an electric current under the influence of electric field, the occurrence of heat flow in the presence of temperature difference, generation of diffusive flow at a concentration gradient, and appearance of polarizability (magnetization) in the external

S. N. Burmistrov, *Physical Kinetics*,
https://doi.org/10.1007/978-981-19-1649-6_1

electric (magnetic) field. The corresponding responses are coefficients of electrical conductivity, thermal conductivity, diffusion, and dielectric (magnetic) susceptibility.

The second group of examples includes the study of collective excitations in the physical system, for example, sound waves in a condensed medium, plasma oscillations in a plasma, and spin waves in magnets. Other research examples include the determination of correlation functions, relaxation times in a non-equilibrium medium, and rates of phase transitions depending on the external conditions.

As a rule, the statistical description of a condensed medium is sufficient with introducing a single-particle distribution function $n(t, \boldsymbol{r}, \boldsymbol{p})$, time-dependent t and defined in the phase space of coordinates $\boldsymbol{r}$ and momenta $\boldsymbol{p}$. Let

$$d\tau = \frac{d\boldsymbol{p}\, d\boldsymbol{r}}{(2\pi\hbar)^d} = \frac{d^d p\, d^d r}{(2\pi\hbar)^d}$$

be the number of the possible states in the phase space element $d\boldsymbol{p}\, d\boldsymbol{r}$. Here d is the dimensionality of the usual geometrical space. The product

$$dN(t, \boldsymbol{r}, \boldsymbol{p}) = n(t, \boldsymbol{r}, \boldsymbol{p})d\tau$$

determines the average number of particles within the given intervals $d\boldsymbol{p}$ and $d\boldsymbol{r}$ at the time moment t, i.e. within $\boldsymbol{p} \div \boldsymbol{p} + d\boldsymbol{p}$ and $\boldsymbol{r} \div \boldsymbol{r} + d\boldsymbol{r}$.

Provided that the state of a particle in the medium is characterized not only by the coordinate or momentum but also by additional degrees of freedom, e.g. particle has spin $\boldsymbol{\sigma}$ or rotational moment (internal angular momentum $\boldsymbol{l}$), all these variables should be involved into a set of variables which the one-particle probability density function n obeys. For the spin 1/2 electrons, it is convenient to represent the probability density function $n(t, \boldsymbol{r}, \boldsymbol{p}, \boldsymbol{\sigma}) = n_{\sigma}(t, \boldsymbol{r}, \boldsymbol{p})$ as a matrix 2×2. The integral,

$$\int n(t, \boldsymbol{r}, \boldsymbol{p}) d\tau_{\boldsymbol{p}} \quad \text{or} \quad \mathrm{Sp}_{\sigma} \int n_{\sigma}(t, \boldsymbol{r}, \boldsymbol{p}) d\tau_{\boldsymbol{p}} \quad \text{where} \quad d\tau_{\boldsymbol{p}} = \frac{d^d p}{(2\pi\hbar)^d},$$

gives the density of spatial particle distribution at the time moment t.

The Boltzmann transport equation

$$\frac{\partial n}{\partial t} + \boldsymbol{v} \cdot \frac{\partial n}{\partial \boldsymbol{r}} + \boldsymbol{F} \cdot \frac{\partial n}{\partial \boldsymbol{p}} = \mathrm{St}\,[n]$$

describes the evolution of one-particle probability density function $n = n(t, \boldsymbol{r}, \boldsymbol{p})$. Here, we denote the particle velocity and the force acting upon the particle of energy $\varepsilon(\boldsymbol{p}, \boldsymbol{r}) = \varepsilon(\boldsymbol{p}) + U(\boldsymbol{r})$, respectively,

$$\boldsymbol{v} = \frac{\partial \varepsilon}{\partial \boldsymbol{p}} \quad \text{and} \quad \boldsymbol{F} = -\frac{\partial \varepsilon}{\partial \boldsymbol{r}}.$$

The right-hand side $\mathrm{St}\,[n]$[1] of the transport equation is called the *collision term* or *collision integral* and determines the rate of varying the probability density function. The case $\mathrm{St}\,[n] = 0$ is referred to as the *collisionless* one and the transport equation reduces to the *Liouville equation*.

The Boltzmann transport equation is employed to describe the kinetic phenomena in gases, plasmas, solids, quantum fluids, and crystals as well as to calculate the transport coefficients in these media. Determining the specific form for the collision term represents a unique problem in each field of physics.

The transport equation and the collision term should not contradict the conservation laws, in particular, for mass, momentum, and energy. In the differential form, the conservation law for some physical quantity must have a universal form. The derivative of the conserving quantity should equal the opposite sign divergence of some flux. Accordingly, the conservation law of mass is expressed with the aid of *continuity equation*:

$$\frac{\partial \rho}{\partial t} + \mathrm{div}\,\boldsymbol{j} = 0.$$

Here are the mass density ρ and *momentum flux* (momentum of unit volume) $\boldsymbol{j}$. The conservation law of momentum yields the equation

$$\frac{\partial j_i}{\partial t} + \frac{\partial \Pi_{ik}}{\partial x_k} = 0$$

where Π_{ik} is the *momentum flux density tensor*. The law of energy conservation reads

$$\frac{\partial E}{\partial t} + \mathrm{div}\,\boldsymbol{Q} = 0.$$

Here E is the energy of unit volume and $\boldsymbol{Q}$ is the *energy flux density*.

The collision term should satisfy the following relations:

$$\int \mathrm{St}\,[n] d\tau_p = 0, \quad \int \boldsymbol{p}\,\mathrm{St}\,[n] d\tau_p = 0, \quad \int \varepsilon\,\mathrm{St}\,[n] d\tau_p = 0$$

in order to enforce the fulfilment of conservation laws for mass, momentum, and energy.

The *relaxation time approximation* or *τ-approximation* is one of the simplest approximations for treating small perturbations of the physical system under external disturbance. The approximation is based on the point that the collision term must identically vanish, i.e. $\mathrm{St}\,[n_0] = 0$, for the state of complete thermodynamical equilibrium with the equilibrium distribution function n_0. For small imbalance $\delta n = n - n_0$ of distribution function n from equilibrium n_0, one can expand the collision term in linear order in the deviation

[1] Abbreviation from German Stoßzahlansatz.

$$\mathrm{St}\,[n] = -\frac{n - n_0}{\tau} \quad \text{and} \quad \frac{1}{\tau} = -\left(\frac{\delta}{\delta n}\mathrm{St}\,[n]\right)_{n=n_0},$$

where τ is the *relaxation time*. In the general case, the relaxation time depends on the momentum (energy) of a particle, $\tau = \tau(\boldsymbol{p})$, and is determined with the concrete interaction and scattering of particles in the medium.

As another approximation, we can mention the approximation of *local equilibrium* which implies the particular thermodynamical equilibrium in every small separate volume of the medium. Then the state of the medium on the whole can be characterized with temperature $T = T(\boldsymbol{r}, t)$, chemical potential $\mu = \mu(\boldsymbol{r}, t)$, and (or) other thermodynamical variables depending on spatial coordinate $\boldsymbol{r}$ and time t. The *locally equilibrium distribution function* coincides with the thermodynamically equilibrium distribution function $n_0(\varepsilon_{\boldsymbol{p}};\ T, \mu)$ and is an implicit function of thermodynamical parameters $T = T(\boldsymbol{r}, t)$ and $\mu = \mu(\boldsymbol{r}, t)$. The approximation of locally equilibrium distribution function does not allow us to take the dissipative and irreversible processes in the medium into account.

1.2 The Generalized Susceptibility and the Linear Response

Let q be some generalized coordinate and $F(t)$ be perturbative time-dependent force. The linear response to small external impact $U = -qF(t)$ can be expressed with the aid of the *linear response function* or *generalized susceptibility* $\alpha(t)$ as

$$q(t) = \int\limits_{-\infty}^{t} \alpha(t - t')F(t')\,dt' = \int\limits_{0}^{\infty} \alpha(t')F(t - t')\,dt'.$$

Here $\alpha(t)$ depends on the properties of the physical system perturbed. Due to *principle of causality* the magnitude of thermodynamical quantity $q(t)$ can only be governed by the previous time points $t' \leqslant t$ unlike the subsequent time points $t' > t$.

With introducing the function of *retarded response* or *retarded* susceptibility

$$\alpha_R(t) = \alpha(t)\vartheta(t),$$

where $\vartheta(t)$ is the Heaviside step function, it is convenient to expand integration over the whole range of time variation from $-\infty$ to $+\infty$ as

$$q(t) = \int\limits_{-\infty}^{+\infty} \alpha_R(t - t')F(t')\,dt'.$$

Applying the property of convolution for the relation between the Fourier transforms gives

$$q(\omega) = \alpha_R(\omega)F(\omega) \quad \text{where} \quad \alpha_R(\omega) = \int_{-\infty}^{\infty} \alpha_R(t)e^{i\omega t}dt = \int_{0}^{\infty} \alpha(t)e^{i\omega t}dt.$$

On the analogy we can introduce the *advanced response* α_A according to $\alpha_A(t) = \alpha(t)\vartheta(-t)$.

Note a number of properties for the retarded response $\alpha_R(\omega)$. In general, $\alpha_R(\omega) = \alpha'_R(\omega) + i\alpha''_R(\omega)$ is a complex function with the even real part and odd imaginary part as a function of frequency ω:

$$\alpha'_R(-\omega) = \alpha'_R(\omega) \quad \text{and} \quad \alpha''_R(-\omega) = -\alpha''_R(\omega).$$

As a function of complex variable $\omega = \omega' + i\omega''$, the retarded response α_R is analytical in the upper half-plane Π^+ $(\omega'' > 0)$ since the integral

$$\alpha_R = \int_{0}^{\infty} \alpha(t)e^{i\omega' t}e^{-\omega'' t}dt$$

converges absolutely in the entire range of values $\omega'' > 0$ provided that $\alpha(t)$ does not grow as $t \to \infty$. On the contrary, the retarded response $\alpha_A(\omega)$ is analytical in the lower half-plane Π^- $(\omega'' < 0)$.

For the positive segment of imaginary semi-axis $\omega'' > 0$, the retarded response α_R is real, i.e. $\alpha^*(i\omega'') = \alpha(i\omega'')$. The advanced response α_A has the analogous property of realness but for the negative segment of imaginary semi-axis $\omega'' < 0$.

For the region of large $|\omega| \to \infty$ frequencies, one may expect that the magnitude of the linear response vanishes $\alpha_R \to 0$. The point is that the physical system, perturbed by the force at frequencies much higher than the maximum resonant frequency in the system, does not have time to respond to the alternating oscillations of the external perturbation. This physical argument, augmented with the analyticity of retarded response $\alpha_R(\omega)$ in the upper half-plane Π^+, allows us to formulate the *dispersion Kramers-Kronig relations*:

$$\alpha_R(\omega) = -\frac{i}{\pi}\mathcal{P}\int_{-\infty}^{\infty} \frac{\alpha_R(\omega')}{\omega' - \omega}d\omega' = -\frac{i}{\pi}\int_{-\infty}^{\infty} \frac{\alpha_R(\omega') - \alpha_R(\omega)}{\omega' - \omega}d\omega'.$$

Here symbol $\mathcal{P}$ denotes the Cauchy principal value. Separating this relation into the real and imaginary parts, we have two formulas

$$\alpha'_R(\omega) = \frac{1}{\pi}\mathcal{P}\int_{-\infty}^{\infty} \frac{\alpha''_R(\omega')}{\omega' - \omega} d\omega' = \frac{1}{\pi}\int_{-\infty}^{\infty} \frac{\alpha''_R(\omega') - \alpha''_R(\omega)}{\omega' - \omega} d\omega',$$

$$\alpha''_R(\omega) = -\frac{1}{\pi}\mathcal{P}\int_{-\infty}^{\infty} \frac{\alpha'_R(\omega')}{\omega' - \omega} d\omega' = -\frac{1}{\pi}\int_{-\infty}^{\infty} \frac{\alpha'_R(\omega') - \alpha'_R(\omega)}{\omega' - \omega} d\omega'.$$

The real α'_R and imaginary α''_R parts of linear response are not independent. The full function α_R can be reconstructed if either the imaginary or the real part is known singly.

1.3 The Fluctuation-Dissipation Theorem

For studying the fluctuations and correlations (interplay) between the random variables $q(t)$ given at the various time moments, the *correlation functions* or *correlators* are usually introduced. Let generalized coordinate $q(t)$ represent some random quantity. The correlation function or correlator of nth order is defined according to

$$K_n(t_1, t_2, \ldots, t_n) = \langle q(t_1)q(t_2)\ldots q(t_n)\rangle.$$

Here the angle brackets denote some procedure of time averaging. As a rule, we restrict ourselves with the correlators of first and second orders. The first-order correlator is obviously the simple average of quantity $q(t)$.

If the random process is uniform in time, the first-order correlator $\langle q(t)\rangle$ is independent of time and represents the average magnitude $\langle q(0)\rangle$. In this case, the second-order correlator proves to be dependent on the relative time difference $t = t_1 - t_2$ alone:

$$K_2(t_1, t_2) = K_2(t_1 - t_2) = \langle q(t)q(0)\rangle.$$

The correlation of deviations or fluctuations $\delta q(t) = q(t) - \langle q(t)\rangle$ from the average magnitudes $\langle q(t)\rangle$ is usually of more interest and the following correlator

$$K(t) = \langle \delta q(t)\delta q(0)\rangle.$$

is analyzed. We call the Fourier transform of correlator $K(t)$ as the *spectral density* $K(\omega)$ of fluctuations:

$$K(\omega) \equiv \langle \delta q(t)\delta q(0)\rangle_\omega = \int_{-\infty}^{\infty} \langle \delta q(t)\delta q(0)\rangle e^{i\omega t} dt.$$

For the thermodynamically equilibrium systems at temperature T, the *fluctuation-dissipation theorem* (FDT) is valid. The theorem allows us to set the relation between the spectral density $K(\omega)$ of the fluctuating variable $\delta q(t)$ and the imaginary part $\alpha''_R(\omega)$ of the corresponding generalized susceptibility α_R, namely,

$$K(\omega) = \hbar\alpha''_R(\omega)\,\coth\frac{\hbar\omega}{2T}\,.$$

The total mean square of the fluctuating random quantity $q(t)$ is given by the integral

$$\langle(\delta q)^2\rangle = \int_{-\infty}^{\infty} K(\omega)\frac{d\omega}{2\pi} = \int_{-\infty}^{\infty} \hbar\alpha''_R(\omega)\,\coth\left(\frac{\hbar\omega}{2T}\right)\frac{d\omega}{2\pi}\,.$$

In the classical limit $T \gg \hbar\omega$, we obtain

$$K(\omega) = \frac{2T}{\omega}\alpha''_R(\omega) \quad \text{and} \quad \langle(\delta q)^2\rangle = \frac{T}{\pi}\int_{-\infty}^{\infty}\frac{\alpha''_R(\omega)}{\omega}d\omega$$

and in the quantum $T = 0$ limit we have

$$K(\omega) = \hbar|\alpha''_R(\omega)| \quad \text{and} \quad \langle(\delta q)^2\rangle = \hbar\int_{-\infty}^{\infty}|\alpha''_R(\omega)|\frac{d\omega}{2\pi}\,.$$

In the latter, we took the odd property $\alpha''_R(\omega)$ into account.

1.4 The Onsager Reciprocal Relations

The response of a thermodynamically equilibrium system to the external force impact entails the appearance of a number of flows absent in the equilibrium system as a reaction to perturbation. For example, the pressure difference results in the mass flow, the temperature difference leads to heat flow, and electric potential difference emerges an electric current. The same type of flow can occur under influence of different forces (perturbations). For example, in liquid mixtures, the same impurity flow can be induced with an existence of one of the following reasons: concentration gradient (diffusion), pressure gradient (barodiffusion), or temperature gradient (thermodiffusion or Soret effect).

For the weak perturbations, the flows, which emerge, are linearly proportional to the external forces. The *transport coefficients* relate the corresponding flows to the perturbations. The *Onsager reciprocal relations* or *principle of symmetry of kinetic coefficients* express the symmetry interrelation between the transport coefficients related to the different pairs of forces and flows.

Let $\boldsymbol{q} = (q_1, q_2, \dots q_n)$ represent a set of thermodynamical variables, e.g. energy, volume, and the number of particles, which specify the entropy $S = S(q_1, q_2, \dots q_n)$ of the physical system. In the thermodynamically equilibrium system the entropy is maximum. Let its maximum correspond to the equilibrium values $\boldsymbol{q}_0 = (q_{10}, q_{20}, \dots q_{n0})$. For the small deviations from equilibrium $\Delta\boldsymbol{q} = \boldsymbol{q} - \boldsymbol{q}_0 = (q_1 - q_{10}, q_2 - q_{20}, \dots q_n - q_{n0})$, the entropy variation ΔS represents the quadratic form

$$\Delta S = -\frac{1}{2}\beta_{ik}\Delta q_i \Delta q_k\,, \quad \beta_{ik} = -\left.\frac{\partial^2 S}{\partial q_i \partial q_k}\right|_{\boldsymbol{q}=\boldsymbol{q}_0},$$

where $\beta_{ik} = \beta_{ki}$ is a positive definite quadratic form.

As a result of perturbation and relaxation to equilibrium, the thermodynamical variables $q_i = q_i(t)$ are time dependent. Provided that the *quasistationary approximation* is valid, the variation rate $\dot{q}_i(t)$ of every quantity $q_i(t)$ is determined with a single set of thermodynamical variables $\boldsymbol{q} = (q_1, q_2, \dots q_n)$ under consideration and can be a certain function of these variables alone:

$$\dot{q}_i = \dot{q}_i(q_1, q_2, \dots q_n).$$

As long as the system remains in the state close to the complete thermodynamical equilibrium, it is possible to decompose the functions $\dot{q}_i(q_1, q_2, \dots q_n)$ in a series in deviations $\Delta\boldsymbol{q} = \boldsymbol{q} - \boldsymbol{q}_0$ and restrict ourselves with the linear terms. Taking into account that in equilibrium $\dot{q}_i(q_{10}, q_{20}, \dots q_{n0}) = 0$, we obtain the following decomposition:

$$\dot{q}_i = -\lambda_{ik}\Delta q_k\,, \quad \lambda_{ik} = -\left.\frac{\partial \dot{q}_i}{\partial q_k}\right|_{\boldsymbol{q}=\boldsymbol{q}_0},$$

where i, k run from 1 to n. The inverse magnitudes λ_{ik}^{-1} play a role of relaxation times for various processes of setting the complete thermodynamical equilibrium.

To give the *canonical* form to these relations, variables Q_i or *thermodynamic forces*, conjugated to the initial variables q_i, are introduced according to

$$Q_i = -\frac{\partial S}{\partial q_i}.$$

For example, the variables conjugated to energy E, volume V, and number of particles N are the inverse temperature $1/T$ and the temperature-divided pressure P/T and chemical potential μ/T, respectively. In the thermodynamically equilibrium state, in which $\boldsymbol{q} = \boldsymbol{q}_0$ and entropy is maximum, we have all $Q_i = 0$. Near the equilibrium, we can write

$$Q_i = \beta_{ik}\Delta q_k \quad \text{or} \quad \Delta q_i = \beta_{il}^{-1} Q_l,$$

where β_{il}^{-1} is the inverse matrix. Expressing the variation rate[2] of the initial thermodynamic variables $\dot{q}_i$ in terms of conjugated quantities, we obtain the form required

$$\dot{q}_i = -L_{ik}Q_k \,.$$

Here new coefficients $L_{ik} = \lambda_{il}\beta_{lk}^{-1}$ are usually called the *kinetic* or *transport coefficients*.

The *principle of symmetry of kinetic coefficients* or *Onsager principle* states the following relationship:

$$L_{ik} = L_{ki} \,.$$

This formula implies that if the time-reversal symmetry of thermodynamic variable requires changing its sign, e.g. for magnetic field $\boldsymbol{H}$ or angular velocity $\boldsymbol{\Omega}$, then it reads as

$$L_{ik}(\boldsymbol{H}) = L_{ki}(-\boldsymbol{H}) \quad \text{and} \quad L_{ik}(\boldsymbol{\Omega}) = L_{ki}(-\boldsymbol{\Omega}).$$

Non-equilibrium processes and relaxation in the condensed medium entail the growth of internal energy $E = E(S)$ which we consider as a function of entropy S. Differentiating the energy with respect to time, we obtain $\dot{E} = T\dot{S}$. For the rate of entropy variation, we find straightforwardly that

$$\dot{S} = \frac{\partial S}{\partial q_i}\dot{q}_i = -Q_i\dot{q}_i = J_iQ_i = L_{ik}Q_iQ_k = \frac{R}{T},$$

where we introduce the *Rayleigh dissipation function* R according to

$$\frac{R}{T} = J_iQ_i = L_{ik}Q_iQ_k = \lambda_{ik}Q_i\Delta q_k = \beta_{li}\beta_{mk}L_{lm}\Delta q_i\Delta q_k \,.$$

The energy dissipation function R determines the heat generation power $dE/dt = R$ as a result of the dissipative processes occurring in the non-equilibrium and relaxing system. Since the approach to equilibrium is accompanied with the entropy growth, the energy dissipation function R must be positive definite. The requirement of positivity for the energy dissipation function R, being a quadratic form, imposes certain requirements on kinetic coefficients L_{ik}. Thus, if the lack of equilibrium is associated with the fluctuation of single quantity, say q_1, we find $R/T = L_{11}Q_1^2 = \beta_1^2L_{11}(\Delta q_1)^2$. The latter means the necessary requirement $L_{11} > 0$.

[2] The variation rate of variables q_i taken with the opposite sign is usually referred to as the *flux* $J_i = -\dot{q}_i$ and $J_i = L_{ik}Q_k$.

1.5 The Lindblad Master Equation

The possible states of quantum system are usually classified into the pure quantum states and the mixed quantum states. The *pure quantum state* is a state described unambiguously and completely with the wave function or state vector, i.e. complete set of quantum numbers related to the given state. The state which is not the pure one refers to the *mixed quantum state*. In this sense, the mixed state[3] is defined as a probabilistic (statistical) mixture[4] of pure states. The mathematical description of the mixed state is realized with introducing the *density matrix* defined[5] as the following Hermitian operator:

$$\hat{\rho} = \sum_n w_n |n\rangle\langle n|.$$

Here w_n is the probability to find the system in the pure quantum state with vector state $|n\rangle$ and a sum is taken over all possible states. Since a sum of all probabilities w_n $(0 \leqslant w_n \leqslant 1)$ equals unity, i.e. $\sum_n w_n = 1$, the trace of density matrix satisfies an equality $\operatorname{tr}\hat{\rho} = 1$.

The method of density matrix allows us to describe the pure and mixed states at one time. The simple criterium of verification is the following. For the pure states, $\operatorname{tr}\hat{\rho}^2 = \operatorname{tr}\hat{\rho} = 1$ and for the mixed ones, $\operatorname{tr}\hat{\rho}^2 < 1$. The average value for the physical operator $\hat{A}$ is defined by the following formula:

$$\langle \hat{A} \rangle = \sum_n w_n \langle n|\hat{A}|n\rangle = \operatorname{tr}(\hat{\rho}\hat{A}).$$

In the coordinate representation, when the quantum state $|n\rangle$ is described in terms of wave function $\psi_n(x)$, the density matrix is determined with the same sum over all possible states of quantum system

$$\rho(x, x') = \sum_n w_n \psi_n(x)\psi_n^*(x'), \quad \sum_n w_n = 1, \quad w_n \geqslant 0.$$

The quantum system interacting with a heat reservoir or thermostat is frequently called the *open* or *dissipative quantum system*. Within the framework of quantum

[3] The unpolarized photon beam (natural light) can serve as an example of the mixed state.

[4] The mixed state should be discerned from the *superposition* of states when the wave functions are summed over various pure states in contrast to summing the probabilities to find the system in one or another pure state.

[5] Let us recall that $|\psi\rangle\langle\psi|$ implies a matrix (tensor) product of columns multiplied by rows, i.e.

$$\begin{pmatrix} \psi_1 \\ \psi_2 \\ \vdots \\ \psi_N \end{pmatrix} (\psi_1^*, \psi_2^*, \dots \psi_N^*) = \begin{pmatrix} \psi_1\psi_1^* & \psi_1\psi_2^* & \dots & \psi_1\psi_N^* \\ \psi_2\psi_1^* & \psi_2\psi_2^* & \dots & \psi_2\psi_N^* \\ \dots & \dots & \dots & \dots \\ \psi_N\psi_1^* & \psi_N\psi_2^* & \dots & \psi_N\psi_N^* \end{pmatrix}.$$

mechanics it is usually said that the dynamics of the closed system is coherent and conserves the amplitude and phase. In the real quantum systems, the phase and coherence do not conserve due to coupling with the heat reservoir. The open quantum system is frequently represented as a system with the small number of degrees of freedom[6] with the discrete energy spectrum. The thermostat interacting with the quantum system has, as a rule, the macroscopically large number of degrees of freedom and continuous energy spectrum.

To describe the dynamics of such open quantum system, characterized by Hamiltonian $\hat{H}$, the *Lindblad master equation*[7] is employed. For explaining and deriving the equation, the *Markov approximation* for random processes, consisting in the complete neglect of any possible memory effects, is used. This approximation implies that the typical time scales of relaxation in the heat reservoir are much shorter than the relaxation time in the quantum system. The Lindblad equation is a simplest master equation[8] for the density matrix $\hat{\rho}(t)$ describing the dissipative evolution of the density matrix:

$$\frac{d\hat{\rho}}{dt} = -\frac{i}{\hbar}[\hat{H}, \hat{\rho}] + \frac{1}{2}\left([\hat{L}\hat{\rho}, \hat{L}^{+}] + [\hat{L}, \hat{\rho}\hat{L}^{+}]\right).$$

Another common representation reads

$$\frac{d\hat{\rho}}{dt} = -\frac{i}{\hbar}[\hat{H}, \hat{\rho}] + \hat{L}\hat{\rho}\hat{L}^{+} - \frac{1}{2}\{\hat{L}^{+}\hat{L}, \hat{\rho}\},$$

where [...] is commutator and {...} is anticommutator. The operator $\hat{L}$ is entirely determined with the interaction of quantum system with the heat reservoir. (equation $\dot{\hat{\rho}} = -(i/\hbar)[\hat{H}, \hat{\rho}]$ is commonly called the *Liouville-von Neumann equation.*) The terms with operator $\hat{L}$, also called the *Lindbladian,* are responsible for energy dissipation and decoherence in the dynamics of quantum system.

The Lindblad equation conserves the self-adjoint property and the trace of density matrix. In fact, from the relations

$$\operatorname{tr}[\hat{H}, \hat{\rho}] = \operatorname{tr}(\hat{H}\hat{\rho}) - \operatorname{tr}(\hat{\rho}\hat{H}) = 0,$$
$$\operatorname{tr}(\hat{L}\hat{\rho}\hat{L}^{+}) = \frac{1}{2}\operatorname{tr}(\hat{L}^{+}\hat{L}\hat{\rho}) + \frac{1}{2}\operatorname{tr}(\hat{\rho}\hat{L}^{+}\hat{L})$$

it follows that

$$\frac{\partial}{\partial t}\left(\operatorname{tr}\hat{\rho}\right) = 0 \quad \text{or} \quad \operatorname{tr}\hat{\rho}(t) = \text{const}.$$

That is, the trace of density matrix remains time-invariant.

---

[6] The simplest object is a two-level system with the density matrix 2×2.

[7] Less commonly it is also called the Gorini-Kossakowski-Sudarshan-Lindblad (GKSL) equation.

[8] In some sense, one may say that the Lindblad equation is a quantum analog of the Boltzmann transport equation, the density matrix $\hat{\rho}(t)$ playing a role of probability density function $n(\boldsymbol{r}, \boldsymbol{p}, t)$.

The Lindblad master equation is used in quantum optics and lasers for describing the physical phenomena, for example, such as spontaneous emission of excited atoms or dynamics of microresonators with the energy losses. Here the electromagnetic field interacting with an excited atom plays a role of heat reservoir. The Lindblad operator[9] $\hat{L}$ will be responsible for absorbing or emitting a photon with the electromagnetic field. For describing several dissipative and decoherence processes, the above Lindblad equation can be generalized with introducing several Lindblad operators $\hat{L}_k$.

Let density matrix be diagonal

$$\rho_{ik} = \delta_{ik} P_i$$

where $P_i = \rho_{ii}$ is the probability that the system stays in the state i corresponding to the diagonal element ii of density matrix. In this case, the Lindblad equation reduces to the *master equation*

$$\frac{\partial P_i}{\partial t} = \sum_k L_{ik} P_k L_{ik}^{+} - \frac{1}{2} \sum_k \left(L_{ki}^{+} L_{ki} P_i + P_i L_{ki}^{+} L_{ki} \right) = \sum_k (W_{ik} P_k - W_{ki} P_i),$$

where $W_{ik} = |L_{ik}|^2$ is the transition probability per unit time from the state k of the system to its state i.

[9] In quantum optics, it is said as a *jump operator* as well.

Chapter 2
Problems

2.1 The Langevin Equation and the Fluctuation-Dissipation Theorem

1. Using the fluctuation-dissipation theorem, find the mean-squared displacement $R(t)$ in the thermodynamically equilibrium medium with temperature T for a particle of mass $m = 1$. The one-dimensional motion of the particle is governed by the Langevin equation

$$\ddot{x}(t) + \gamma \dot{x}(t) = f(t),$$

where $\gamma > 0$ is the friction coefficient associated with the mobility $\mu = 1/\gamma$ and $f(t)$ is a random force.

2. The particle of mass m oscillates in the one-dimensional harmonic trap of frequency ω_0. The trap is in the thermodynamically equilibrium medium with temperature T. The equation of particle motion has the Langevin form

$$m\ddot{x}(t) + \eta \dot{x}(t) + m\omega_0^2 x(t) = F(t),$$

where $\eta = m\gamma$ is the friction coefficient and $F(t)$ is a random force. Due to fluctuations the particle position $x(t)$ experiences the random deviations from central position $x = 0$. Using the fluctuation-dissipation theorem, find the mean square $\langle x^2 \rangle$ of the particle position.

2.2 The Boltzmann Transport Equation in the τ-Approximation: The Thermoelectric Effects in Conductors

3. Find the collision integral for scattering an electron with nonmagnetic elastic impurities in the Born approximation. Using the Boltzmann transport equation, calculate the collision time, transport collision time, and determine the electrical

S. N. Burmistrov, *Physical Kinetics*,
https://doi.org/10.1007/978-981-19-1649-6_2

conductivity. Put the electron spectrum equal to $\varepsilon(\boldsymbol{p}) = \boldsymbol{p}^2/2m$ and impurity concentration equal to n_i.

Consider the following potentials of impurity-electron interaction:
(a) point-like potential $u(\boldsymbol{r}) = u_0\delta(\boldsymbol{r})$;
(b) Coulomb screened potential $u(\boldsymbol{r}) = (Ze^2/r)\exp(-\varkappa r)$.

4. As a simplest model to describe the thermoelectrical phenomena, one may use the Boltzmann transport equation in the time relaxation approximation (τ-approximation) and the isotropic model of free electrons with dispersion $\varepsilon(\boldsymbol{p}) = \boldsymbol{p}^2/2m$. For simplicity, the collision time $\tau = \tau(\varepsilon)$ depends on the electron energy alone.

(a) Find the expressions for electric conductivity σ, Seebeck coefficient S (also known as thermopower), Peltier coefficient Π, and heat conductivity $\varkappa$.

(b) Determine the energy flux $\boldsymbol{Q}$, entropy flux $\boldsymbol{F}$, and energy dissipation function R. Show that the requirement for the positive determinacy of dissipative function results in the positive magnitudes of electric conductivity σ and thermal conductivity $\varkappa$.

5. Using the results of the previous problem, calculate electric conductivity σ, Seebeck coefficient S, Peltier coefficient Π, and thermal conductivity $\varkappa$ for the two cases: (a) metal (degenerate electron gas) and (b) semiconductor (non-degenerated electron gas).

Find the *figure of merit* ZT for the thermoelectric phenomena

$$ZT = \frac{S^2\sigma T}{\varkappa}$$

and the *Lorenz number* L

$$L = \frac{\varkappa}{\sigma T}.$$

Suppose that the scattering time τ has a power-like dependence on energy, i.e. $\tau(\varepsilon) \sim \varepsilon^{r-1/2}$.

6. One of mechanisms of heat conduction in metals is associated with the presence of impurities and the scattering of conduction electrons with those impurities. Determine the impurity contribution to the thermal conductivity of superconductor, using the τ-approximation of the Boltzmann transport equation. Suppose that the mean free path of electron excitations in the superconducting state remains the same as in the normal state $l = (n_i\sigma)^{-1}$. Here n_i is the impurity concentration and σ is the cross section of electron scattering with impurity. Neglect the possible dependence of chemical potential μ as a function of temperature.

2.3 Galvanomagnetic Effects in Metals

7. As a simplest model for describing the thermogalvanic phenomena in magnetic field, the Boltzmann transport equation can be used in the approximation of time relaxation (τ-approximation). For simplicity, consider the isotropic free electron model with dispersion $\varepsilon(\boldsymbol{p}) = \boldsymbol{p}^2/2m$. The collision time depends on the electron energy $\tau = \tau(\varepsilon)$ alone. The magnetic field H is sufficiently small $(eH/mc)\tau = \Omega\tau \ll 1$ in order to be restricted with the linear approximation in magnetic field.

(a) Find the expressions for the electric current density $\boldsymbol{j}$ and for the energy dissipation flux $\boldsymbol{q}$.

(b) Calculate the Hall R, Nernst N, and Righi-Leduc L coefficients for the case of a metal (degenerate electron gas).

8. Find the conductivity tensor $\sigma_{\alpha\beta}$ of a metal in the constant uniform magnetic field H. The electron distribution function $n(\boldsymbol{p})$ is governed with the Boltzmann transport equation in the approximation of relaxation time $\tau = \tau(\varepsilon)$. The electron spectrum $\varepsilon(\boldsymbol{p}) = \boldsymbol{p}^2/2m$ is simple and isotropic.

2.4 Kinetic Phenomena in a Metal and Normal Fermi Liquid

9. Determine the collision term for the degenerate Fermi liquid in the Born approximation. The matrix element of the fermion-fermion interaction is equal to $U(\boldsymbol{q})$ where $\boldsymbol{q}$ is the transferred momentum in the course of the fermion-fermion collision.

Find the behavior of relaxation time τ as a function of energy and temperature in the region of the energy and temperature small as compared with the Fermi energy ε_F.

10. Determine the collision term in the case of two-dimensional degenerate Fermi gas in the Born approximation. The matrix element of the fermion-fermion interaction is equal to $U(\boldsymbol{q})$ where $\boldsymbol{q}$ is the transferred momentum in the course of the fermion-fermion collision.

Find the behavior of relaxation time τ as a function of energy and temperature in the region of the energy and temperature small as compared with the Fermi energy ε_F.

11. The non-equilibrium states of Fermi liquid are described with the density distribution function $n = n(\boldsymbol{r}, \boldsymbol{p}, t)$. At zero temperature or at sufficiently low temperatures, the collisions between quasiparticles are so rare that we can neglect the collisions completely. In the lack of collisions between particles, we can employ the Liouville theorem about vanishing the total time derivative of distribution function, i.e.

$$\frac{dn}{dt} = \frac{\partial n}{\partial t} + \{H, n\} = 0,$$

where $\{H, n\}$ is the Poisson brackets for Hamiltonian H and distribution function n. Using the following transport equation:

$$\frac{\partial n}{\partial t} + \frac{\partial H}{\partial \boldsymbol{p}}\frac{\partial n}{\partial \boldsymbol{r}} - \frac{\partial H}{\partial \boldsymbol{r}}\frac{\partial n}{\partial \boldsymbol{p}} = 0,$$

consider the small oscillations of distribution function at $T = 0$.

Find the condition when the propagation of undamped waves, called *zero-sound*, is possible. Suppose the function of interaction for quasiparticles to be independent of momenta $f(\boldsymbol{p}, \boldsymbol{p}') = f_0$.

12. Find the dispersion and absorption of low- and high-frequency sound in isotropic normal Fermi liquid at low temperatures. For calculating, use the collision term $\mathrm{St}\,[n_{\boldsymbol{p}}]$ in the approximation of effective relaxation time $\tau = \tau(T)$ and in the representation which allows us to take the conservation laws of the total particle number, momentum, and energy into account:

$$\mathrm{St}\,[n_{\boldsymbol{p}}] = -\frac{1}{\tau}\big[\delta n_{\boldsymbol{p}} - \langle \delta n_{\boldsymbol{p}}\rangle - 3\langle \delta n_{\boldsymbol{p}} \cos\theta\rangle \cos\theta\big].$$

Here $\delta n_{\boldsymbol{p}} = n_{\boldsymbol{p}} - n_0(\varepsilon_{\boldsymbol{p}})$ is the deviation of the quasiparticle distribution function from equilibrium $n_0(\varepsilon_{\boldsymbol{p}})$. The angle brackets denote averaging over all possible directions of momentum vector $\boldsymbol{p}$

$$\langle \cdots \rangle = \int \cdots \frac{d\Omega_{\boldsymbol{p}}}{4\pi}, \qquad \Omega_{\boldsymbol{p}} = (\theta, \varphi).$$

The polar and azimuthal angles are equal to θ and φ, respectively. For simplicity, the interaction function of quasiparticles is independent of momenta and corresponds to repulsion $f(\boldsymbol{p}, \boldsymbol{p}') = f_0 > 0$.

2.5 Electron-Phonon Scattering in a Metal: The Transport Coefficients

13. As a rule, the linearized transport equation reduces to non-uniform integral equation as

$$\int d^3p'\, W(\boldsymbol{p}, \boldsymbol{p}')[\Phi(\boldsymbol{p}) - \Phi(\boldsymbol{p}')] = f(\boldsymbol{p})$$

with the definite-positive symmetric kernel $W(\boldsymbol{p}, \boldsymbol{p}') = W(\boldsymbol{p}', \boldsymbol{p})$ and $W(\boldsymbol{p}, \boldsymbol{p}') \geqslant 0$. Function $\Phi(\boldsymbol{p})$ is unknown and the given function $f(\boldsymbol{p})$ depends on the external fields. The region of integration represents the whole momentum $\boldsymbol{p}$-space.

(a) Show that the integral equation can be derived from the requirement of absolute extremum for the following functional:

$$F[\Phi(\boldsymbol{p})] = \frac{1}{2}\iint W(\boldsymbol{p}, \boldsymbol{p}')[\Phi(\boldsymbol{p}) - \Phi(\boldsymbol{p}')]^2\, d^3p\, d^3p' - \int d^3p\, f(\boldsymbol{p})\Phi(\boldsymbol{p}).$$

(b) Prove the following variational principle which can be used for searching the solution of integral equation. The minimum value of a ratio

$$R[\Phi(\boldsymbol{p})] = \frac{\iint W(\boldsymbol{p}, \boldsymbol{p}')[\Phi(\boldsymbol{p}) - \Phi(\boldsymbol{p}')]^2\, d^3p\, d^3p'}{\left[\int f(\boldsymbol{p})\Phi(\boldsymbol{p})\, d^3p\right]^2}$$

is delivered by function $\Phi(\boldsymbol{p})$ being the solution of integral equation. With the aid of choosing the reasonable trial functions the variational principle allows us to obtain some approximation to the genuine solution of integral equation.

14. In the ideal periodic crystal, an electron experiences no scattering. The thermal oscillations of atoms in the crystal lattice beside the equilibrium sites can be described as a set of elementary oscillation quanta called *phonons*. The thermal oscillations of atoms violate the strict periodicity and result in the electron scattering. The scattering is usually considered as a process of absorbing or emitting the phonons.

Find the electrical conductivity σ of a metal and the transport scattering time τ_{tr} of an electron interacting with the thermal oscillations of crystalline lattice. Neglect the discreteness of ionic lattice and treat the crystalline lattice as a continuous medium.

The indication. Consider the simple model of a metal representing the isotropic Fermi liquid with the quadratic electron dispersion $\varepsilon_p = p^2/2m$. As an additional simplification, put that there is only a single acoustic branch of phonons with the acoustic dispersion $\omega_k = sk$ where s is the sound velocity. The phonon spectrum is limited with the maximal wave vector $k_D \sim \pi/a$ meaning the lack of oscillations with the wavelength smaller than interatomic distance a. When solving the problem, take into account that the electron concentration $n \sim 1/a^3$ in a metal and Fermi momentum $p_F \sim \hbar/a$. Accordingly, treat the magnitudes p_F and $\hbar k_D$ to be the same order of magnitude, neglecting any quantitative difference between them. Use also that a ratio of sound velocity to the Fermi velocity $s/v_F \sim \sqrt{m/M} \ll 1$ is small as a square root from a ratio of electron mass m to ion mass M.

Choose the *Bloch approximation* or neglect the *effect of phonon drifting with electrons*, namely, possible deviation of phonon distribution function $N_{\boldsymbol{k}}$ from the equilibrium Planck function.

15. As a rule, the metals have a high thermal conductivity due to heat transfer with electrons. The thermal conductivity is limited by the scattering of electrons with the ion lattice oscillations, i.e. phonons. For simplest model of a metal, one can take the model from the previous problem. In other words, consider the metal as an isotropic Fermi liquid with the electron dispersion law $\varepsilon_p = p^2/2m$ and single acoustic branch of phonon dispersion $\omega_k = sk$ where s is the sound velocity.

Calculate the thermal conductivity $\varkappa$ and the corresponding electron-phonon scattering time $\tau_\varkappa(T)$, using the following trial function as a solution of the linearized transport equation:

$$n_p = n_0(\varepsilon_p) + \delta n_p, \quad \delta n_p = n_0'(\varepsilon_p)\Phi_p,$$

where

$$\Phi_p = \Phi_p^{(0)}\tau_{\varkappa}(T), \quad \Phi_p^{(0)} = \frac{\varepsilon_p - \mu}{T}(\boldsymbol{v}_p \nabla T).$$

Here $n_0(\varepsilon_p)$ is the equilibrium electron distribution function, μ is the electron chemical potential, $\boldsymbol{v}_p = \boldsymbol{p}/m$ is the electron velocity, and T is the temperature of a metal.

As in the previous problem, for solving the problem, use the *Bloch approximation* when the possible deviation of phonon distribution function N_k from the equilibrium Planck function or the *phonon drifting effect* is neglected. In general, the phonon drifting effect is essential in the region of temperatures lower as compared with the Debye temperature. Employ the results of the previous problem for the collision term and matrix element of electron-phonon interaction.

16. Find the low-temperature Seebeck coefficient S for a metal, using the free electron model with dispersion $\varepsilon_p = p^2/2m$. Suppose that phonons in a metal are represented by the single acoustic branch with sound velocity s and dispersion $\omega = sk$. When solving the problem, employ the conservation law for the total momentum of the electron and phonon subsystems and its consequence for the following combination of electron and phonon collision terms:

$$\int \boldsymbol{p}\, \mathrm{St}_e[n, N]\frac{2d^3p}{(2\pi\hbar)^3} + \int \hbar\boldsymbol{k}\, \mathrm{St}_{ph}[N, n]\frac{d^3k}{(2\pi)^3} = 0.$$

Here the integration is performed over the whole electron momenta and phonon wave vectors.

2.6 The Electron Scattering with Magnetic Impurities: The Kondo Effect

17. The *s-d* exchange model is often used to describe the scattering process of conduction electrons in a metal with nonzero spin magnetic impurities. In the *s-d* model, the exchange interaction between the conduction electron and all the magnetic impurity spins is written as a sum over the total number of impurities located randomly at points $\boldsymbol{r} = \boldsymbol{R}_a$

$$U(\boldsymbol{r}) = \sum_a u(\boldsymbol{r} - \boldsymbol{R}_a) \quad \text{and} \quad u(\boldsymbol{r}) = J(\boldsymbol{\sigma}\boldsymbol{S})\,\delta(\boldsymbol{r}).$$

Here $\boldsymbol{\sigma} = (\sigma^x, \sigma^y, \sigma^z)$ is the Pauli matrices, $\boldsymbol{S}_a$ is the spin of impurity at the site $\boldsymbol{r} = \boldsymbol{R}_a$, and J is the exchange integral usually small as compared with the Fermi energy ε_F divided by the electron density n, i.e. $J \ll \varepsilon_F/n$.

Find the amplitude and electron scattering probability in the first and second Born approximation, regarding the electron spectrum equal to $\varepsilon_p = \boldsymbol{p}^2/2m$ and magnetic impurity concentration equal to n_i. In the second Born approximation, it is necessary to take the Pauli principle into account in the course of analyzing a sum over the intermediate states in the second-order terms. The magnetic impurity concentration is small $n_i \ll n$ and thus a possible correlation between the impurity spins can be neglected.

Calculate the temperature contribution to electrical resistivity $\rho(T)$ of a metal with the accuracy to terms J^3 in the exchange constant as a result of conduction electron scattering at magnetic impurities (*Kondo effect*).

2.7 Weak Localization Effects in a Metal with Impurities

18. The weak localization effects in the conductivity of a metal with impurities can be described with the aid of the Boltzmann transport equation

$$\frac{\partial n}{\partial t} + \boldsymbol{v}\frac{\partial n}{\partial \boldsymbol{r}} + e\boldsymbol{E}\frac{\partial n}{\partial \boldsymbol{p}} = \mathrm{St}\,[n],$$

using the following self-consistent approximation for the collision term.

For the random and chaotic impurity sites, the motion of electron, treating as a classical particle, has a diffusive character with the diffusion coefficient $D_0 = v_F l/3$ where v_F is the Fermi velocity and l is the mean free path. On the other hand, electron as a quantum wave-like particle can demonstrate the quantum interference effects in its motion under multiple elastic scattering with nonmagnetic impurity. Such scattering conserves the phase coherence of wave function.

The simultaneous manifestation of classical and quantum character in the elastic scattering of electron with impurity can phenomenologically be described with introducing two terms into the collision integral

$$\mathrm{St}\,[n] = -\frac{n(\boldsymbol{r},\boldsymbol{p},t) - n_0(\varepsilon_{\boldsymbol{p}})}{\tau} + \int\limits_{-\infty}^{t} dt'\,\alpha(t-t')\big[n(\boldsymbol{r},-\boldsymbol{p},t) - n_0(\varepsilon_{\boldsymbol{p}})\big].$$

The first term corresponds to an ordinary τ-approximation of relaxation time which has a sense of mean free time $\tau = l/v_F$, i.e. average time between collisions. The second and nonlocal term, depending on the distribution function with the opposite sign of momentum $\boldsymbol{p}$, simulates the interference of geometrically identical and self-intersecting paths traveled with an electron in the opposite directions (Fig. 2.1).

The integrand kernel α, responsible for the quantum diffusion effects of an electron in the random impurity field, is given by the following response function of diffusive structure:

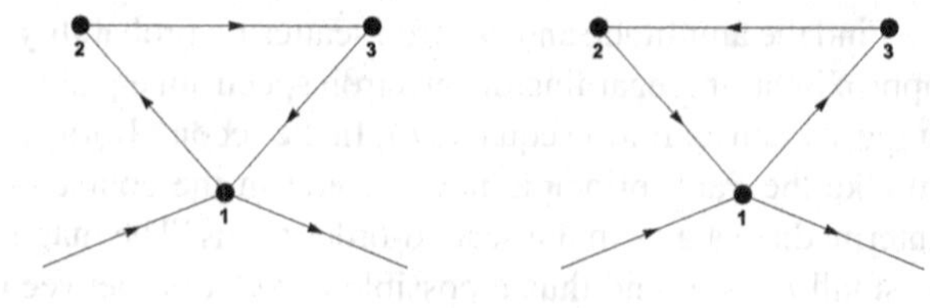

Fig. 2.1 An example is given of simplest electron paths with self-intersecting. The path with the clockwise 1–2–3–1 walk and counterclockwise 1–3–2–1 walk. The impurity atoms are shown with the solid circles

$$\alpha(t) = \frac{2n_i|V|^2}{\hbar^2}\int \frac{d^3q}{(2\pi)^3} e^{-(D_0q^2+\tau_\varphi^{-1})t}, \quad (t \geqslant 0).$$

Here n_i is the impurity concentration, V is the Fourier transform of the point-like electron-impurity interaction $U(\boldsymbol{r}) = V\delta(\boldsymbol{r})$, and τ_φ is the time of inelastic electron scattering which does not conserve the phase of wave function. Usually the inelastic scattering is associated with the electron-electron and electron-phonon interactions and $\tau_\varphi \gg \tau$. The typical region of integrating over wave vector q is approximately limited with the following interval: $l_\varphi^{-1} \lesssim q \lesssim l^{-1}$, where $l_\varphi = v_F\tau_\varphi$ is the electron mean free path with respect to inelastic scattering. The mean free time τ, expressed via the magnitude of electron-impurity interaction, is given by the formula

$$\hbar\tau^{-1} = \pi n_i\nu(\varepsilon_F)|V|^2,$$

$\nu(\varepsilon_F)$ being the density of states at the Fermi surface.

Find the conductivity $\sigma(\omega)$ in the alternating uniform electric field $\boldsymbol{E}$ of frequency ω and determine the quantum correction $\delta\sigma$ to the classical Drude formula for conductivity. The electron spectrum is isotropic and equal to $\varepsilon(\boldsymbol{p}) = \boldsymbol{p}^2/2m$.

19. The Einstein relation $\sigma(0) = e^2\nu(\varepsilon_F)D(0)$ is a connection between conductivity σ and diffusion coefficient D. Assuming that for small frequencies, the relation $\sigma(\omega)/\sigma_0 = D(\omega)/D_0$ remains[1] still valid, let us rewrite the equation obtained for determining the quantum correction to conductivity as

$$\frac{D(\omega)}{D_0} = \left(1 - i\omega\tau + \frac{2n_i|V|^2\tau}{\hbar^2}\int \frac{d^3q}{(2\pi)^3}\frac{1}{-i\omega + D_0q^2 + \tau_\varphi^{-1}}\right)^{-1}.$$

The self-consistent generalization, suggested by *P. Wölfle* and *D. Vollhardt* to describe the *Anderson localization* in an impurity-disordered metal, consists of replacing the trial diffusion coefficient $D_0 = v_F l/3$ in the integrand with its genuine magnitude $D(\omega)$. This assumes that the equation for quantum correction holds for its form at the noticeable values of parameter $\lambda = \hbar/(\pi p_F l) \sim 1$. So,

$$\frac{D(\omega)}{D_0} = \left(1 - i\omega\tau + \pi\lambda^2\frac{2\pi^2 l^3}{\tau}\int \frac{d^3q}{(2\pi)^3}\frac{1}{-i\omega + D(\omega)q^2 + \tau_\varphi^{-1}}\right)^{-1}.$$

[1] The Einstein relation is strictly valid for $\omega = 0$ alone.

Determine the critical value for parameter λ_c when the diffusion coefficient vanishes and the metal-insulator transition occurs. Consider the static $\omega = 0$ and zero $T = 0$ temperature case, assuming the lack of inelastic scattering $\tau_\varphi^{-1} = 0$.

2.8 Ballistic Electron Transport in the Mesoscopic Systems

20. A typical mesoscopic system with the quantum microcontact (Fig. 2.2) has two macroscopic conducting regions or electron reservoirs connected with an electrically conducting thin channel with the typical width W and longitudinal length L not exceeding the mean free path l. The voltage equal to V is applied across the channel.

Find *conductance* G at zero temperature, connecting the current I through the quantum contact and the voltage V across the contact according to

$$I = GV.$$

It is supposed that the electron motion in the microcontact channel is collisionless $L \ll l$ and, in addition, is semiclassical, i.e. $W \gg \lambda_F$ where $\lambda_F = \hbar/p_F$ is the Fermi wavelength of electron. The spectrum of electrons in the channel

$$\varepsilon_\lambda = \varepsilon_n + p^2/2m, \quad \lambda = (n, p)$$

is quantized in the transverse direction and classified by the quantum number n ($n = 1, 2, 3, \ldots$) which is also called the *subband* or *quantum channel*. In the longitudinal microcontact channel, the electrons travel freely and are characterized with momentum p. The electron reservoirs are always in the thermal equilibrium. Neglect the possible reflection of electrons from the terminals of the microcontact.

21. The typical mesoscopic system with the quantum microcontact (Fig. 2.3), considered in the previous problem, represents two macroscopic metal regions (electron reservoirs) connected with the conducting thin channel of width W and length

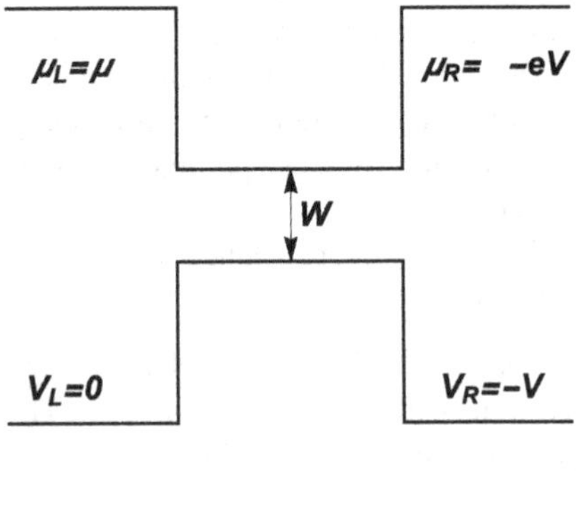

Fig. 2.2 A scheme of microcontact with width W

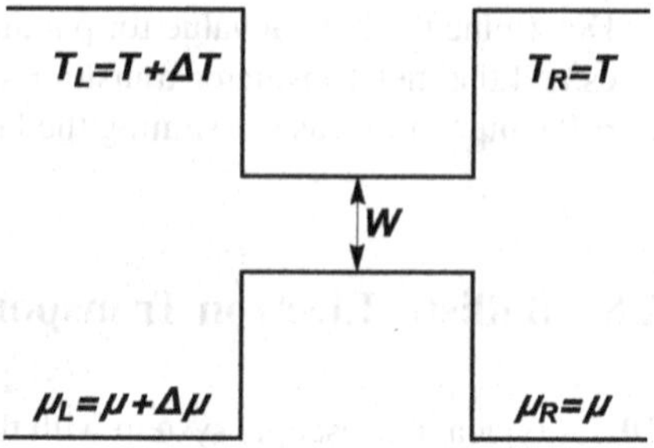

Fig. 2.3 A scheme of microcontact with width W

L much smaller than the electron mean free path l. The electron reservoirs have a small difference in temperatures $\Delta T \ll T$ and chemical potentials $\Delta\mu \ll \mu$.

Find the energy flux Q from one reservoir to the other under ballistic and collisionless $L \ll l$ motion of electrons in the microcontact. The semiclassical approximation $W \gg \lambda_F$, $\lambda_F = \hbar/p_F$ being the Fermi wavelength, is assumed to be valid. The electron spectrum in the microcontact

$$\varepsilon_\lambda = \varepsilon_n + p^2/2m, \quad \lambda = (n, p)$$

is quantized in the transverse direction and classified by the quantum number n ($n = 1, 2, 3, \ldots$) called the *subband* or *quantum conducting channel*. In the longitudinal direction, the electrons travel freely and are characterized with momentum p. The electron reservoirs are always in the thermal equilibrium. Neglect the possible reflection of electrons from the terminals of the microcontact.

Temperature T is assumed to be small as compared both with the energy spacing of subbands and with the chemical potential μ, i.e. $T \ll \Delta\varepsilon_n = \varepsilon_{n+1} - \varepsilon_n$ and $T \ll \mu$. For simplicity, the microcontact to be treated as flat.

2.9 The Rarified Bose and Fermi Gases

22. The flat wall travels at velocity V across the rarified ideal gas of Bose particles. The temperature of gas T is lower than that of the Bose-Einstein condensation temperature T_c. The mass of particles is m. The gas is so rarified that the collisions between particles can completely be neglected, i.e. mean free path is large. The collisions of Bose particles with the wall are assumed to have an elastic mirror character.

Determine the resistance force f to the wall motion, emerging in addition to the pressure of ideal Bose gas.

23. Let the flat wall travel at velocity V across the rarified degenerate ideal gas of fermions with mass m and density n. The gas is so rarified that the collisions between particles can completely be neglected, i.e. mean free path is large.

Assuming the *full accommodation*, determine the resistance force f to the wall motion, emerging in addition to the pressure of ideal fermion gas. Under full accommodation the particles reflected from the wall reach the complete thermal equilibrium with the wall at the same temperature T as the gas has. The velocity V of the wall is small compared with the Fermi velocity v_F. The same temperature T for the gas and the wall is also small compared with the Fermi energy ϵ_F.

2.10 The Hopping Mott Conductivity and the Coulomb Gap

24. Doping an impurity atom of the donor type to an ideal pure semiconductor, e.g. doping silicon with phosphorus atoms, results in the formation of a hydrogen-like bound state of an electron with localization radius a and energy ε. The wave function of such electron state concentrates beside an impurity atom and decays exponentially far away from the impurity site according to $\psi(r) \sim \exp(-r/a)$. Due to random and disordered location of impurity atoms in the crystalline lattice of semiconductor the binding energy of electron with the impurity has, in general, random magnitude as well. For sufficiently small impurity concentration $na^3 \ll 1$, the overlapping of electron wave functions corresponding to different impurity sites is not large. Thus, the electron is involved into the process of electric charge transfer only due to hopping from one impurity site to another, varying its energy as a difference between the binding energies at various sites. The similar mechanisms of charge transfer in the disordered electron systems are frequently called the *hopping conductivity*.

Find the low-temperature behavior of conductivity $\sigma(T)$ within exponential accuracy, using the following assumptions:
(a) Electron jumps are only essential to the impurity sites of energies ε_λ lying within sufficiently small vicinity Δ of Fermi level μ so that $|\mu - \varepsilon_\lambda| \leqslant \Delta \ll \mu$. The temperature is low enough $\Delta \gg T$ and the typical energy variations are much larger than the temperature.
(b) Put the density of states in the vicinity Δ of Fermi level to be constant and equal to $g(\mu)$.
(c) Localization radius a is independent of energy ε_λ.
(d) Suppose an existence of thermal equilibrium due to emitting and absorbing the phonons. The interaction with phonons ensures the fulfilment of energy conservation law for hopping the electrons from one level to another. For simplification, put that phonons are represented with the single acoustic branch of frequency $\omega_k = sk$ and consider only one-phonon processes owing to sufficiently low temperature as compared with the maximum phonon energy $\hbar\omega_D$.

25. Determine the exponent p in the *Mott law*

$$\rho(T) \sim \exp\bigl(T_0/T\bigr)^p$$

for the resistivity of thin (two-dimensional) film of doped semiconductor under assumptions of the previous problem about the hopping conductivity mechanism.

26. In two previous problems, the hopping mechanism of conductivity is analyzed for weakly doped semiconductor. It was assumed that the electrons, localized at impurity sites, do not interact with each other and the density of electron states $g(\varepsilon)$ is constant near the Fermi level μ. An existence of the Coulomb interaction between electrons at various impurity sites breaks this property and results in vanishing the density of states $g(\varepsilon)$ at point $\varepsilon = \mu$. This phenomenon $g(\mu) = 0$, called the *Coulomb gap*, disturbs essentially the temperature behavior of conductivity in a doped semiconductor.

Find the exponent p in the Mott law

$$\rho(T) \sim \exp\bigl(T_0/T\bigr)^p$$

for the resistivity of weakly doped semiconductor at low temperatures if the density of electron states near the Fermi level follows the law

$$g(\varepsilon) = g_0\left(\frac{|\varepsilon - \mu|}{\varepsilon_0}\right)^{\beta},$$

where $\beta = 2$ in the space $d = 3$ and $\beta = 1$ in the space $d = 2$.

2.11 Kinetics of Phonons and Thermal Conductivity in Dielectrics

27. In a solid dielectric, the thermal vibrations of crystalline lattice atoms beside the equilibrium sites can be described as a set of vibrational quanta or phonons which as a whole compose a rarified gas of elementary excitations. In a crystal, the elementary cell consists of ν atoms. The complete spectrum of vibrations has 3ν branches or, correspondingly, 3ν types of phonons. The whole spectrum subdivides into three acoustic branches, one being longitudinal and two being transverse, and into the other $3(\nu - 1)$ optical branches of vibrations.

Determine the thermal conductivity $\varkappa$ in a dielectric, using the transport equation for the phonon distribution function $N_{\boldsymbol{k}}(t, \boldsymbol{r})$ in the approximation of relaxation time τ. For simplicity, put that the dielectric has the cubic symmetry and the phonons are represented by the single acoustic branch with dispersion $\omega = \omega(\boldsymbol{k})$.

2.12 The Hydrodynamics of Normal and Superfluid Liquids: Sound Oscillations and Energy Dissipation

28. Derive the hydrodynamical equations for normal isotropic fluid within the framework of phenomenological approach using the thermodynamic identities and conservation laws of mass, momentum, energy, and entropy growth. Find the dissipative Rayleigh function R and obtain the equation of motion for viscous fluid (*Navier-Stokes equation*).

29. Using the results of the previous problem for the dissipative function R, find the sound absorption coefficient $\alpha(\omega)$ of frequency ω in normal isotropic fluid provided that the first coefficient of viscosity (*shear viscosity*) η, second coefficient of viscosity (*bulk viscosity*) ζ, and thermal conductivity $\varkappa$ are given.

30. Derive the equations of two-fluid hydrodynamics for superfluid isotropic liquid, not involving the dissipative effects into consideration. For the derivation, use the conservation laws of mass, momentum, energy, entropy, and Galilean transformation.

The following notions for superfluid liquid should be taken into consideration. In superfluid liquid, the two independent flows are possible at the same time. The first is a superfluid flow with velocity $\boldsymbol{v}_s$ and superfluid component density ρ_s. The second is a normal flow with $\boldsymbol{v}_n$ and normal component density ρ_n. A sum of densities of the superfluid and normal components yields the total density of liquid $\rho = \rho_n + \rho_s$. In addition, the superfluid flow is potential and irrotational, i.e. curl $\boldsymbol{v}_s = 0$.

31. In superfluid liquid there may exist two types of sound oscillations called the *first* and *second* sounds. Find the velocities of first and second sounds, using the hydrodynamic equations found in the previous problem. For simplicity, take the volumetric coefficient of thermal expansion α_V as zero since it is negligibly small at sufficiently low temperatures.

32. One of the properties of superfluid helium is its ability to flow through thin capillaries. At sufficiently low temperatures, the mean free path of elementary excitations starts to exceed the capillary diameter. Under such conditions the normal component of a superfluid slows completely down due to viscous interaction with the capillary walls. The velocity of normal component becomes zero $\boldsymbol{v}_n = 0$. However, there can propagate a sound wave in the superfluid component since the latter does not interact with the capillary walls. Such sound wave is referred to as the *fourth sound*.

Find the fourth sound velocity u_4, using the hydrodynamic equations of superfluid liquid. For simplicity, put the volumetric coefficient of thermal expansion α_V equal to zero. The normal component density is ρ_n and superfluid component density is ρ_s. The velocities of first and second sounds are u_1 and u_2, respectively.

2.13 Kapitza Resistance and Kinetic Phenomena at the Superfluid-Solid Helium Interface

33. The thermal boundary resistance or *Kapitza resistance* for the contact of two dielectrics, e.g. between the solid body and the superfluid helium, appears at low temperature due to various physical characteristics of the adjacent media.

Determine the Kapitza resistance R_K relating the constant heat flow

$$\Delta Q = \frac{\Delta T}{R_K}$$

with the temperature jump ΔT between superfluid helium and solid body. The temperatures of superfluid helium and solid body are T and $T' = T - \Delta T$, respectively. The heat exchange between two dielectric media occurs due to transition of phonons from one media to the other.

For simplicity, assume the reflection and transition coefficients of each phonon to be independent of temperatures at the both interfacial sides and the presence of other phonons. Take also that the phonons in each of the adjacent media are represented with the longitudinal acoustic mode alone. The acoustic modes in the media have the same sound velocities $u = u'$ but the densities of media are different and equal to ρ and ρ', respectively. The both temperatures T and T' are much smaller as compared with the corresponding Debye temperatures.

34. The crystalline (solid) phase of helium isotope ^{4}He can be in the thermodynamical equilibrium with its superfluid phase at sufficiently low temperatures down to absolute zero. At zero temperature, the entropy of the both phases vanishes and elementary excitations (phonons) are completely frozen out. Thus, in first approximation, we can neglect the latent heat of phase transition, presence of elementary excitations, and normal component in the superfluid phase. Under such conditions the melting and crystallization processes at the crystal-liquid interface occur in the dissipationless manner. The phase conversion rate is so large that the phase equilibrium conditions at the liquid-solid interface are always fulfilled, entailing the equality of chemical potentials, pressures, and temperatures. As a result, the undamped oscillations, called the *crystallization waves*, can propagate along the superfluid-crystal ^{4}He interface, representing the periodic processes of melting and crystallization.

Find the spectrum $\omega = \omega(k)$ of small amplitude crystallization waves under the following simplifying assumptions. The helium crystal is an isotropic solid being immobile. The superfluid phase is incompressible and the surface tension α between the phases is completely isotropic, being independent of the direction of the normal to the crystal surface with respect to crystallographic axes. The density of solid phase equals ρ' and that of superfluid phase is ρ. Neglect the presence of normal component and its small density in the superfluid phase. Due to weak temperature dependence the chemical potentials $\mu' = \mu'(P)$ and $\mu = \mu(P)$ of both phases can be regarded to be dependent on pressure P alone.

35. In the previous problem, it is assumed that the melting and growth of ^{4}He crystal takes place under total absence of any dissipative processes and the phase conversion rate is very large, entailing the phase equilibrium as an equality of chemical potentials μ and μ' at the interface. Strictly speaking, this is wrong. The melting and growth of solid phase are accompanied with, though small, but finite energy dissipation. This results in arising a finite difference in the chemical potentials at the liquid-solid ^{4}He interface.

If the deviation from phase equilibrium is small, the difference in chemical potentials μ and μ' can be related with the solid phase growth rate $\dot{\zeta}$ in the linear approximation according to

$$\dot{\zeta} = K(\mu - \mu').$$

Here K is the *kinetic growth coefficient*.

Using the approximations and representations of the previous problem, find the dispersion and damping of crystallization waves as a function of kinetic growth coefficient K. In ^{4}He the kinetic growth coefficient in essence depends on the temperature alone, i.e. $K = K(T)$.

Determine the energy and dissipative Rayleigh function for the superfluid-crystal ^{4}He interface, taking the dispersion equation and mechanical analogy into account. Show that the sign of kinetic growth coefficient K is positive.

36. In quantum liquids and solids, as a rule, the mean free path of elementary excitations increases drastically as the temperature lowers. The mean free path becomes so large that exceeds all the rest typical lengths in the system as, for example, size of a crystal or wavelength of crystallization waves. Under such conditions, called *ballistic* or *collisionless*, in order to describe a condensed medium, we can employ the picture of elementary excitations which can be characterized by the distribution function $n(\boldsymbol{p})$. In superfluid ^{4}He at low $T \lesssim 0.5$ K temperatures, the main type of excitations is phonons with dispersion $\varepsilon = up$, u being the sound velocity.

Find *kinetic growth coefficient* $K(T)$ of flat superfluid-solid ^{4}He interface, assuming for simplicity that the phonons from superfluid phase experience the mirror reflection with the probability of unity from the superfluid-crystal interface. Do not consider elementary excitations in the solid crystalline phase. The densities of superfluid and solid phases are ρ and ρ', respectively. Neglect a difference in the densities of superfluid component ρ_s and liquid ρ.

37. To study the kinetic properties of interface between two phases, it is necessary to disturb the phase equilibrium, for example, shifting the pressure from its equilibrium magnitude with the next observation of the interface response. For this purpose, the sound wave incident onto the interface can be used to induce the pressure modulation δP at the interface.

Find the acoustic reflection r and transmission τ coefficients determined as ratios of reflected δP_r and transmitted δP_τ pressure amplitudes to the pressure amplitude δP_0 of the incident sound wave. The sound wave of frequency ω propagates across the flat interface from the first phase to the second. The densities and sound velocities of the phase are equal to ρ_1, ρ_2, and u_1, u_2, respectively. The phase conversion rate or kinetic properties of interface can be characterized by the kinetic growth coefficient ξ connecting the mass flux J across the interface with the difference between chemical potentials $\mu_1 = \mu_1(P)$ and $\mu_2 = \mu_2(P)$ according to

$$J = \xi\left(\frac{\rho_1\rho_2}{\rho_1 - \rho_2}\right)^2(\mu_1 - \mu_2).$$

(In the general case, the kinetic growth coefficient can depend on frequency $\xi = \xi_\omega$.) For simplicity, suppose that the both phases are normal liquids and coexist at pressure $P = P_0$.

What fraction of the incident sound wave energy will be lost as a result of reflection and transmission at the interface?

2.14 Collisionless Plasma: The Permittivity and the Longitudinal and Transverse Oscillations

38. Find the space-frequency dispersion of permittivity $\varepsilon_{\alpha\beta}(\omega, \boldsymbol{k})$ in the *collisionless* isotropic plasma when the collisions between plasma particles play no significant role. Separate the permittivity $\varepsilon_{\alpha\beta}(\omega, \boldsymbol{k})$ into the *longitudinal* and *transverse* components.

When solving, use the Boltzmann transport and Maxwell equations. As a first approximation, neglect completely the motion of positive ions due to strong inequality between the ion and electron masses $M \gg m$.

39. Find the *Debye screening radius* r_D and the frequencies of longitudinal and transverse plasma oscillations. Use the results of the previous problem for the permittivity of collisionless electron plasma.

40. Find the space-frequency dispersion of permittivity for a thin metallic film, treating it as a two-dimensional layer of collisionless electron plasma. Find the dispersion $\omega = \omega(\boldsymbol{k})$ of longitudinal plasma oscillations (*plasmons*).

41. In graphene, which is a graphite single layer, the electron energy dependence on momentum $\boldsymbol{p} = (p_x, p_y)$ is described with the massless Dirac spectrum $\varepsilon_{\boldsymbol{p}} = v\boldsymbol{\sigma} \cdot \boldsymbol{p}$ where v is the electron velocity and $\boldsymbol{\sigma} = (\sigma_x, \sigma_y)$ are the Pauli matrices. The electron spectrum reads $\varepsilon_{\boldsymbol{p}} = \pm v|\boldsymbol{p}|$. The density of states $g(\varepsilon)$ is determined by the integral

$$g(\varepsilon) = \nu \int \frac{d^2 p}{(2\pi\hbar)^2} \big[\delta(\varepsilon - vp) + \delta(\varepsilon + vp)\big] = \nu \frac{|\varepsilon|}{2\pi\hbar^2 v^2},$$

where $\nu = 2 \times 2$ is the multiplicity of spectrum degeneracy in the electron spin and two electron subbands.

Using this model and the results of previous problem, find the longitudinal permittivity $\varepsilon_l(\omega, k)$ and dispersion $\omega = \omega(\boldsymbol{k})$ of longitudinal plasma oscillations (*plasmons*) in graphene.

2.15 Kinetics of First-Order Phase Transitions: Melting and Solidification

42 Find the law which the ice thickness follows in an unlimited and deep lake (Fig. 2.4) under the following freezing conditions. The air temperature T_0 above the lake is kept constant and lower than the water freezing temperature T_m. The temperature of

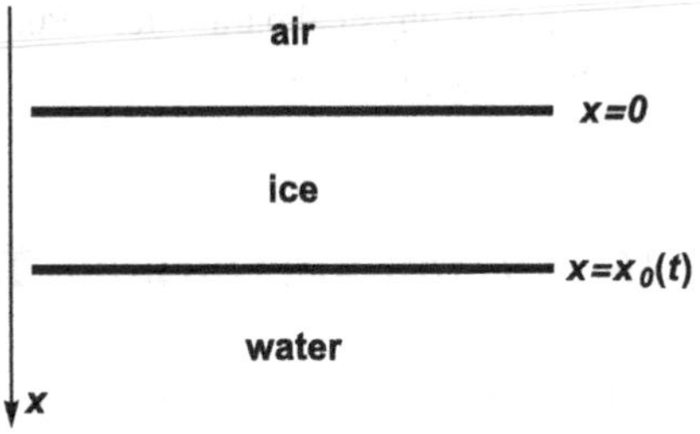

Fig. 2.4 The position of the phases and interphase boundaries

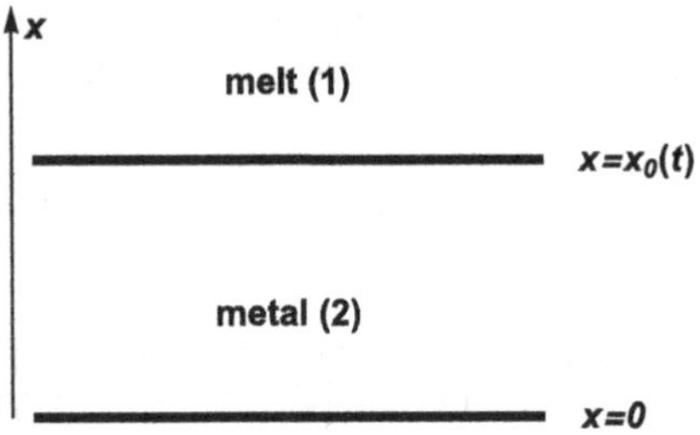

Fig. 2.5 The positions of the metal, melt, and their boundaries

water, due to its high (infinite) thermal conductivity, retains constant over all water bulk and equals the freezing temperature T_m. The specific heat capacity of ice is C_p at constant pressure, $\varkappa$ is the thermal conductivity of ice, and L is the specific melting heat of ice. Neglect any possible dependence of physical parameters on the temperature. Put the densities of ice and water to be the same and equal to ρ.

43. Find the law which the thickness of a metal follows and increases under the homogeneous crystallization in the unlimited volume of the melt (Fig. 2.5). The conditions are the following. The temperature of the furnace bottom below the melt is smaller than the crystallization temperature T_c and is kept invariable and equal to T_0. The temperature of the melt at the upper furnace side with respect to the metal boundary is also kept constant and equal to T_l. The latter temperature is not smaller than the crystallization one T_c. The specific heat capacities of the melt and metal are C_1 and C_2 at constant pressure, respectively. The specific crystallization heat of the melt equals L.

Take a difference of the melt density ρ_1 and metal density ρ_2 into account. (For definiteness, $\rho_1 < \rho_2$.) Do not consider a possible temperature dependence of physical parameters. The rate of the melt growth is proposed to be sufficiently small.

44. Find the initial rate of solidifying the liquid in the region $x > 0$ provided that the liquid remains at the crystallization temperature T_0 from the initial time moment and later on. Let us denote $\varkappa$ as a thermal conductivity of solid phase. Then, thermal diffusivity χ is the thermal conductivity divided by density and specific heat capacity at constant pressure, ρ is the solid phase density, and L is the specific latent heat of crystallization.

Solidification occurs under conditions when the heat flux supplied from the flat bottom at plane $x = 0$ is constant and equal to Q. Neglect a possible temperature dependence of physical parameters. Put the densities of liquid and solid phases to be the same and equal to ρ.

2.16 Macroscopic Quantum Tunneling

45. Let us turn to the *Caldeira-Leggett model*. The particle of mass M is in the potential $U_0(q)$ and interacts with the external medium or thermostat. The medium is an infinite set $\{\alpha\}$ of phonons, i.e. oscillators with frequencies ω_α. The Hamiltonian of medium is the energy of oscillators

$$H_{\text{m}} = \sum_\alpha \left(\frac{m\dot{x}_\alpha^2}{2} + \frac{m\omega_\alpha^2 x_\alpha^2}{2} \right)$$

and each force acting on the particle is proportional to the deviation of oscillator x_α from the equilibrium position $x_\alpha = 0$. The particle-medium coupling reads

$$U_{\text{int}} = \sum_\alpha q C_\alpha x_\alpha,$$

where C_α is the coupling constant. The spectral density of phonon oscillations is introduced according to

$$J(\Omega) = \frac{\pi}{2} \sum_\alpha \frac{C_\alpha^2}{m\omega_\alpha} \delta(\Omega - \omega_\alpha).$$

Eliminate the medium variables x_α, supposing that the phonon spectrum has an arbitrarily small frequencies and $J(\Omega) = \eta\Omega$ $(\eta > 0)$. Next, derive the effective equation of particle motion and determine the friction coefficient, random force, and random force correlator. Neglect the influence of the particle on the medium and assume that the medium is always in the thermodynamic equilibrium at temperature T.

46. The particle of mass M is in the potential $U_0(q)$ and interacts with the external medium or thermostat. The medium is an infinite set $\{\alpha\}$ of phonons, i.e. oscillators with frequencies ω_α. The Hamiltonian of medium is the energy of oscillators

$$H_{\text{m}} = \sum_\alpha \left(\frac{m\dot{x}_\alpha^2}{2} + \frac{m\omega_\alpha^2 x_\alpha^2}{2} \right).$$

The potential of particle-medium coupling, in general, is nonlinear and equals

$$U_{\text{int}} = \sum_\alpha \gamma(q) C_\alpha x_\alpha,$$

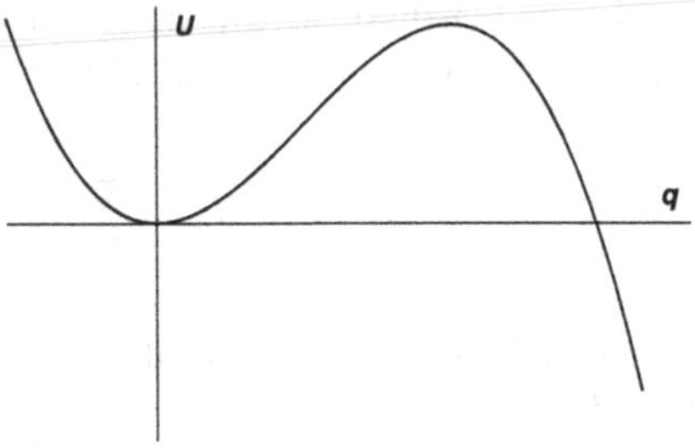

Fig. 2.6 The shape of potential

where C_α is the coupling constant. The spectral density of oscillations in the medium is determined as

$$J(\Omega) = \frac{\pi}{2}\sum_\alpha \frac{C_\alpha^2}{m\omega_\alpha}\delta(\Omega - \omega_\alpha).$$

Supposing that $J(\Omega) = \eta\Omega$ $(\eta > 0)$, eliminate the variables of the medium x_α and find the effective equation of particle motion in the medium. Determine also the renormalization of potential $U_0(q)$ and friction coefficient $\mu(q)$. This problem generalizes the previous problem to the case of nonlinear particle-medium coupling.

47. Let potential barrier separate the metastable state at the local energy minimum from the state with the lower energy (Fig. 2.6). The particle overcomes the potential barrier with the aid of quantum tunneling at low temperatures. On the contrary, the barrier is overcome with the aid of the classical and thermally activated process at high temperatures. Let potential barrier satisfy the semiclassical condition and the probability of escaping from the potential well be so small that the particle is always in the thermodynamic equilibrium with temperature T as long as the particle remains in the classically accessible region of potential.

Find the dependence of decay rate $\Gamma = \Gamma(T)$ of metastable state within the exponential accuracy. Express the answer via extremal value of *effective (Euclidean) action* S_{eff} determined for the particle motion in the *inverted potential* $\tilde{U}(q) = -U(q)$ on the imagine time axis τ. Determine the crossover temperature T_q from the quantum regime of metastable state decay to the classical regime of thermal activation. The particle mass equals M.

48. As a result of interaction between the particle and the external dissipative medium, the particle motion is subjected to the following equation:

$$M\ddot{q} + \mu(q)\dot{q} + U'(q) = f(t),$$

the friction force being proportional to the particle velocity $\dot{q}$ and the friction coefficient being $\mu(q)$. The particle is in the metastable state separated from the ground state of smaller energy with a potential barrier. The friction force affects the probability of tunneling across the potential barrier as well as the decay rate of metastable state.

The external dissipative medium can be described as an infinite set $\{\alpha\}$ of phonons, i.e. oscillators of frequencies ω_α. The corresponding Hamiltonian reads

$$H_{\mathrm{m}} = \sum_\alpha \left(\frac{m\dot{x}_\alpha^2}{2} + \frac{m\omega_\alpha^2 x_\alpha^2}{2} \right).$$

The particle of mass M is in the potential $U_0(q)$ and interacts with the medium according to

$$U_{\mathrm{int}} = \sum_\alpha \gamma(q) C_\alpha x_\alpha,$$

where C_α is the coupling constant. The spectral density of oscillations is defined as

$$J(\Omega) = \frac{\pi}{2} \sum_\alpha \frac{C_\alpha^2}{m\omega_\alpha} \delta(\Omega - \omega_\alpha) = \eta\Omega \quad (\eta > 0).$$

Let particle be in the potential well in the metastable state separated with the potential barrier from the ground state. It is known that the friction force affects a probability of tunneling across the potential barrier and, correspondingly, decay rate of metastable state. The tunneling provided that the energy dissipation exists in the physical system is called the *macroscopic quantum tunneling*.

Using the results of the previous problems and the representations about quantum tunneling as a particle motion on imaginary time τ, determine the effective action $S_{\mathrm{eff}}[q(\tau)]$ and find the decay probability $\Gamma(T)$ of metastable state in the dissipative medium within the exponential accuracy. Estimate the thermal-quantum crossover temperature T_q.

The potential barrier satisfies the semiclassical condition. The probability of escaping from the potential well is so small that the particle in the well is in the thermodynamic equilibrium with the medium at temperature T.

49. The particle of mass M is in the potential well in the shape of cubic parabola

$$U(q) = \frac{1}{2} M\omega^2 q^2 \left(1 - \frac{q}{q_c}\right).$$

The friction coefficient μ is so large that $\mu \gg M\omega$ and the friction forces are much stronger as compared with the inertia forces. The particle motion in the classically accessible region is an aperiodic damping governed by the equation

$$\mu\dot{q} + U'(q) = 0.$$

Find the decay rate $\Gamma(T)$ of metastable state as a function of temperature.

2.17 Macroscopic Quantum Nucleation

50. The first-order phase transition in a condensed medium occurs via inception of the stable phase nuclei in the metastable phase. The energetically unfavorable effect of the interphase boundary is compensated with the gain in the bulk energy for a sufficiently large nucleus. Due to thermal heterophase fluctuations the critical nucleus appears, which is able to overcome the potential barrier.

Near absolute zero of temperature the classical mechanism of thermal activation becomes inefficient. The decay of metastable condensed medium occurs due to quantum heterogeneous fluctuations. This implies the tunneling mechanism for new phase nucleus to overcome the potential barrier separating the stable state of the medium from the metastable state. The quantum phenomena which accompany such decay of metastable condensed medium can be classified as *macroscopic quantum nucleation*.

The physical systems which have first-order phase transitions down to zero temperature are, first of all, helium systems. Here we can mention the following phenomena: crystallization of overpressurized liquid helium, cavitation of gas bubbles in liquid under negative pressures, and phase separation of supersaturated liquid and solid ^{3}He-^{4}He mixtures. There exist more exotic examples as a collapse of metastable condensate in Bose gas with attraction or phase transition of deconfinement type with nucleating the quark matter from nuclear matter in the neutron star core.

So, consider the metastable liquid of density ρ with the spherical nucleus of new stable phase. The nucleus has radius R, density $\rho' = \rho + \Delta\rho$, and the following energy:

$$U(R) = 4\pi\sigma R^2 - \frac{4\pi R^3}{3}\rho'\Delta\mu = 4\pi\sigma R^2\left(1 - \frac{R}{R_c}\right).$$

Here σ is the surface tension, $\Delta\mu > 0$ is the difference of thermodynamic potentials responsible for the non-equilibrium degree of metastable phase, and $R_c \sim 1/\Delta\mu$ is the critical nucleus radius.

Determine the kinetic energy $K(R, \dot{R})$ of a nucleus, its effective mass $M(R)$, and find the decay rate of metastable phase $\Gamma(T) \sim \exp(-A)$ as a function of temperature T within the exponential accuracy.

When solving, use the approximations of macroscopic nucleus, its quasi stationary growth, and dissipationless medium.

(a) Assume that the metastable and stable phases are sufficiently close to phase equilibrium $\Delta\mu = 0$. This entails that the critical nucleus radius R_c is much larger than the interatomic distances. Thus, the critical nucleus is a macroscopic formation and contains the large number of particles.

(b) The liquid phase is incompressible $\rho = \text{const}$. This is a good approximation for the nucleus growth rate $\dot{R}(t)$ small as compared with the sound velocity.

(c) Neglect the energy dissipation processes of viscosity and heat conduction originating from the nucleus growth of new phase.

51. The previous problem deals with the phenomenon of *macroscopic quantum nucleation* in first-order phase transitions. In the example of metastable liquid decay, the simplifying approximation is an assumption about the energy dissipationless character for the nucleus growth of new stable phase. Within this approximation the total nucleus energy $E = K + U$ conserves, being a sum of kinetic $K(R, \dot{R})$ and potential $U(R)$ energies. In general, this is not so. The nucleus growth of new phase is accompanied by dissipating the nucleus energy due to existence of viscosity (internal friction) and heat conduction in the metastable medium. Viscosity or internal friction results from the inhomogeneous distribution of liquid velocity beside the nucleus. The heat conduction is due to releasing or absorbing the latent heat $L(T)$ of phase transition at the nucleus boundary.

Consider the effect of viscosity and heat conduction on the growth of spherical nucleus, assuming the approximations of macroscopic nucleus and its quasistationary (slow) nucleus growth. Find the equation of nucleus growth and determine how viscosity and heat conduction will change the behavior for decay probability $\Gamma(T)$ of metastable liquid state. Analyze two growth regimes:

(a) *Hydrodynamical* one when the critical nucleus radius R_c is much larger than the mean free path $l(T)$ of excitations in the medium.

(b) *Ballistic* one when the critical nucleus radius R_c is much smaller than the mean free path $l(T)$ of excitations. The shear viscosity and thermal conductivity are equal to $\eta(T)$ and $\varkappa(T)$, respectively.

2.18 Open Quantum Systems: The Lindblad Equation

52. The two-level system[2] is the simplest quantum system and has only two states. One is the ground state with the state vector $|g\rangle$ and the other is the excited state with the state vector $|e\rangle$. The Hamiltonian of two-level system can be expressed in terms of the Pauli matrix σ_3 as

$$\hat{H} = \frac{1}{2}\hbar\omega\sigma_3 = \begin{pmatrix} \frac{1}{2}\hbar\omega & 0 \\ 0 & -\frac{1}{2}\hbar\omega \end{pmatrix} \quad (\omega > 0).$$

Here $\hbar\omega$ is the spacing between the excited $|e\rangle$ and ground $|g\rangle$ energy levels, and ω is the transition frequency between the levels. The matrices

$$\sigma_+ = \begin{pmatrix} 0 & 1 \\ 0 & 0 \end{pmatrix} \quad \text{and} \quad \sigma_- = \begin{pmatrix} 0 & 0 \\ 1 & 0 \end{pmatrix}$$

have the following commutators with Hamiltonian $\hat{H}$:

[2] For example, an atom in which two levels interact effectively with the electric field or the 1/2 spin particle interacting with the magnetic field.

$$[\hat{H}, \sigma_+] = \hbar\omega \quad \text{and} \quad [\hat{H}, \sigma_-] = -\hbar\omega.$$

Correspondingly, operators σ_+ and σ_- change the energy of the system by the magnitude $\pm\hbar\omega$ and are responsible for excitation (photon absorption) or for relaxation to the ground state (phonon emission).

Find the time evolution of density matrix

$$\hat{\rho}(t) = \begin{pmatrix} \rho_{ee}(t) & \rho_{eg}(t) \\ \rho_{ge}(t) & \rho_{ee}(t) \end{pmatrix} \qquad (\rho_{ge} = \rho_{eg}^*)$$

for the Lindblad operator $\hat{L}_- = \sqrt{\gamma}\sigma_-$ ($\gamma > 0$). The initial population of excited level is $\rho_{ee}(0)$ at $t = 0$ and, accordingly, the ground level population equals $\rho_{gg}(0) = 1 - \rho_{ee}(0)$.

53. Analyze the time evolution of the two-level system given in the previous problem if the Lindblad operator has the following form: $\hat{L}_+ = \sqrt{\gamma}\sigma_+$ ($\gamma > 0$). The initial population of the ground level is $\rho_{gg}(0)$ at $t = 0$ and, accordingly, the excited level population equals $\rho_{ee}(0) = 1 - \rho_{gg}(0)$.

54. Consider the time evolution of the two-level system from the two previous problems for the following diagonal Lindblad operator:

$$\hat{L}_3 = \begin{pmatrix} \gamma_e & 0 \\ 0 & -\gamma_g \end{pmatrix}.$$

The initial level populations are known for time $t = 0$.

55. The two-level system is subjected to the stochastic process in which the phases of wave states fluctuate randomly in time. Let initial state vector

$$|\psi(0)\rangle = a|e\rangle + b|g\rangle \quad (|a|^2 + |b|^2 = 1)$$

cross over to the state with the vector

$$|\psi(t)\rangle = ae^{i\theta_e(t)}|e\rangle + be^{i\omega t}e^{i\theta_g(t)}|g\rangle,$$

at $t > 0$. Here the phases θ_e and θ_g are the random fluctuating quantities which obey the probabilistic distribution of diffusive character with the following probability density $P(\theta_e, \theta_g)$:

$$P(\theta_e, \theta_g)d\theta_e\, d\theta_g = \frac{d\theta_e}{\sqrt{2\pi\lambda_e t}}\frac{d\theta_g}{\sqrt{2\pi\lambda_e t}}e^{-\frac{\theta_e^2}{2\lambda_e t}}e^{-\frac{\theta_g^2}{2\lambda_g t}} \quad (-\infty < \theta_{e,g} < \infty).$$

The frequency ω is the frequency of transitions between the levels.

Determine the time behavior of density matrix $\hat{\rho}(t)$ and show that the evolution of density matrix can be described with the Lindblad-type equation.

2.19 Elements of Diagrammatic Keldysh Technique for Non-equilibrium Systems

56. One of the systematic and practical methods for deriving the quantum transport equation is to use the *Keldysh contour*. In the plane of imaginary time t the Keldysh contour, starting from point $t_0 = -\infty + i\beta$ and ending at point $t_1 = -\infty$, can be drawn as a sequence of three branches $C = C_\tau + C_- + C_+$ (Fig. 2.7). Here $\beta = 1/T$ is the inverse temperature.

The line C_τ leaves from point $t_0 = -\infty + i\beta$ and arrives at point $t = -\infty + i\delta$. From this point the line C_- runs along the real-time axis t and arrives at point $t = +\infty$. Then the contour turns in the opposite direction to the end point $t_1 = -\infty$, running under real-time axis t and constituting the branch C_+ of total contour C. The Keldysh contour is augmented with ordering in time. The direction of time at the contour is shown with the arrows in Fig. 2.7. The time moment t at the branch C_- is always earlier than the time moment t' at the branch C_+. (In this and next problems, we imply the system of units in which the Planck constant equals unity $\hbar = 1$.)

(a) For an electron of energy $\varepsilon = \varepsilon_p - \mu$, find the Green function $G(t, t')$ satisfying the following equation:

$$\left(i\frac{\partial}{\partial t} - \varepsilon\right)G(t, t') = \delta(t - t').$$

Here the time moments t and t' run along the Keldysh contour C. The Green function for electron as a particle related to the Fermi statistics obeys the *condition of antiperiodicity*

$$G(t_1\,, t) = -G(t_0\,, t).$$

(b) Consider separately the values of the Green function $G(t, t')$ when the time moments t and t' belong to the same and different branches of the Keldysh contour C.

57. The Green function $G(t, t')$ for the Bose particle of energy $\varepsilon = \varepsilon_p - \mu$ on the Keldysh contour C is governed by the same equation as for the electron Green function

$$\left(i\frac{\partial}{\partial t} - \varepsilon\right)G(t, t') = \delta(t - t').$$

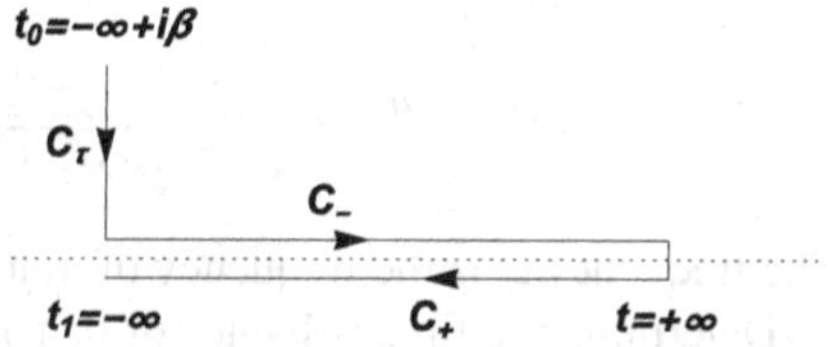

Fig. 2.7 The Keldysh contour

The time moments t and t' vary along the Keldysh contour C. The distinction of Bose statistics from the Fermi one is the *periodicity condition* alone. The following boundary condition is put for the Bose particles:

$$G(t_1, t) = +G(t_0, t).$$

Find the Green function $G(t, t')$ for a Bose particle on the Keldysh contour C and its values when the time moments t and t' refer to the same and different branches of contour C.

58. The Green contour function $\hat{G}$ in the Keldysh space can be realized in terms of the following 2×2 matrix:

$$\hat{G} = \begin{pmatrix} G^{--} & G^{-+} \\ G^{+-} & G^{++} \end{pmatrix}.$$

The matrix components are not linearly independent and connected with the retarded G^R, advanced G^A, and anomalous G^K Green functions according to the following identities:

$$G^R = G^{--} - G^{-+} = G^{+-} - G^{++},$$
$$G^A = G^{--} - G^{+-} = G^{-+} - G^{++},$$
$$G^K = G^{-+} + G^{+-} = G^{--} + G^{++}.$$

The contour Green function $\hat{G}$ can be transformed to the *triangular* or *antidiagonal form* with the aid of rotation in the Keldysh space

$$R^{-1}\hat{G}R \to \hat{G} = \begin{pmatrix} 0 & G^A \\ G^R & G^K \end{pmatrix}.$$

Augmenting the rotation R with the reflection matrix P, one can transform the above matrix to *diagonal Larkin-Ovchinnikov form*

$$R^{-1}P\hat{G}R \to \hat{G} = \begin{pmatrix} G^R & G^K \\ 0 & G^A \end{pmatrix}.$$

Find the transformation matrices R and P.

59. The phonon Green function $D_{\omega_0}(t, t')$ for the phonon dispersion $\omega_0 = \omega_0(\mathbf{k})$, $\mathbf{k}$ being wave vector, satisfies the equation

$$\left(-\frac{\partial^2}{\partial t^2} - \omega_0^2(\mathbf{k})\right)D_{\omega_0}(t, t') = \omega_0^2(\mathbf{k})\delta(t - t').$$

The time moments t and t' vary along the Keldysh C (Fig. 2.7). The phonons as spinless particles are governed with the Bose statistics, resulting in the following condition of periodicity for the Green function $D_{\omega_0}(t, t')$:

$$D_{\omega_0}(t_0, t') = D_{\omega_0}(t_1, t').$$

This condition fulfils identically for an arbitrary dependence on the phonon frequency $\omega_0 = \omega_0(\boldsymbol{k})$ and connects the values of the Green function at the initial t_0 and end t_1 points of Keldysh contour C.

Find the phonon Green function $D_{\omega_0}(t, t')$ on the Keldysh contour C.

60. In the external field of potential $U(t)$, the Green function $G(t, t')$ for an electron with energy ξ satisfies the following equation:

$$\left(i\frac{\partial}{\partial t} - \xi - U(t)\right)G(t, t') = \delta(t - t'),$$

where both time moments t and t' are the points of the Keldysh contour C (Fig. 2.7).

(a) Calculate first-order correction $\delta G(t, t')$ to the free electron Green function $G_0(t, t') = G_0(t - t')$ in the perturbing potential $U(t)$. Expand the correction $\delta G(t, t')$ into the Fourier integral over frequency ω.

(b) Determine the perturbation of electron density distribution $\delta n(t)$ under influence of potential $U(t)$, using the relation

$$n(t) = -2iG^{-+}(t, t).$$

The relation connects the distribution of particle density $n(t)$ with the Green function component $G^{-+}(t, t')$. Coefficient 2 takes the electron spin components into account.

Decompose perturbation $\delta n(t)$ into the Fourier integral over frequency ω and obtain relation $\delta n(\omega) = \chi(\omega)U(\omega)$. Express the *susceptibility* or *linear response function* $\chi(\omega)$ in terms of retarded G_0^R and advanced G_0^A Green functions, assuming the thermal equilibrium of electrons with temperature T.

Write the expression for the *Matsubara* susceptibility $\chi^M(\omega_n) = \chi(\omega = i\omega_n)$ at even Matsubara frequencies $\omega_n = 2\pi nT$ $(n = 0, \pm 1, \pm 2 \ldots)$. Use the frequency Matsubara function $\mathfrak{G}(\varepsilon_k)$ determined for fermions at the odd Matsubara frequencies $\varepsilon_k = \pi T(2k + 1)$.

Chapter 3
Solutions of Problems

3.1 The Langevin Equation and the Fluctuation-Dissipation Theorem

1. Let us represent the solution of the linear equation in the form of convolution

$$x(t) = \int\limits_{-\infty}^{t} \alpha(t-\tau) f(\tau)\, d\tau \equiv \int\limits_{-\infty}^{\infty} \alpha_R(t-\tau) f(\tau)\, d\tau, \quad \alpha_R(t) = \alpha(t)\vartheta(t).$$

Function α is called the *susceptibility* or linear response to force f. We restrict the upper limit of integration with time moment t in order to satisfy the principle of causality. The motion of a particle at time moment t can only be affected with the forces that acted in the preceding time moments $\tau \leqslant t$. Function $\alpha_R(t)$, equal identically to zero at $t < 0$, is referred as to the *retarded response*.

The Fourier transform of retarded response

$$\alpha_R(\omega) = \int_{-\infty}^{\infty} \alpha_R(t) e^{i\omega t}\, dt = \int\limits_{0}^{\infty} \alpha_R(t) e^{i\omega t}\, dt,$$

as a function of complex variable $\omega = \omega' + i\omega''$ is analytical in the upper half-plane Π^+ ($\omega'' > 0$). In fact, $\alpha_R(\omega)$ has no singularities since the integral

$$\int\limits_{0}^{\infty} \alpha_R(t) e^{i\omega' t} e^{-\omega'' t}\, dt$$

converges absolutely for $\omega'' > 0$.

S. N. Burmistrov, *Physical Kinetics*,
https://doi.org/10.1007/978-981-19-1649-6_3

Let us find α_R using the transformation to the Fourier representation

$$x(t) = \int_{-\infty}^{\infty} x(\omega)e^{-i\omega t}\frac{d\omega}{2\pi} \quad \text{and} \quad f(t) = \int_{-\infty}^{\infty} f(\omega)e^{-i\omega t}\frac{d\omega}{2\pi}.$$

We have according to the equation of motion

$$(-\omega^2 - i\gamma\omega)x(\omega) = f(\omega) \quad \text{and} \quad \alpha_R^{-1}(\omega)x(\omega) = f(\omega).$$

Hence we obtain

$$\alpha_R(\omega) = -\frac{1}{(\omega + i\delta)(\omega + i\gamma)}.$$

Here we have added the infinitely small quantity $\delta = +0$ to the factor ω in the denominator in order to provide analyticity of retarded response in the upper half-plane Π^+. Performing the inverse Fourier transform, we find

$$\alpha_R(t) = -\int_{-\infty}^{\infty} \frac{e^{-i\omega t}}{(\omega + i\delta)(\omega + i\gamma)}\frac{d\omega}{2\pi}\bigg|_{\delta=+0} = \frac{1 - e^{-\gamma t}}{\gamma}\vartheta(t).$$

Let force $f(t)$ be random quantity and consider the mean displacement of a particle or *first-order correlator*

$$\langle x(t)\rangle = \langle \int \alpha_R(t-\tau)f(\tau)\,d\tau\rangle = \int \alpha_R(t-\tau)\langle f(\tau)\rangle\,d\tau,$$

where $\langle ...\rangle$ denotes averaging over random distribution of force $f(t)$. It is usually convenient to choose $\langle f(t)\rangle = 0$ and then $\langle x(t)\rangle = 0$.

Let us turn to the *second-order correlator* of displacements $G(t_2, t_1) = \langle x(t_2)x(t_1)\rangle$

$$G(t_2, t_1) = \iint \alpha_R(t_2 - \tau_2)\alpha_R(t_1 - \tau_1)\langle f(\tau_2)f(\tau_1)\rangle\,d\tau_2\,d\tau_1$$

and set *time force correlator* $K(\tau_2, \tau_1) = \langle f(\tau_2)f(\tau_1)\rangle$. As a rule, one deals with a uniform temporal correlation $K(\tau_2, \tau_1) = K(\tau_2 - \tau_1)$ dependent on the relative time difference alone. This results in $G(t_2, t_1) = G(t_2 - t_1)$. The Fourier transform for the correlator of displacements or *spectral density* of correlator is given by the formula

$$G(\omega) \equiv \langle x(t)x(0)\rangle_\omega = \int \langle x(t)x(0)\rangle e^{i\omega t}\,dt.$$

The Fourier transform of convolution reduces to the product of Fourier transforms

$$\langle x(t)x(0)\rangle_\omega = \alpha_R(\omega)\alpha_R(-\omega)K(\omega) = |\alpha_R(\omega)|^2 K(\omega),$$

$K(\omega)$ being the Fourier transform or spectral density of the force correlator $K(\tau)$.

In the thermodynamically equilibrium medium there is unambiguous relation between the linear response function or susceptibility with the correlator of random force. The relation can be expressed with the aid of *fluctuation-dissipation theorem* according to

$$K(\omega) = \frac{\hbar\alpha_R''(\omega)}{|\alpha_R(\omega)|^2}\coth\frac{\hbar\omega}{2T} = -\mathrm{Im}\left(\frac{1}{\alpha_R}\right)\hbar\coth\frac{\hbar\omega}{2T}.$$

Here $\alpha_R''(\omega)$ is the imaginary part of response function $\alpha_R = \alpha_R' + i\alpha_R''$. This yields the following expression for the Fourier transform or spectral density of the random force correlator:

$$K(\omega) = \gamma\hbar\omega\coth\frac{\hbar\omega}{2T} = \begin{cases} 2\gamma T, & \text{classical limit, } \hbar = 0 \\ \gamma\hbar|\omega|, & \text{quantum limit, } T = 0 \end{cases}.$$

In the time representation the force correlator $K(t)$ equals

$$K(t) = \begin{cases} 2\gamma T\delta(t), & \text{classical limit, } \hbar = 0 \\ -\gamma\hbar/(\pi t^2), & \text{quantum limit, } T = 0 \end{cases}$$

and the random force proves to be uncorrelated in the classical limit and correlated in the quantum limit.

Let us consider the mean-squared displacement $R^2(t)$ determined as an average

$$\begin{aligned} R^2(t) &= \langle\big(x(t) - x(0)\big)^2\rangle = \langle x^2(t)\rangle - 2\langle x(t)x(0)\rangle + \langle x(0)x(0)\rangle \\ &= 2[\langle x(0)x(0)\rangle - \langle x(t)x(0)\rangle] = 2\big[G(0) - G(t)\big]. \end{aligned}$$

Using that

$$G(t) = \int \langle x(t)x(0)\rangle_\omega e^{-i\omega t}\frac{d\omega}{2\pi} = \int_{-\infty}^{\infty}\frac{d\omega}{2\pi}\hbar\alpha_R''(\omega)\coth\frac{\hbar\omega}{2T}\,e^{-i\omega t},$$

we find the mean-squared displacement

$$\begin{aligned} R^2(t) &= 2\int_{-\infty}^{\infty}\frac{d\omega}{2\pi}\,\frac{\hbar\gamma\omega}{\omega^2+\gamma^2}\coth\frac{\hbar\omega}{2T}\,\frac{1-e^{-i\omega t}}{\omega^2+\delta^2} \\ &= 2\int_0^{\infty}\frac{d\omega}{\pi}\,\frac{\hbar\gamma}{\omega^2+\gamma^2}\coth\frac{\hbar\omega}{2T}\,\frac{1-\cos\omega t}{\omega}. \end{aligned}$$

In the first integral we can go over to limit $\delta = 0$ since the nominator in the fraction as $\omega \to 0$ is proportional to ω^2 and point $\omega = 0$ in the denominator becomes non-singular.

Note that there are three-time parameters in the problem: current time t, mean free time $\tau_0 = \gamma^{-1}$, and inverse temperature $\hbar/T$. At first, we consider the limiting cases. Let us start from the classical limit $\hbar = 0$. Calculating the second integral

$$\int_0^\infty \frac{d\omega}{\pi} \frac{2\gamma T}{\omega^2+\gamma^2} \frac{1-\cos\omega t}{\omega^2} = \lim_{\delta\to 0} \int_{-\infty}^{\infty} \frac{d\omega}{2\pi} \frac{2\gamma T}{\omega^2+\gamma^2} \frac{1-\cos\omega t}{\omega^2+\delta^2}$$

with the aid of residue theorem and then taking the limit $\delta \to 0$, we arrive at

$$R^2_{\text{cl}}(t) = \frac{2T}{\gamma}\left(t - \frac{1-e^{-\gamma t}}{\gamma}\right) = \begin{cases} Tt^2 = v_T^2 t^2, & t \ll \tau_0 = \gamma^{-1} \\ \frac{2T}{\gamma}t = 2Dt, & t \gg \tau_0 = \gamma^{-1} \end{cases}.$$

For small time $t \ll \tau_0$, the particle motion has a form of free collisionless motion $R(t) \sim v_T t$ at the average thermal velocity v_T. For large time, the particle motion after a few of collisions becomes *diffusive* $R(t) \sim \sqrt{t}$. The diffusion coefficient D and mobility $\mu = \gamma^{-1}$ are connected with the *Einstein relation* $D = T\mu$.

Let us turn now to the quantum limit or zero temperature $T = 0$ case. The mean-squared displacement equals

$$R^2(t) = 2\int_0^\infty \frac{d\omega}{\pi} \frac{\hbar\gamma}{\omega^2+\gamma^2} \frac{1-\cos\omega t}{\omega} = \frac{2\hbar}{\pi\gamma}\int_0^\infty \frac{dx}{1+x^2}\frac{1-\cos(\gamma t x)}{x}.$$

It is readily to estimate the answer according to

$$R^2(t) \sim \frac{2\hbar}{\pi\gamma}\int_0^{(\gamma t)^{-1}} dx\, \frac{\gamma^2 t^2 x}{2(1+x^2)} + \frac{2\hbar}{\pi\gamma}\int_{(\gamma t)^{-1}}^{\infty} \frac{dx}{1+x^2}\frac{1}{x}.$$

Here we involve that the cosine function can be expanded in the region $x \lesssim (\gamma t)^{-1}$ and it gives no essential contribution to the integral in region $x \gtrsim (\gamma t)^{-1}$ due to rapid oscillations. Hence we have

$$R^2(t) \sim \frac{\hbar}{\pi\gamma}\left[\frac{\gamma^2 t^2}{2}\ln\left(1+\frac{1}{\gamma^2 t^2}\right) + \ln\left(1+\gamma^2 t^2\right)\right] \approx \begin{cases} \frac{\hbar\gamma t^2}{\pi}\ln\frac{1}{\gamma t}, & \gamma t \ll 1 \\ \frac{2\hbar}{\pi\gamma}\ln(\gamma t), & \gamma t \gg 1 \end{cases}.$$

The exact magnitude of the mean-squared displacement is given by the formula

$$R^2(t) = \frac{2\hbar}{\pi\gamma}\left(\ln(\gamma t) + \gamma_E - \frac{e^{-\gamma t}\text{Ei}\,(\gamma t) + e^{\gamma t}\text{Ei}\,(-\gamma t)}{2}\right), \quad \gamma_E = 0.577\ldots,$$

where Ei (x) is the exponential integral function defined with the Cauchy principle value

$$\mathrm{Ei}\,(x) = \mathcal{P}\int_{-\infty}^{x} \frac{e^t}{t}dt,$$

and $\gamma_E = 0.577\ldots$ is the Euler-Mascheroni constant. Let us write more exact asymptotic formulas

$$R^2(t) = \frac{2\hbar}{\pi\gamma}\begin{cases} \frac{\gamma^2 t^2}{2}\left(\ln\frac{1}{\gamma t} + \frac{3}{2} - \gamma_E\right) + \ldots, & t \ll \gamma^{-1} = \tau_0 \\ \ln(\gamma t) + \gamma_E + \frac{1}{\gamma^2 t^2} + \ldots, & t \gg \gamma^{-1} = \tau_0 \end{cases}.$$

In the quantum limit at $T = 0$ the behavior of the mean-squared displacement differs in kind from that in the classical limit. For the long time limit, the diffusive behavior is absent and the diffusion coefficient $D = 0$ in the correspondence with the Einstein relation $D = T\mu$. From the formal point of view one may imagine that the diffusion coefficient depends on time as $D(t) = \hbar\ln(\gamma t)/(\pi\gamma t) \to 0$ with $t \to \infty$.

For the short time interval $t \ll \tau_0$, if no attention is paid to the weak logarithmic dependence, we may believe in spite of zero temperature that the particle has nonzero typical velocity of the order of $v_T \approx \sqrt{\hbar\gamma/\pi}$. This results from the influence of random force.

Let us turn to analyzing the case of an arbitrary temperature $T \neq 0$. Here we have three parameters with the dimensionality of time: t, $\tau_0 = \gamma^{-1}$, and $\hbar/T$. To study the limiting behavior at $t \to 0$, it is convenient to employ the general expression for the mean-squared displacement

$$R^2(t) = 2\int_{-\infty}^{\infty} \frac{d\omega}{2\pi}\,\frac{\hbar\gamma\omega}{\omega^2+\gamma^2}\coth\frac{\hbar\omega}{2T}\,\frac{1-e^{-i\omega t}}{\omega^2+\delta^2}\Big|_{\delta\to 0}.$$

The latter can be estimated with the aid of the residue theorem

$$R^2(t) = \frac{2T}{\gamma}t - \frac{\hbar}{\gamma}\frac{1-e^{-\gamma t}}{\tan\,(\hbar\gamma/2T)} + \sum_{n=1}^{\infty}\frac{4\gamma T}{\gamma^2-\omega_n^2}\,\frac{1-e^{-\omega_n t}}{\omega_n}, \quad \omega_n = \frac{2\pi n T}{\hbar}.$$

The frequencies $\omega_n = 2\pi n T/\hbar$, n being the integer, are commonly classified as *Matsubara frequencies*.

Using the expansion of the cotangent function

$$\cot x = \frac{1}{x} + \sum_{n=1}^{\infty}\frac{2x}{x^2-\pi^2 n^2}$$

into the Laurent series at $x = \hbar\gamma/2T$, we can represent $R^2(t)$ as

$$R^2(t) = R^2_{\rm cl}(t) - 4T\sum_{n=1}^{\infty}\left(\frac{1-e^{-\gamma t}}{\gamma^2-\omega_n^2} - \frac{\gamma}{\omega_n}\frac{1-e^{-\omega_n t}}{\gamma^2-\omega_n^2}\right).$$

Here $R^2_{\rm cl}(t)$ implies the displacement in the classical limit. Hence, one can see that nonzero temperature in the long time limit

$$t \gg \begin{cases} \frac{1}{\gamma}, & 2\pi T \gg \hbar\gamma \\ \frac{\hbar}{2\pi T}\ln\frac{\hbar\gamma}{2\pi T}, & 2\pi T \ll \hbar\gamma \end{cases} \text{(or approximately)} \sim \max\left\{\frac{1}{\gamma}, \frac{\hbar}{2\pi T}\right\}$$

leads however to breaking the quantum correlations down. The diffusive behavior of a particle recovers with conserving the Einstein relation between diffusion coefficient and mobility $D = T\mu$. In other words, the broadening of Brownian particle energy, as a result of both collisions and temperature is essential for developing the diffusive motion of a particle in the thermodynamically equilibrium medium.

Let us write the answer for the mean-squared displacement $R^2(t)$ at short times

$$R^2(t) \sim \frac{\hbar\gamma}{\pi}t^2 \ln\frac{1}{t\max\{\gamma, 2\pi T/\hbar\}}, \qquad t \ll \min\left\{\frac{1}{\gamma}, \frac{\hbar}{2\pi T}\right\}.$$

2. The Fourier transform of linear response or susceptibility is equal to

$$\alpha(\omega) = \alpha'(\omega) + i\alpha''(\omega) = -\frac{1}{m(\omega^2-\omega_0^2+i\gamma\omega)}.$$

Using the fluctuation-dissipation theorem, we find the Fourier transform of correlator

$$\langle x(t)x(0)\rangle_\omega = \alpha''(\omega)\hbar\coth\frac{\hbar\omega}{2T} = \frac{1}{m}\frac{\gamma\hbar\omega}{(\omega^2-\omega_0^2)^2+\gamma^2\omega^2}\coth\frac{\hbar\omega}{2T}.$$

Then the mean square of the particle position in the harmonic trap is given by the integral

$$\langle x^2\rangle = \frac{1}{m}\int_{-\infty}^{\infty}\frac{d\omega}{2\pi}\frac{\gamma\hbar\omega}{(\omega^2-\omega_0^2)^2+\gamma^2\omega^2}\coth\frac{\hbar\omega}{2T}.$$

The case of thermodynamically equilibrium and frictionless oscillator, i.e. $\gamma = 0$, is trivial. At $\gamma \to 0$ the close vicinity of points $\omega = \pm\omega_0$ contributes only to the integral. This allows us to simplify the integrand and arrive at the obvious answer well-known from the courses of statistical physics

$$\langle x^2\rangle = \frac{1}{m}\int_{-\infty}^{\infty}\frac{d\omega}{2\pi}\frac{\gamma\hbar\omega_0\coth(\hbar\omega_0/2T)}{4\omega_0^2(\omega-(\pm\omega_0))^2+\gamma^2\omega_0^2} = \frac{\hbar}{2m\omega_0}\coth\frac{\hbar\omega_0}{2T}.$$

Let us turn now to other limiting cases. We start from the classical limit $\hbar = 0$. Then the integral is readily calculated and we find the following expression for the mean square of particle position:

$$\langle x^2 \rangle = \frac{1}{m} \int_{-\infty}^{\infty} \frac{d\omega}{2\pi} \frac{2\gamma T}{(\omega^2 - \omega_0^2)^2 + \gamma^2 \omega^2} = \frac{T}{m\omega_0^2} = T\alpha(0).$$

The answer is independent of friction coefficient $\eta = m\gamma$ and can be expressed via the magnitude of susceptibility $\alpha(0)$ at zero frequency. One can also say that the mean potential energy of particle $m\omega_0^2 \langle x^2 \rangle /2$ equals $T/2$.

In the quantum limit of zero temperature $T = 0$ the mean square of particle position has nonzero magnitude and equals the integral

$$\langle x^2 \rangle = \frac{1}{\pi m} \int_{0}^{\infty} \frac{\gamma \hbar \omega \, d\omega}{(\omega^2 - \omega_0^2)^2 + \gamma^2 \omega^2} = \frac{\hbar \gamma}{2\pi m \omega_0^2} \int_0^{\infty} \frac{dx}{(x-1)^2 + x\gamma^2/\omega_0^2}.$$

Calculating the integral yields

$$\langle x^2 \rangle = \frac{\hbar}{\pi m \omega_0} \frac{1}{\sqrt{|a-4|}} \begin{cases} \frac{\pi}{2} - \arcsin \frac{a-2}{2}, & a \leqslant 4, \\ \ln \left[\frac{a-2}{2} + \sqrt{\frac{(a-2)^2}{4} - 1} \right], & a \geqslant 4, \end{cases} \quad a = (\gamma/\omega_0)^2.$$

The mean square of particle position $\langle x^2 \rangle$ decreases gradually with increasing the energy dissipation in the system, i.e. with the growth of coefficient γ (left Fig. 3.1). Let us give the limiting expressions for $\langle x^2 \rangle$ and its magnitude at $\gamma = 2\omega_0$:

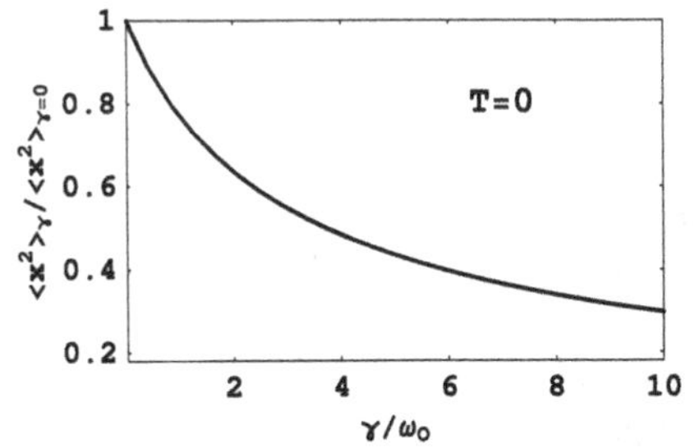

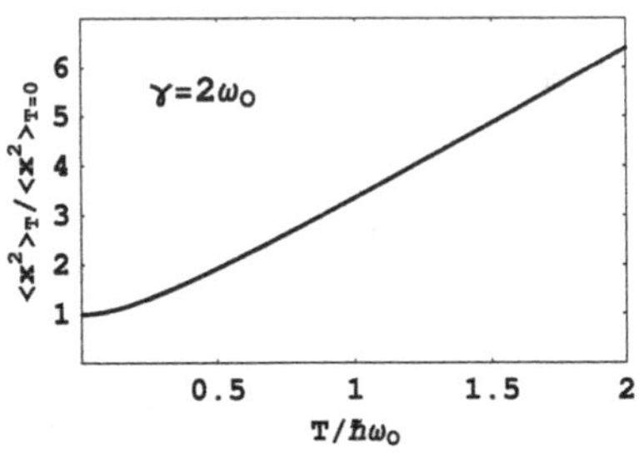

Fig. 3.1 Left: The decrease of the mean square of particle position as a result of the growth of friction coefficient at zero temperature. Right: The increase of the mean square of particle position with the temperature growth

$$\langle x^2\rangle = \begin{cases} \frac{\hbar}{2m\omega_0}\left(1-\frac{\gamma}{\pi\omega_0}+\ldots\right), & \gamma \ll 2\omega_0, \\ \hbar/(\pi m\omega_0), & \gamma = 2\omega_0, \\ \frac{2\hbar}{\pi m\gamma}\ln(\gamma/\omega_0)+\ldots, & \gamma \gg 2\omega_0. \end{cases}$$

In the lack of energy dissipation $\gamma = 0$ we have an ordinary quantum-mechanical answer $m\omega_0^2\langle x^2\rangle/2 = \hbar\omega_0/4$ for the average potential energy. For the strongly dissipative and aperiodic $\gamma \gg \omega_0$ character of motion, the average potential energy is much less and equals $\sim (\hbar\omega_0^2/\gamma)\ln(\gamma/\omega_0)$.

For high temperatures, the correction to the classical value can readily be obtained from the expansion

$$\coth x = 1/x + x/3 + \ldots \quad \text{for } x \ll 1.$$

In fact, assuming $T \gg \hbar\omega_0$, we can find

$$\begin{aligned} \langle x^2\rangle &\approx \frac{1}{m}\int_{-\infty}^{\infty}\frac{d\omega}{2\pi}\,\frac{\gamma\hbar\omega}{(\omega^2-\omega_0^2)^2+\gamma^2\omega^2}\left(\frac{2T}{\hbar\omega}+\frac{1}{3}\frac{\hbar\omega}{2T}\right) \\ &= \frac{T}{m\omega_0^2}\left(1+\frac{\hbar^2\omega_0^2}{12T^2}\right). \end{aligned}$$

The next terms of expansion in the inverse temperature will depend on coefficient γ as well.

The low temperature expansion can be found with integrating the initial expression for $\langle x^2\rangle$ by parts. So, introducing the function

$$F(\omega) = \frac{1}{m}\int_0^{\omega}\frac{d\omega}{\pi}\,\frac{\gamma\hbar\omega}{(\omega^2-\omega_0^2)^2+\gamma^2\omega^2}$$

and replacing $x = \hbar\omega/2T$, we have

$$\begin{aligned} \langle x^2\rangle &= F(\omega)\coth\frac{\hbar\omega}{2T}\Big|_0^{\infty} + \int_0^{\infty}\frac{F(\omega)}{\sinh^2(\hbar\omega/2T)}\,d\Big(\frac{\hbar\omega}{2T}\Big) \\ &= \langle x^2\rangle_{T=0} + \int_0^{\infty}\frac{F(2Tx/\hbar)}{\sinh^2 x}\,dx. \end{aligned}$$

Expanding the integrand F in the Taylor series and taking into account that function $F(0)$ and its odd derivatives vanishes at zero, we arrive at the following two first terms of low temperature $T \ll \hbar\omega_0$ expansion:

$$
\begin{aligned}
\langle x^2 \rangle - \langle x^2 \rangle_{T=0} &= \left(\frac{2T}{\hbar}\right)^2 \frac{F''(0)}{2} \int_0^\infty \frac{x^2\,dx}{\sinh^2 x} + \left(\frac{2T}{\hbar}\right)^4 \frac{F''''(0)}{24} \int_0^\infty \frac{x^4\,dx}{\sinh^2 x} \\
&= \left(\frac{2\pi T}{\hbar}\right)^2 \frac{F''(0)}{12} + \left(\frac{2\pi T}{\hbar}\right)^4 \frac{F''''(0)}{720} \\
&= \frac{2\hbar\gamma}{3\pi m\omega_0^2} \left(\frac{\pi T}{\hbar\omega_0}\right)^2 \left(1 + \frac{2\omega_0^2 - \gamma^2}{5\omega_0^2} \left(\frac{\pi T}{\hbar\omega_0}\right)^2\right).
\end{aligned}
$$

As an example, in the right-hand Fig. 3.1 the temperature behavior is shown for the mean square of particle position $\langle x^2 \rangle$ at $\gamma = 2\omega_0$.

3.2 The Boltzmann Transport Equation in the τ-Approximation: The Thermoelectric Effects in Conductors

3. Let f_p be distribution function of electrons with the fixed spin projection. The scattering processes vary the number of electrons with momentum $\boldsymbol{p}$. Due to the Pauli principle the scattering of electron can occur only provided that the initial state with momentum $\boldsymbol{p}$ is occupied by the electron and the final state with momentum $\boldsymbol{p}'$ is free. Thus the scattering probability $W_{\boldsymbol{p}\,\boldsymbol{p}'}$ or probability of electron transition from state $\boldsymbol{p}'$ to the state $\boldsymbol{p}$ should be written for the injecting term in the collision integral St $[f]$ as follows:

$$
(1 - f_{\boldsymbol{p}})W_{\boldsymbol{p}\,\boldsymbol{p}'} f_{\boldsymbol{p}'}\,.
$$

As it concerns the ejecting term, the probability $W_{\boldsymbol{p}'\boldsymbol{p}}$ of the scattering from the state $\boldsymbol{p}$ to state $\boldsymbol{p}'$ should be written as

$$
(1 - f_{\boldsymbol{p}'})W_{\boldsymbol{p}'\boldsymbol{p}} f_{\boldsymbol{p}}\,.
$$

Summing the probabilities over an arbitrary momentum $\boldsymbol{p}'$ and volume V, we obtain the collision integral as a difference of the injecting and ejecting terms

$$
\mathrm{St}\,[f] = \int \left[(1 - f_{\boldsymbol{p}})W_{\boldsymbol{p}\,\boldsymbol{p}'} f_{\boldsymbol{p}'} - (1 - f_{\boldsymbol{p}'})W_{\boldsymbol{p}'\boldsymbol{p}} f_{\boldsymbol{p}}\right] \frac{V\,d^3p'}{(2\pi\hbar)^3}\,.
$$

Let $u(\boldsymbol{r})$ be interaction potential of electron with a separate impurity. Then we can represent the interaction potential $U(\boldsymbol{r})$ of electron with the total amount of impurities located at points $\boldsymbol{R}_a$ as a following sum:

$$
U(\boldsymbol{r}) = \sum_a u(\boldsymbol{r} - \boldsymbol{R}_a)\,.
$$

In the Born approximation the elastic transition probability reduced to both unit time and unit volume is given by the formula

$$W_{p'p} = \frac{2\pi}{\hbar}|U_{p'p}|^2\delta(\varepsilon_{p'} - \varepsilon_p).$$

The matrix element $U_{p'p}$ in the wave functions of electron free motion with momentum $\boldsymbol{p}$

$$\psi_p(\boldsymbol{r}) = \frac{e^{i\boldsymbol{pr}/\hbar}}{\sqrt{V}},$$

factor $V^{-1/2}$ implying the normalization per one electron in volume V, has the following form:

$$U_{p'p} = \int \psi^*_{p'}(\boldsymbol{r})U(\boldsymbol{r})\psi_p(\boldsymbol{r})\, d^3r = \frac{u_{p'p}}{V}\sum_a e^{-i(\boldsymbol{p}'-\boldsymbol{p})\boldsymbol{R}_a/\hbar}.$$

Here $u_{p'p}$ represents the Fourier transform of interaction potential $u(\boldsymbol{r})$, depending on the momentum difference $\boldsymbol{p}' - \boldsymbol{p}$

$$u_{p'p} = u(\boldsymbol{p}' - \boldsymbol{p}) = \int u(\boldsymbol{r})e^{-i(\boldsymbol{p}'-\boldsymbol{p})\boldsymbol{r}/\hbar}\, d^3r.$$

Next, calculating the square of modulus $|U_{p'p}|^2$, we take into account that impurities are uniformly and randomly distributed over the bulk. Accordingly, we separate the double sum over impurity sites into two parts. The first contains the same sites and the second does the different sites, i.e.

$$|U_{p'p}|^2 = \frac{|u_{p'p}|^2}{V^2}\left(\sum_{a=b}1 + \sum_{a\neq b} e^{-i(\boldsymbol{p}'-\boldsymbol{p})(\boldsymbol{R}_a-\boldsymbol{R}_b)/\hbar}\right).$$

The second sum consists of the macroscopically large number of oscillating sign-alternating functions with the random and nonzero argument $(\boldsymbol{R}_a - \boldsymbol{R}_b)$. After averaging out, this sum vanishes, not contributing to the final answer. The first sum over the same impurity sites yields the total number of impurities $N_i = n_i V$. The result of averaging over impurity sites reduces to

$$|U_{p'p}|^2_{\text{av}} = \frac{N_i}{V^2}|u_{p'p}|^2 = \frac{N_i}{V^2}|u_{p'-p}|^2.$$

Accordingly, the scattering probability or disorder scattering matrix element $W_{pp'}$ in the collision integral reads

$$W^{(\text{av})}_{pp'} = \frac{2\pi}{\hbar}n_i\frac{|u_{p'-p}|^2}{V}\delta(\varepsilon_{p'} - \varepsilon_p).$$

For the centrally symmetrical interaction potential $u(\boldsymbol{r}) = u(-\boldsymbol{r})$, it is obvious that

$$W^{(av)}_{\boldsymbol{p}\,\boldsymbol{p}'} = W^{(av)}_{\boldsymbol{p}'\boldsymbol{p}}\,.$$

The collision integral simplifies

$$\begin{aligned}\mathrm{St}\,[f] &= \int W^{(av)}_{\boldsymbol{p}\,\boldsymbol{p}'}(f_{\boldsymbol{p}'} - f_{\boldsymbol{p}})\frac{V d^3 p'}{(2\pi\hbar)^3} \\ &= \frac{2\pi}{\hbar}n_i \int |u_{\boldsymbol{p}'-\boldsymbol{p}}|^2 \delta(\varepsilon_{\boldsymbol{p}'} - \varepsilon_{\boldsymbol{p}})(f_{\boldsymbol{p}'} - f_{\boldsymbol{p}})\frac{d^3 p'}{(2\pi\hbar)^3}\,.\end{aligned}$$

Let us write the transport Boltzmann equation in the homogeneous constant field $\boldsymbol{E}$

$$e\boldsymbol{E}\cdot\frac{\partial f_{\boldsymbol{p}}}{\partial \boldsymbol{p}} = \mathrm{St}\,[f]$$

which we solve in the linear approximation $f(\boldsymbol{p}) = f_0(\varepsilon_{\boldsymbol{p}}) + f_1(\boldsymbol{p})$ in the electric field $\boldsymbol{E}$. Then we have

$$e\boldsymbol{E}\cdot\frac{\partial f_0(\varepsilon_{\boldsymbol{p}})}{\partial \boldsymbol{p}} = \frac{2\pi}{\hbar}n_i \int |u_{\boldsymbol{p}'-\boldsymbol{p}}|^2 \delta(\varepsilon_{\boldsymbol{p}'} - \varepsilon_{\boldsymbol{p}})\big[f_1(\boldsymbol{p}') - f_1(\boldsymbol{p})\big]\frac{d^3 p'}{(2\pi\hbar)^3}\,.$$

For the equilibrium distribution function $f_0(\varepsilon_{\boldsymbol{p}})$, it is evident that the collision integral vanishes $\mathrm{St}\,[f_0] = 0$ due to equality

$$\delta(\varepsilon_{\boldsymbol{p}'} - \varepsilon_{\boldsymbol{p}})\big(f_0(\varepsilon_{\boldsymbol{p}'}) - f_0(\varepsilon_{\boldsymbol{p}})\big) = 0.$$

Taking into account that $\partial f_0/\partial \boldsymbol{p} = \boldsymbol{v}\partial f_0/\partial\varepsilon$, where $\boldsymbol{v} = \partial\varepsilon/\partial\boldsymbol{p}$ is the electron velocity, we seek for the transport equation as

$$f_1(\boldsymbol{p}) = \boldsymbol{v}\boldsymbol{E}\chi(\varepsilon).$$

Then we have after substitution

$$\begin{aligned} e\boldsymbol{v}\boldsymbol{E}\frac{\partial f_0}{\partial\varepsilon} &= \frac{2\pi n_i}{\hbar}\int |u_{\boldsymbol{p}'-\boldsymbol{p}}|^2\delta(\varepsilon_{\boldsymbol{p}'} - \varepsilon_{\boldsymbol{p}})\big[\boldsymbol{v}'\boldsymbol{E}\chi(\varepsilon') - \boldsymbol{v}\boldsymbol{E}\chi(\varepsilon)\big]\frac{p'^2\,dp'\,d\Omega}{(2\pi\hbar)^3} \\ &= \frac{\pi n_i}{\hbar}\int |u(\vartheta)|^2\delta(\varepsilon' - \varepsilon)\big[\boldsymbol{v}'\boldsymbol{E}\chi(\varepsilon') - \boldsymbol{v}\boldsymbol{E}\chi(\varepsilon)\big]\nu(\varepsilon')\,d\varepsilon'\frac{d\Omega}{4\pi} \\ &= \frac{\pi n_i}{\hbar}\nu(\varepsilon)\chi(\varepsilon)\int |u(\vartheta)|^2\big[\boldsymbol{v}'\boldsymbol{E} - \boldsymbol{v}\boldsymbol{E}\big]\frac{d\Omega}{4\pi}\,, \quad \nu(\varepsilon) = \frac{mp(\varepsilon)}{\pi^2\hbar^3} = \frac{m\sqrt{2m\varepsilon}}{\pi^2\hbar^3}\,.\end{aligned}$$

Here $\nu(\varepsilon)$ is the density of states. For brevity, we denote $u(\vartheta) \equiv u(2p\sin(\vartheta/2))$, where ϑ is the angle between the vectors $\boldsymbol{p}'$ and $\boldsymbol{p}$. In addition, we recall also that for elastic scattering, $|\boldsymbol{p}'| = |\boldsymbol{p}|$.

In order to integrate over solid angle Ω, we direct the polar axis along vector $\boldsymbol{p}$ and then introduce the spherical angles ϑ and φ determining the direction of vectors $\boldsymbol{p}'$ and the spherical angles ϑ_1 and φ_1 determining the direction of electric field $\boldsymbol{E}$. Then we have for angle α between two directions of vectors $\boldsymbol{p}'$ and $\boldsymbol{E}$

$$\cos\alpha = \cos\vartheta\cos\vartheta_1 + \sin\vartheta\sin\vartheta_1\cos(\varphi - \varphi_1).$$

Substituting this value for a dot product yields

$$evE\frac{\partial f_0}{\partial\varepsilon}\cos\vartheta_1 = \frac{\pi n_i}{\hbar}\nu(\varepsilon)\chi(\varepsilon)\frac{1}{4\pi}\int\limits_0^{\pi}\sin\vartheta\, d\vartheta\int\limits_0^{2\pi}d\varphi|u(\vartheta)|^2 vE$$
$$\times\left[\cos\vartheta\cos\vartheta_1 + \sin\vartheta\sin\vartheta_1\cos(\varphi-\varphi_1) - \cos\theta_1\right].$$

After integrating over angle φ, the middle term $\cos(\varphi - \varphi_1)$ gives no contribution to the answer. Reducing the common factor $evE\cos\vartheta_1$ on the left- and right-hand sides of equation, we find $\chi(\varepsilon)$ and $f_1(\boldsymbol{p})$

$$\chi(\varepsilon) = -e\tau_{tr}(\varepsilon)\frac{\partial f_0(\varepsilon)}{\partial\epsilon}, \quad f_1(\boldsymbol{p}) = e(\boldsymbol{v}\boldsymbol{E})\tau_{tr}(\varepsilon)\left(-\frac{\partial f_0(\varepsilon)}{\partial\epsilon}\right).$$

Here we have introduced the transport scattering time τ_{tr}, defining it according to the relation

$$\frac{1}{\tau_{tr}(\varepsilon)} = \frac{\pi n_i}{\hbar}\nu(\varepsilon)\int|u(\vartheta)|^2(1-\cos\theta)\frac{d\Omega}{4\pi}.$$

The collision integral can be represented as

$$\mathrm{St}[f] = -\frac{f - f_0}{\tau_{tr}}.$$

Such representation is referred to as the relaxation time (or *tau*) approximation. An ordinary collision time is defined with the following formula:

$$\frac{1}{\tau(\varepsilon)} = \frac{\pi n_i}{\hbar}\nu(\varepsilon)\int|u(\vartheta)|^2\frac{d\Omega}{4\pi}.$$

To calculate electric conductivity σ, we write the expression for the current

$$\boldsymbol{j} = \int e\boldsymbol{v}f(\boldsymbol{p})\frac{2d^3p}{(2\pi\hbar)^3} = \int e\boldsymbol{v}f_0(\boldsymbol{p})\frac{2d^3p}{(2\pi\hbar)^3} + \int e\boldsymbol{v}f_1(\boldsymbol{p})\frac{2d^3p}{(2\pi\hbar)^3}.$$

The first integral is independent of electric field and, obviously, vanishes. For the second integral, we find

$$\boldsymbol{j} = e^2\int \boldsymbol{v}(\boldsymbol{v}\boldsymbol{E})\tau_{tr}(\varepsilon)\Big(-\frac{\partial f_0}{\partial \varepsilon}\Big)\frac{2d^3p}{(2\pi\hbar)^3} = \boldsymbol{E}\frac{e^2}{3}\int v^2\tau_{tr}(\varepsilon)\Big(-\frac{\partial f_0}{\partial \varepsilon}\Big)\frac{2d^3p}{(2\pi\hbar)^3}.$$

Determining the conductivity σ with relation $\boldsymbol{j} = \sigma\boldsymbol{E}$ and involving that $\varepsilon = mv^2/2$, we have

$$\sigma = \frac{2e^2}{3m}\int \varepsilon\tau_{tr}(\varepsilon)\Big(-\frac{\partial f_0}{\partial \varepsilon}\Big)\frac{2d^3p}{(2\pi\hbar)^3} \sim \frac{1}{n_i}.$$

Accordingly, the impurity resistivity σ^{-1} is proportional to the concentration of impurities.

If time $\tau_{tr}(\varepsilon)$ is energy independent, it can be drawn out the integral sign. Then, for simplifying the answer, we use the relation

$$\int \varepsilon\Big(-\frac{\partial f_0}{\partial \varepsilon}\Big)\frac{2d^3p}{(2\pi\hbar)^3} = \frac{3}{2}\int f_0(\varepsilon)\frac{2d^3p}{(2\pi\hbar)^3} = \frac{3}{2}n$$

which is derived by integrating parts. As a result, the conductivity σ and resistivity ρ are given with the formulas

$$\sigma = \frac{ne^2}{m}\tau_{tr}\,, \qquad \rho = \sigma^{-1} = \frac{m}{ne^2}\frac{1}{\tau_{tr}} \sim n_i,$$

n being the electron concentration. Such form of the answer is also conserved at zero temperature since $f_0'(\varepsilon) = -\delta(\varepsilon - \varepsilon_F)$ in this case. The time $\tau_{tr} = \tau_{tr}(\varepsilon_F)$ can also be drawn out the integral sign, the energy ε being equal to the Fermi energy. In any case the impurity resistivity will be proportional to the concentration of impurities.

(a) Fourier transform $u(q)$ of point-like potential $u(\boldsymbol{r}) = u_0\delta(\boldsymbol{r})$ is independent of momentum, i.e. $u(q) = u_0$. The transport scattering time and ordinary scattering time are simply the same

$$\tau_{tr}^{-1} = \tau^{-1} = \pi n_i\nu(\varepsilon)|u_0|^2/\hbar.$$

(b) Fourier transform $u(q)$ of screened Coulomb potential equals

$$u(q) = \frac{4\pi Ze^2}{(q/\hbar)^2 + \varkappa^2} = \frac{4\pi Ze^2\hbar^2}{q^2 + (\hbar\varkappa)^2}\,.$$

For the next integration over angles, it is convenient to introduce the variable $q = 2p\sin\theta/2$ and, correspondingly, $d\Omega/4\pi = q\,dq/(2p^2)$. As a result, we have

$$\frac{\hbar}{\tau} = \pi n_i \nu \int\limits_0^{2p} \frac{q\,dq}{2p^2}\left(\frac{4\pi Ze^2\hbar^2}{q^2+(\hbar\varkappa)^2}\right)^2,$$

$$\frac{\hbar}{\tau_{tr}} = \pi n_i \nu \int\limits_0^{2p} \frac{q\,dq}{2p^2}\,\frac{q^2}{2p^2}\left(\frac{4\pi Ze^2\hbar^2}{q^2+(\hbar\varkappa)^2}\right)^2.$$

Then, substituting $x = (q/2p)^2$ gives

$$\frac{\hbar}{\tau} = \pi n_i \nu \left(\frac{\pi Ze^2\hbar^2}{p^2}\right)^2 \int\limits_0^1 \frac{dx}{(x+\hbar^2\varkappa^2/4p^2)^2} = \pi n_i \nu \frac{(4\pi Ze^2\hbar^2)^2}{\varkappa^2(\hbar^2\varkappa^2+4p^2)},$$

$$\frac{\hbar}{\tau_{tr}} = \pi n_i \nu \left(\frac{\pi Ze^2\hbar^2}{p^2}\right)^2 \int\limits_0^1 \frac{2x\,dx}{(x+\hbar^2\varkappa^2/4p^2)^2} = 2\pi n_i \nu \left(\frac{\pi Ze^2\hbar^2}{p^2}\right)^2 \times \left[\ln\frac{\hbar^2\varkappa^2+4p^2}{\hbar^2\varkappa^2} - \frac{4p^2}{\hbar^2\varkappa^2+4p^2}\right].$$

The ratio of transport time to the ordinary time equals

$$\frac{\tau}{\tau_{tr}} = 2a\left((1+a)\ln\frac{1+a}{a} - 1\right) \approx \begin{cases} 1, & \hbar\varkappa \gg p \\ (\hbar\varkappa/p)^2\ln(2p/\hbar\varkappa), & \hbar\varkappa \ll p \end{cases}.$$

Here magnitude a equals $a = (\hbar\varkappa/2p)^2$. At $\hbar\varkappa \ll p$ or when the typical radius of the potential influence exceeds noticeably the impact parameter $\varkappa^{-1} \gg \hbar/p$, the transport time is significantly larger as compared with the ordinary scattering time $\tau_{tr} \gg \tau$. From the physical point of view this property is associated with domination of the small-angular scattering $\theta \sim \hbar\varkappa/p \ll 1$ and, correspondingly, with the noticeably smaller contribution to $1/\tau_{tr}$ due to inequality $1 - \cos\theta \approx \theta^2/2 \ll 1$.

4. Let conductor be placed in the constant electric field $\boldsymbol{E}$ and constant temperature gradient ∇T. Let us write the Boltzmann equation for the stationary distribution function $f = f(\boldsymbol{r}, \boldsymbol{p})$ in the τ-approximation

$$\boldsymbol{v}\frac{\partial f}{\partial \boldsymbol{r}} + e\boldsymbol{E}\frac{\partial f}{\partial \boldsymbol{p}} = -\frac{f-f_0}{\tau(\varepsilon)}.$$

Here $f_0(\varepsilon)$ is the equilibrium distribution function with the following properties:

$$f_0(\varepsilon) = f_0\left(\frac{\varepsilon-\mu}{T}\right), \quad \frac{\partial f_0}{\partial \mu} = -\frac{\partial f_0}{\partial \varepsilon}, \quad \frac{\partial f_0}{\partial T} = -\frac{\varepsilon-\mu}{T}\frac{\partial f_0}{\partial \varepsilon}, \quad \varepsilon = \frac{p^2}{2m},$$

where $\mu = \mu(T)$ is the chemical potential depending on temperature. In the linear approximation in $\boldsymbol{E}$ and ∇T we have the equation

$$\boldsymbol{v}\frac{\partial f_0}{\partial \boldsymbol{r}} + e\boldsymbol{E}\frac{\partial f_0}{\partial \boldsymbol{p}} = -\frac{f - f_0}{\tau(\varepsilon)}$$

for finding the distribution function f. Next, the variation of distribution function reads $\delta f = f - f_0$

$$\delta f = e(\boldsymbol{v}\boldsymbol{E}')\tau(-\frac{\partial f_0}{\partial \varepsilon}) - (\boldsymbol{v}\nabla T)\frac{\varepsilon - \mu}{T}\tau(-\frac{\partial f_0}{\partial \varepsilon}), \quad \boldsymbol{E}' = \boldsymbol{E} - \frac{\nabla \mu}{e}.$$

For brevity, we have introduced the field $\boldsymbol{E}'$ as a magnitude of deviation from the equilibrium condition in the external electric potential $\Phi(\boldsymbol{r})$

$$e\Phi(\boldsymbol{r}) + \mu = \text{const.}$$

Let us define electric current density $\boldsymbol{j}$ and dissipative heat flow $\boldsymbol{q}$ according to the following integrals:

$$\boldsymbol{j} = \int e\boldsymbol{v} f \, d\Gamma_p \quad \text{and} \quad \boldsymbol{q} = \int \boldsymbol{v}(\varepsilon - \mu) f \, d\Gamma_p,$$

where $d\Gamma_p = 2d^3p/(2\pi\hbar)^3$ is the volume element in the p-space with taking two spin projection into account. Due to isotropy of electron spectrum the equilibrium part of distribution functions gives no contribution to the current and heat flow. Thus we have

$$\boldsymbol{j} = \int e^2\boldsymbol{v}(\boldsymbol{v}\boldsymbol{E}')\tau(-\frac{\partial f_0}{\partial \varepsilon})\, d\Gamma_p - \int e\boldsymbol{v}(\boldsymbol{v}\nabla T)\frac{\varepsilon - \mu}{T}\tau(-\frac{\partial f_0}{\partial \varepsilon})\, d\Gamma_p,$$

$$\boldsymbol{q} = \int e\boldsymbol{v}(\boldsymbol{v}\boldsymbol{E}')(\varepsilon - \mu)\tau(-\frac{\partial f_0}{\partial \varepsilon})\, d\Gamma_p - \int \boldsymbol{v}(\boldsymbol{v}\nabla T)\frac{(\varepsilon - \mu)^2}{T}\tau(-\frac{\partial f_0}{\partial \varepsilon})\, d\Gamma_p.$$

The isotropy of electron spectrum allows us to simplify integration over angles according to

$$\int \boldsymbol{v}(\boldsymbol{v}\boldsymbol{A})\, d\Gamma_p = \frac{\boldsymbol{A}}{3}\int v^2\, d\Gamma_p,$$

and to obtain the following relations:

$$\boldsymbol{j} = \frac{\boldsymbol{E}'}{3}\int e^2v^2\tau(-\frac{\partial f_0}{\partial \varepsilon})\, d\Gamma_p - \frac{\nabla T}{3}\int ev^2\frac{\varepsilon - \mu}{T}\tau(-\frac{\partial f_0}{\partial \varepsilon})\, d\Gamma_p,$$

$$\boldsymbol{q} = \frac{\boldsymbol{E}'}{3}\int ev^2(\varepsilon - \mu)\tau(-\frac{\partial f_0}{\partial \varepsilon})\, d\Gamma_p - \frac{\nabla T}{3}\int v^2\frac{(\varepsilon - \mu)^2}{T}\tau(-\frac{\partial f_0}{\partial \varepsilon})\, d\Gamma_p.$$

With the aid of *kinetic coefficients* these equations express the linear relation between dissipative fluxes $\boldsymbol{j}, \boldsymbol{q}$ and small perturbations $\boldsymbol{E}'$, ∇T of the equilibrium state in the system

$$\begin{aligned} \boldsymbol{j} &= \sigma(\boldsymbol{E}' - S\nabla T) \\ \boldsymbol{q} &= \sigma\Pi\boldsymbol{E}' - \varkappa_E\nabla T \end{aligned} \qquad \text{or} \qquad \begin{aligned} \boldsymbol{E}' &= \tfrac{1}{\sigma}\boldsymbol{j} + S\nabla T \\ \boldsymbol{q} &= \Pi\boldsymbol{j} - \varkappa\nabla T \end{aligned} \qquad \text{and} \quad \varkappa_E = \varkappa + \sigma\Pi S.$$

The coefficient σ is referred to as *electrical conductivity*, S is the *Seebeck coefficient* or *thermopower*, and Π is the *Peltier coefficient*. The *thermal conductivity* $\varkappa$ is determined under the lack of electric current. It is easy to see that the cross-coefficients S and Π are connected with the *second Thomson relation*

$$\Pi = ST,$$

representing the specific case of the *Onsager reciprocal relations*.

The following laws and effects are a consequence of these relations. For $\nabla T = 0$, we have $\nabla\mu = 0$, entailing the *Ohm's law* $\boldsymbol{j} = \sigma\boldsymbol{E}$. For $\boldsymbol{j} = 0$, we arrive at the *Fourier's heat conduction law* $\boldsymbol{q} = -\varkappa\nabla T$. The *contact potential difference* or voltage may be generated at the junction between two different metals. In fact, at $\boldsymbol{j} = 0$ and $\nabla T = 0$ one has $\boldsymbol{E}' = 0$ or $\nabla\mu/e = \boldsymbol{E} = -\nabla\Phi$. If these two metals have various chemical potentials μ_1 and μ_2, the contact potential difference $\Phi_{12} = \Phi_2 - \Phi_1 = -(\mu_2 - \mu_1)/e$ can be found using $\nabla\Phi = \nabla\mu/e$.

The coupling between electrical current and heat flow leads to *thermoelectric effects*. If two ends of the open conductor are maintained at various temperatures T_1 and T_2, the potential difference or thermopower appears between the ends of the conductor. In fact, at $\boldsymbol{j} = 0$, we have $\boldsymbol{E}' = S\nabla T$ and

$$-\nabla\Phi = \boldsymbol{E} = S\nabla T + \nabla\mu/e = (S + e^{-1}\partial\mu/\partial T)\nabla T = S'\nabla T.$$

The potential difference $\Phi_{12} = \Phi_2 - \Phi_1 = -S'(T_2 - T_1)$ which originates represents the *Seebeck effect*.

The *Peltier effect* implies an appearance of the heat flow linear in the electric current in the absence of temperature gradient. At $\nabla T = 0$ there is an additional heat flow $\boldsymbol{q} = \Pi\boldsymbol{j}$, heat flow direction depending on the direction of electric current.

(b) In the differential form the energy conservation law reads

$$\frac{\partial\mathcal{E}}{\partial t} + \operatorname{div}\boldsymbol{Q} = 0,$$

where $\mathcal{E}$ is the energy per the unit volume and $\boldsymbol{Q}$ is the energy flux. The equation determining the production rate of entropy S in the unit volume can be written as

$$\frac{\partial S}{\partial t} + \operatorname{div}\boldsymbol{F} = \frac{R}{T}.$$

Here $\boldsymbol{F}$ is the entropy flow and R is the dissipative Rayleigh function. The latter determines the entropy production rate and must always be positive definite $R \geqslant 0$. Let us employ the following identity for energy $\mathcal{E}$:

$$d\mathcal{E} = T\,dS + \mu\,d\rho,$$

assuming the density ρ to be constant for simplicity. Next, we seek for the relation between the time derivatives

$$\dot{\mathcal{E}} = T\dot{S} \quad \text{and} \quad T\dot{S} = -\operatorname{div}\boldsymbol{Q}.$$

Let us define the energy flux according to

$$\boldsymbol{Q}(\boldsymbol{r}) = \int \boldsymbol{v}\Big(\varepsilon_p + e\Phi(\boldsymbol{r})\Big) f\,d\Gamma_p = \int \boldsymbol{v}\Big(\varepsilon_p - \mu(\boldsymbol{r})\Big) f\,d\Gamma_p$$
$$+ \int \boldsymbol{v}\Big(\Phi(\boldsymbol{r}) + \frac{\mu(\boldsymbol{r})}{e}\Big) ef\,d\Gamma_p = \boldsymbol{q} + \Big(\Phi(\boldsymbol{r}) + \frac{\mu(\boldsymbol{r})}{e}\Big)\boldsymbol{j}.$$

Here $\boldsymbol{q}$ is the dissipative heat flow, $\boldsymbol{j}$ is the electric current density, and $\Phi(\boldsymbol{r})$ is the electric potential. Then one has

$$-\dot{S} = \frac{1}{T}\operatorname{div}\boldsymbol{Q} = \frac{1}{T}\operatorname{div}\boldsymbol{q} + \frac{1}{T}\operatorname{div}\Big[\big(\Phi + \frac{\mu}{e}\big)\boldsymbol{j}\Big]$$
$$= \operatorname{div}\big(\frac{\boldsymbol{q}}{T}\big) - \boldsymbol{q}\cdot\nabla\frac{1}{T} + \frac{\Phi + \mu/e}{T}\operatorname{div}\boldsymbol{j} + \frac{1}{T}\,\boldsymbol{j}\cdot\nabla\big(\Phi + \frac{\mu}{e}\big).$$

The continuity equation for the electric current means $\operatorname{div}\boldsymbol{j} = 0$ for the time-independent charge density. Then we obtain

$$\dot{S} = -\operatorname{div}\big(\frac{\boldsymbol{q}}{T}\big) - \frac{\boldsymbol{q}\cdot\nabla T}{T^2} + \frac{\boldsymbol{j}}{T}\big(\boldsymbol{E} - \frac{\nabla\mu}{e}\big) = -\operatorname{div}\big(\frac{\boldsymbol{q}}{T}\big) - \frac{\boldsymbol{q}\cdot\nabla T}{T^2} + \frac{\boldsymbol{j}\boldsymbol{E}'}{T}.$$

Substituting the relations for $\boldsymbol{q}$ and $\boldsymbol{E}'$ yields

$$\frac{\partial S}{\partial t} + \operatorname{div}\big(\frac{\boldsymbol{q}}{T}\big) = -\frac{\Pi\boldsymbol{j} - \varkappa\nabla T}{T^2}\nabla T + \frac{\boldsymbol{j}}{T}\big(\frac{\boldsymbol{j}}{\sigma} + S\nabla T\big)$$
$$= \varkappa\frac{(\nabla T)^2}{T^2} + \frac{\boldsymbol{j}^2}{\sigma T} + \big(S - \frac{\Pi}{T}\big)\frac{\boldsymbol{j}\cdot\nabla T}{T}.$$

The sign-alternating term $\boldsymbol{j}\cdot\nabla T$ vanishes as a requirement of the Onsager reciprocal principle for the kinetic coefficients $S = \Pi/T$. Finally, we arrive at the equation for the entropy production

$$\frac{\partial S}{\partial t} + \operatorname{div}\boldsymbol{F} = \frac{R}{T}.$$

Here the density entropy flow equals $\boldsymbol{F} = \boldsymbol{q}/T$ and the dissipative function R is given by the expression

$$R = \varkappa\frac{(\nabla T)^2}{T} + \frac{\boldsymbol{j}^2}{\sigma}.$$

The latter has a sense of energy dissipation power per unit volume, resulting from the processes of thermal and electric conductivity. The requirement of positive definiteness for the dissipative function results in the necessity to have the positive values of both the thermal conductivity $\varkappa > 0$ and the electric conductivity $\sigma > 0$.

The rate of varying the energy per unit volume $\mathcal{E}$ can be expressed as

$$\dot{\mathcal{E}} = -\operatorname{div}\boldsymbol{Q} = \frac{\boldsymbol{j}^2}{\sigma} + \operatorname{div}(\varkappa\nabla T) - \mu_T\boldsymbol{j}\cdot\nabla T, \quad \mu_T = T\frac{d}{dT}\left(\frac{\Pi}{T}\right) = T\frac{dS}{dT}.$$

The coefficient μ_T is referred to as *Thomson coefficient*[1] and the last term in a sum represents the *Thomson heat*. Under stationary conditions $\dot{\mathcal{E}} = 0$ we have

$$\operatorname{div}(-\varkappa\nabla T) = \boldsymbol{j}^2/\sigma - \mu_T\boldsymbol{j}\cdot\nabla T$$

and the heat generation is a sum of Joule heat and Thomson heat. Varying the direction of electrical current changes the generation of Thomson heat to its absorption.

5. (a) For the calculations below, it is convenient to introduce the density of states $\nu(\varepsilon)$ and go over to integrating over the energy

$$\nu(\varepsilon) = \frac{m\sqrt{2m\varepsilon}}{\pi^2\hbar^3}, \quad d\Gamma_p = \nu(\varepsilon)\,d\varepsilon.$$

The electrical conductivity is given by the integral

$$\sigma = \frac{1}{3}\int e^2v^2(\varepsilon)\tau(\varepsilon)\nu(\varepsilon)\Big(-\frac{\partial f_0}{\partial\varepsilon}\Big)\,d\varepsilon, \quad -\frac{\partial f_0}{\partial\varepsilon} = \frac{1}{4T}\frac{1}{\cosh^2(\varepsilon-\mu)/2T}.$$

In a metal, as a degenerate Fermi liquid $T \ll \mu$, all the integrals are gained beside the Fermi surface, i.e. in the region $|\varepsilon - \mu| \lesssim T$. Therefore, we can take out the slowly varying functions from the integral sign, thus neglecting the exponentially small terms of the order of $\exp(-\mu/T)$ as compared with unity. So, we obtain

$$\sigma = \frac{1}{3}e^2(v^2\tau\nu)_{\varepsilon=\mu} = (e^2D\nu)_\mu = \frac{ne^2}{m}\tau(\mu), \quad D = \frac{1}{3}v^2\tau.$$

Here D is the diffusion coefficient, ν is the density of states taken at the Fermi surface, and n is the electron concentration.

[1] The formula $\mu_T = TdS/dT$ is also called the first Thomson relation.

In the lack of electric field the thermal conductivity $\varkappa_E$ equals

$$\varkappa_E = \frac{1}{3}\int_0^\infty v^2(\varepsilon)\tau(\varepsilon)\nu(\varepsilon)\frac{(\varepsilon-\mu)^2}{T}\frac{d\varepsilon}{4T\cosh^2\frac{\varepsilon-\mu}{2T}} \approx \frac{1}{3}(v^2\tau\nu)_{\varepsilon=\mu}$$

$$\times\int_{-\infty}^{\infty}\frac{\xi^2}{4T^2}\frac{d\xi}{\cosh^2\frac{\xi}{2T}} = \frac{1}{3}(v^2\tau\nu)_\mu 2T\int_{-\infty}^{\infty}\frac{x^2\,dx}{\cosh^2 x} = \frac{\pi^2 T}{9}(v^2\tau\nu)_\mu = C(T)D(\mu).$$

Here $C(T) = \pi^2 T\nu(\mu)/3$ is the electron heat capacity of a metal and $D(\mu)$ is the diffusion coefficient.

Let us turn now to calculating the Peltier coefficient and consider the integral

$$\sigma\Pi = \frac{1}{3}\int_0^\infty e(v^2\tau\nu)_\varepsilon(\varepsilon-\mu)\frac{d\varepsilon}{4T\cosh^2\frac{\varepsilon-\mu}{2T}} \approx \frac{e}{3}\int_{-\mu}^{\infty}(v^2\tau\nu)_{\xi+\mu}\frac{\xi}{4T}\frac{d\xi}{\cosh^2\frac{\xi}{2T}}.$$

If quantity $(v^2\tau\nu)_\varepsilon = \text{const}$ is ε-independent, the integral for $\sigma\Pi$ is exponentially small. Thus we use the expansion

$$(v^2\tau\nu)_{\xi+\mu} = (v^2\tau\nu)_\mu + (v^2\tau\nu)'_\mu\xi + \dots$$

and the last term yields nonzero power-like contribution. Next, expanding the lower integration limit from $-\mu$ to $-\infty$, we arrive at

$$\sigma\Pi \approx \frac{e}{3}(v^2\tau\nu)'_\mu\int_{-\infty}^{\infty}\frac{\xi^2}{4T}\frac{d\xi}{\cosh^2\frac{\xi}{2T}} = \frac{e}{3}(v^2\tau\nu)'_\mu 2T^2\frac{\pi^2}{6} = \frac{\pi^2}{9}eT^2(v^2\tau\nu)'_\mu.$$

Hence the Peltier coefficient equals

$$\Pi(T) = \frac{\pi^2T^2}{3e}\frac{(v^2\tau\nu)'_\mu}{(v^2\tau\nu)_\mu} = \frac{\pi^2T^2}{3e}\frac{d}{d\mu}\big[\ln(v^2\tau\nu)_\mu\big] = \frac{\pi^2T^2}{3e}\frac{d}{d\mu}\big[\ln\sigma(\mu)\big].$$

The Seebeck coefficient expressed in terms of conductivity is called the *Mott formula*

$$S(T) = \frac{\pi^2 T}{3e}\frac{d}{d\mu}\big[\ln\sigma(\mu)\big].$$

Involving $\tau \sim \varepsilon^{r-1/2}$, we have

$$S(T) = (r+1)\frac{\pi^2}{3e}\frac{T}{\mu}.$$

The magnitude and the sign of the Seebeck and Peltier coefficients are dependent on the scattering mechanism.

In metals one may not discern the thermal conductivities $\varkappa$ and $\varkappa_E$ within accuracy to $T^2/\mu^2 \ll 1$ and put $\varkappa \approx \varkappa_E$. In fact, it is seen that

$$\frac{\varkappa_E - \varkappa}{\varkappa_E} = \frac{S^2\sigma T}{\varkappa_E} \sim \frac{T^2}{\mu^2} \ll 1.$$

The *figure of merit* ZT_E of thermoelectric effects is determined as a ratio of Peltier heat flow $\Pi \boldsymbol{j}$ to the heat flow $-\varkappa\nabla T$ passing due to heat conduction mechanism under zero electric field

$$ZT_E = \left(\frac{\Pi \boldsymbol{j}}{-\varkappa\nabla T}\right)_{\boldsymbol{E}'=0} = \frac{-\Pi\sigma S\nabla T}{-\varkappa\nabla T} = \frac{S^2\sigma T}{\varkappa}.$$

Under the lack of electrical current $\boldsymbol{j} = 0$ the figure of merit ZT can be introduced as

$$ZT = \frac{S^2\sigma T}{\varkappa_E}.$$

The figure of merit ZT of thermoelectricity in metals is very small

$$ZT = \frac{S^2\sigma T}{\varkappa_E} = (r+1)\frac{\pi^2}{3}\frac{T^2}{\mu^2} \ll 1.$$

The smallness for ZT is associated with the large density of electric charge careers in metals.

The *Lorentz number* can readily be calculated

$$L = \frac{\varkappa}{\sigma T} \approx \frac{\varkappa_E}{\sigma T} = \frac{\pi^2}{3e^2}.$$

Provided that the Lorentz number proves to be temperature independent, the *Wiedemann-Frantz law* is said to be valid.

(b) *Case of non-degenerated semiconductor*. For the Boltzmann statistics of charge careers, the equilibrium distribution will be equal to $f_0 = \exp[(\mu - \varepsilon)/T]$. For convenience of calculation, let us introduce electron density n and μ^* equal to a ratio of chemical potential μ to temperature T

$$n = \int_0^\infty f_0(\varepsilon)\nu(\varepsilon)\,d\varepsilon = e^{\frac{\mu}{T}}\int_0^\infty e^{-\frac{\varepsilon}{T}}\nu(\varepsilon)\,d\varepsilon, \quad \mu^* = \frac{\mu}{T} = \ln\left[\frac{n}{2}\left(\frac{2\pi\hbar^2}{mT}\right)^{3/2}\right].$$

Henceforward we take into account that the following holds for the Boltzmann distribution function:

$$-\frac{\partial f_0}{\partial \varepsilon} = \frac{f_0}{T}.$$

The electrical conductivity σ is given by the integral

$$\sigma = \frac{e^2}{3T}\int_0^\infty v^2 \tau \nu f_0\, d\varepsilon = \frac{2ne^2}{3mT}\frac{\int_0^\infty \varepsilon \tau \nu e^{-\frac{\varepsilon}{T}}\, d\varepsilon}{\int_0^\infty \nu e^{-\frac{\varepsilon}{T}}\, d\varepsilon} = \frac{2ne^2}{3mT}\langle \varepsilon\tau \rangle,$$

where, for brevity, the angle brackets mean a ratio of integrals of the following type:

$$\langle \varepsilon^k \rangle = \frac{\int_0^\infty \varepsilon^k f_0 \nu\, d\varepsilon}{\int_0^\infty f_0 \nu\, d\varepsilon} = \frac{\int_0^\infty \varepsilon^k \nu\, e^{-\frac{\varepsilon}{T}}\, d\varepsilon}{\int_0^\infty \nu\, e^{-\frac{\varepsilon}{T}}\, d\varepsilon} = T^k \frac{\Gamma(k+3/2)}{\Gamma(3/2)}$$

and $\Gamma(x)$ is the Gamma function. Recalling that $\tau(\varepsilon) \sim \varepsilon^{r-1/2}$, we find finally

$$\sigma = \frac{ne^2}{m}\tau(T)\frac{\Gamma(r+2)}{\Gamma(5/2)}.$$

Let us turn to determining the Peltier coefficient

$$\sigma\Pi = \frac{e}{3T}\int_0^\infty v^2 \tau \nu (\varepsilon - \mu) f_0\, d\varepsilon = \frac{ne}{3T}\big[\langle \varepsilon v^2 \tau \rangle - \mu \langle v^2 \tau \rangle\big].$$

After the similar calculations we obtain the Peltier and Seebeck coefficients

$$\Pi = \frac{T}{e}(r + 2 - \mu^*) \quad \text{and} \quad S = \frac{1}{e}(r + 2 - \mu^*).$$

The thermal conductivity in zero electric field reads

$$\varkappa_E = \frac{1}{3T^2}\int_0^\infty v^2 \tau \nu (\varepsilon - \mu)^2 f_0\, d\varepsilon = \frac{n}{3T^2}\big[\langle \varepsilon^2 v^2 \tau \rangle - 2\mu \langle \varepsilon v^2 \tau \rangle + \mu^2 \langle v^2 \tau \rangle\big].$$

The simple calculation results in

$$\varkappa_E = \frac{nT}{m}\tau(T)\frac{\Gamma(r+2)}{\Gamma(5/2)}\big[(r+2) + (r+2-\mu^*)^2\big].$$

Then we find the thermal conductivity for zero electric current

$$\varkappa = \varkappa_E - \sigma \Pi S = \frac{nT}{m}\tau(T)\frac{\Gamma(r+2)}{\Gamma(5/2)}(r+2) = \frac{nT}{m}\tau(T)\frac{\Gamma(r+3)}{\Gamma(5/2)}.$$

It is worthwhile to note that the coefficients $\varkappa$ and $\varkappa_E$ are noticeably different in semiconductors.

Unlike metals the Lorentz number L in semiconductors becomes dependent on the type of scattering and equals

$$L = \frac{\varkappa}{\sigma T} = \frac{r+2}{e^2}.$$

The figure of merit ZT in semiconductors is significantly larger than in metals

$$ZT = \frac{S^2\sigma T}{\varkappa_E} = \frac{(r+2-\mu^*)^2}{(r+2)+(r+2-\mu^*)^2}$$

and can reach the magnitudes close to unity.

6. The heat flow $\boldsymbol{q}$ is determined from the relation

$$\boldsymbol{q} = \int \boldsymbol{v}(\varepsilon_{\boldsymbol{p}} - \mu) f(\varepsilon_{\boldsymbol{p}}) \frac{2d^3p}{(2\pi\hbar)^3} = \int \boldsymbol{v}(\varepsilon_{\boldsymbol{p}} - \mu)\delta f(\varepsilon_{\boldsymbol{p}}) \frac{2d^3p}{(2\pi\hbar)^3},$$

where $\delta f = f - f_0$ is the perturbation of distribution function f for electron excitations with the dispersion

$$\varepsilon = \varepsilon(\boldsymbol{p}, \boldsymbol{r}) = \sqrt{\xi_p^2 + \Delta^2(T)} + \mu, \quad \xi_p = p^2/2m - \mu$$

under temperature gradient ∇T and electron-impurity scattering, μ being the chemical potential. Here $\Delta(T)$ is the superconducting gap in the energy spectrum of electron excitations and $f_0 = \left[\exp[(\varepsilon - \mu)/T] + 1\right]^{-1}$ is the equilibrium Fermi distribution function.

In order to find the distribution function, we write the corresponding transport equation

$$\frac{\partial \varepsilon}{\partial \boldsymbol{p}}\frac{\partial f}{\partial \boldsymbol{r}} - \frac{\partial \varepsilon}{\partial \boldsymbol{r}}\frac{\partial f}{\partial \boldsymbol{p}} = -\frac{f - f_0}{\tau_s(\varepsilon)}.$$

Within the linear approximation in the temperature-gradient smallness, we can replace f with f_0 on the left-hand side of equation. Then the initial equation reduces to

$$-\frac{f - f_0}{\tau_s(\varepsilon)} = \frac{\partial \varepsilon}{\partial \boldsymbol{p}}\frac{\partial f_0}{\partial \boldsymbol{r}} - \frac{\partial \varepsilon}{\partial \boldsymbol{r}}\frac{\partial f_0}{\partial \boldsymbol{p}}$$

$$= \left(\boldsymbol{v}\nabla\varepsilon - (\boldsymbol{v}\nabla T)\frac{\varepsilon - \mu}{T}\right)\frac{\partial f_0}{\partial \varepsilon} - (\boldsymbol{v}\nabla\varepsilon)\frac{\partial f_0}{\partial \varepsilon} = (\boldsymbol{v}\nabla T)\frac{\varepsilon - \mu}{T}\left(-\frac{\partial f_0}{\partial \varepsilon}\right).$$

Note that the explicit dependence of electron excitation energy on the superconducting gap $\Delta(T)$ is completely eliminated from the transport equation.

In order to find the parameter $\tau_s(\varepsilon)$ in the superconducting state, we recall that the mean free path l is determined with the particle velocity multiplied by the mean free time. The velocity of electron excitation will be equal to

$$\boldsymbol{v} = \frac{\partial\varepsilon}{\partial\boldsymbol{p}} = \frac{\partial\varepsilon}{\partial\xi}\frac{\partial\xi}{\partial\boldsymbol{p}} = \boldsymbol{v}_n\frac{\xi_p}{\sqrt{\xi_p^2+\Delta^2}},$$

where $\boldsymbol{v}_n = \boldsymbol{p}/m$ is the electron velocity in the normal state. Since the mean free path remains unchanged according to the statement of the problem, the mean free time in the superconducting state becomes equal to

$$\tau_s = \tau_n\frac{\sqrt{\xi_p^2+\Delta^2}}{|\xi_p|} = \tau_n\frac{\varepsilon_p-\mu}{|\xi_p|},$$

where $\tau_n = \tau(\xi_p+\mu)$ is the mean free time of electrons with respect to impurity scattering in the normal state.

As a result, we arrive at the following formulas:

$$\delta f = -\boldsymbol{v}(\boldsymbol{v}\nabla T)\frac{\varepsilon-\mu}{T}\tau_s\left(-\frac{\partial f_0}{\partial\varepsilon}\right) = -\boldsymbol{v}(\boldsymbol{v}\nabla T)\frac{(\varepsilon-\mu)^2}{T|\xi|}\tau_n\left(-\frac{\partial f_0}{\partial\varepsilon}\right),$$

$$\boldsymbol{q} = -\int \boldsymbol{v}(\boldsymbol{v}\nabla T)\frac{(\varepsilon_p-\mu)^3}{T|\xi_p|}\tau_n(\xi_p+\mu)\left(-\frac{\partial f_0}{\partial\varepsilon_p}\right)\frac{2d^3p}{(2\pi\hbar)^3} = -\varkappa_s\nabla T.$$

Due to isotropy of energy spectrum the integration over the solid angle results in the following expression for the thermal conductivity:

$$\varkappa_s = \frac{1}{3}\int v_p^2\frac{(\varepsilon_p-\mu)^3}{T|\xi_p|}\tau_n(\xi_p+\mu)\left(-\frac{\partial f_0}{\partial\varepsilon_p}\right)\frac{2d^3p}{(2\pi\hbar)^3}$$
$$= \frac{1}{3}\int\frac{p^2}{m^2}\frac{\xi_p^2}{(\varepsilon_p-\mu)^2}\frac{(\varepsilon_p-\mu)^3}{T|\xi_p|}\tau_n(\xi_p+\mu)\left(-\frac{\partial f_0}{\partial\varepsilon_p}\right)\frac{8\pi p^2dp}{(2\pi\hbar)^3}.$$

Before the numerical estimate of integral we note first of all that the integral over momentum is gained from the region of momenta close to the Fermi momentum, i.e. $|p-p_F| \ll p_F$ or $|\xi_p| \ll \mu$. In this case one can approximately put $\xi_p = v_F(p-p_F)$, differential $dp = d(p-p_F) = d\xi/v_F$, and finally $p = p_F$ if the dependence on momentum is smooth. So, we obtain

$$\varkappa_s \approx \frac{1}{3}\int\limits_{-\infty}^{\infty}\frac{p_F^2}{m^2}\frac{\xi^2}{E^2}\frac{E^3}{T|\xi|}\tau_n(\mu)\left(-\frac{\partial f_0}{\partial E}\right)\frac{p_F^2}{\pi^2\hbar^3}\frac{d\xi}{v_F}, \quad E = \sqrt{\xi^2+\Delta^2}.$$

Going over to the integration region $\xi > 0$ and variable $E = (\xi^2+\Delta^2)^{1/2}$, we have

$$\varkappa_s = \frac{2}{3}\frac{p_F^2}{m^2}\frac{p_F^2\tau_n(\mu)}{\pi^2\hbar^3 v_F}\frac{1}{T}\int\limits_{\Delta}^{\infty} E^2\Big(-\frac{\partial f_0}{\partial E}\Big)dE \quad \text{where} \quad f_0 = \frac{1}{e^{E/T}+1}.$$

Then for the aim of comparison, we indicate the thermal conductivity in the normal state

$$\varkappa_n(T) = \frac{T}{9}\frac{p_F^2 v_F}{\pi^2\hbar^3}\tau_n(\mu) = C_n(T)D(\mu),$$

where $C_n(T)$ is the electron specific heat of a metal in the normal state and $D(\mu) = v_F^2\tau(\mu)/3$ is the diffusion coefficient. Lastly, we arrive at the final formula

$$\frac{\varkappa_s(T)}{\varkappa_n(T)} = \frac{6}{(\pi T)^2}\int\limits_{\Delta}^{\infty} E^2\Big(-\frac{\partial f_0}{\partial E}\Big)dE$$

$$= \frac{12}{\pi^2}\int\limits_{\Delta/2T}^{\infty}\frac{x^2\,dx}{\cosh^2 x} \approx \begin{cases} 1 - \frac{\Delta^3}{2\pi^2 T^3}, & \Delta \ll T, \\ \frac{6\Delta^2}{(\pi T)^2}e^{-\Delta/T}, & \Delta \gg T. \end{cases}$$

Thus, for the temperatures close to absolute zero, it is possible the situation when the thermal conductivity in the superconducting state can be several orders of magnitude smaller as compared with that in the normal state at the same temperature. This allows one to use superconductors as thermal insulators.

3.3 Galvanomagnetic Effects in Metals

7. (a) Let conductor be under the following conditions. There are small and constant electric field $\boldsymbol{E}$, magnetic field $\boldsymbol{H}$, and temperature gradient ∇T. Let us write the Boltzmann equation for the distribution function $f = f(\boldsymbol{r}, \boldsymbol{p})$ in the τ-approximation

$$\boldsymbol{v}\frac{\partial f}{\partial \boldsymbol{r}} + e\Big(\boldsymbol{E} + \frac{1}{c}[\boldsymbol{v}\times\boldsymbol{H}]\Big)\frac{\partial f}{\partial \boldsymbol{p}} = -\frac{f - f_0}{\tau(\varepsilon)}.$$

Here $f_0(\varepsilon)$ is the equilibrium distribution function with the following properties:

$$f_0(\varepsilon) = f_0\Big(\frac{\varepsilon-\mu}{T}\Big), \quad \frac{\partial f_0}{\partial \mu} = -\frac{\partial f_0}{\partial \varepsilon}, \quad \frac{\partial f_0}{\partial T} = -\frac{\varepsilon-\mu}{T}\frac{\partial f_0}{\partial \varepsilon}, \quad \varepsilon = \frac{p^2}{2m}$$

and $\mu = \mu(T)$ is the chemical potential dependent on the temperature.

Let us represent the non-equilibrium distribution function as

$$f(\boldsymbol{r}, \boldsymbol{p}) = f_0(\varepsilon) + \delta f(\boldsymbol{r}, \boldsymbol{p}).$$

In the linear approximation in disturbance we must consider only the terms proportional to $\boldsymbol{E}$, $\boldsymbol{H}$, and ∇T in δf. In order to determine the linear contribution to δf from $\boldsymbol{E}$ and ∇T, it is sufficient to substitute $f = f_0$ on the left-hand side of the Boltzmann equation. As it concerns the magnetic field term, the situation is more complicated. Because of $\partial f_0/\partial \boldsymbol{p} = \boldsymbol{v} f_0'$ and $[\boldsymbol{v} \times \boldsymbol{H}]\boldsymbol{v} = 0$ the magnetic field contribution simply vanishes in this approximation. Therefore, to consider the magnetic field contribution to δf, one should treat the derivative $\partial(\delta f)/\partial \boldsymbol{p}$ more thoroughly. As a result, we obtain the following equation for determining the correction δf:

$$\boldsymbol{v}\frac{\partial f_0}{\partial \boldsymbol{r}} + e\boldsymbol{E}\frac{\partial f_0}{\partial \boldsymbol{p}} + \frac{e}{c}[\boldsymbol{v} \times \boldsymbol{H}]\Big)\frac{\partial(\delta f)}{\partial \boldsymbol{p}} = -\frac{\delta f}{\tau(\varepsilon)}.$$

Using the above-mentioned relations for the derivatives of function f_0, we arrive at the equation

$$\left(e(\boldsymbol{v}\boldsymbol{E}') - \frac{\varepsilon - \mu}{T}(\boldsymbol{v}\nabla T)\right)\frac{\partial f_0}{\partial \varepsilon} + \frac{e}{c}[\boldsymbol{v} \times \boldsymbol{H}]\frac{\partial(\delta f)}{\partial \boldsymbol{p}} = -\frac{\delta f}{\tau(\varepsilon)}$$

which is necessary to solve. For brevity, we have introduced the quantity $\boldsymbol{E}' = \boldsymbol{E} - \nabla\mu/e$ as a deviation from the equilibrium condition

$$e\Phi(\boldsymbol{r}) + \mu = \text{const}$$

in the external electric potential $\Phi(\boldsymbol{r})$.

The solution of the Boltzmann equation is trialed in the form $\delta f = \boldsymbol{v}\chi(\varepsilon)$ which conserves the structure of free term. After substituting it into equation, we employ the relation

$$m\frac{\partial}{\partial \boldsymbol{p}}(\boldsymbol{v}\chi(\varepsilon_p)) = \frac{\partial}{\partial \boldsymbol{p}}(\boldsymbol{p}\chi(\varepsilon_p)) = \chi(\varepsilon_p) + \boldsymbol{v}(\boldsymbol{p}\frac{\partial\chi(\varepsilon_p)}{\partial\epsilon_p}) = \chi + \boldsymbol{v}(\boldsymbol{p}\chi'),$$

and take $[\boldsymbol{v} \times \boldsymbol{H}]\cdot\boldsymbol{v} = 0$ and $[\boldsymbol{v} \times \boldsymbol{H}]\cdot\chi = \boldsymbol{v}\cdot[\boldsymbol{H} \times \chi]$ into account. Then we have the equation as a result

$$\left(e(\boldsymbol{v}\boldsymbol{E}') - \frac{\varepsilon - \mu}{T}(\boldsymbol{v}\nabla T)\right)\frac{\partial f_0}{\partial \varepsilon} + \frac{e}{mc}\,\boldsymbol{v}\cdot[\boldsymbol{H} \times \chi] = -\frac{\boldsymbol{v}\chi}{\tau(\varepsilon)}.$$

Thus the function χ satisfies the equation

$$\left(e\boldsymbol{E}' - \frac{\varepsilon - \mu}{T}\nabla T\right)\frac{\partial f_0}{\partial \varepsilon} + \frac{e}{mc}[\boldsymbol{H} \times \chi] = -\frac{\chi}{\tau(\varepsilon)}.$$

Hence, in the linear approximation in field H we find χ and δf

$$\chi \approx \Big(e\boldsymbol{E}' - \frac{\varepsilon-\mu}{T}\nabla T - \frac{e\tau}{mc}[\boldsymbol{H}\times(e\boldsymbol{E}' - \frac{\varepsilon-\mu}{T}\nabla T)]\Big)\tau(-\frac{\partial f_0}{\partial\varepsilon}),$$

$$\delta f \approx \boldsymbol{v}\Big(e\boldsymbol{E}' - \frac{\varepsilon-\mu}{T}\nabla T - \frac{e\tau}{mc}[\boldsymbol{H}\times(e\boldsymbol{E}' - \frac{\varepsilon-\mu}{T}\nabla T)]\Big)\tau(-\frac{\partial f_0}{\partial\varepsilon}).$$

The neglect of the second term as compared with the first one entails the criterium of smallness for the magnetic field and linear approximation

$$\Omega\tau \ll 1 \quad \text{or} \quad R_L = v/\Omega \gg l = v\tau; \quad \Omega = eH/mc,$$

Ω being the *cyclotron frequency*. The second inequality means that the *Larmor radius* R_L of electron orbit in the magnetic field is much larger than the mean free path l.

As above, electrical current density $\boldsymbol{j}$ and dissipative heat flow $\boldsymbol{q}$ are presented with the following integrals:

$$\boldsymbol{j} = \int e\boldsymbol{v}f\,d\Gamma_p \quad \text{and} \quad \boldsymbol{q} = \int \boldsymbol{v}(\varepsilon-\mu)f\,d\Gamma_p,$$

where $d\Gamma_p = 2d^3p/(2\pi\hbar)^3$ is the volume element in the p-space augmented with two spin projections. Due to isotropy of electron spectrum, the equilibrium part f_0 of distribution function gives no contribution to the current and heat flow. The spectrum isotropy allows us to simplify the integration over angles according to

$$\int \boldsymbol{v}(\boldsymbol{v}\boldsymbol{A})\,d\Gamma_p = \frac{\boldsymbol{A}}{3}\int v^2\,d\Gamma_p\,.$$

The magnetic field-independent terms lead to the known contributions to current $\boldsymbol{j}$ and heat flow $\boldsymbol{q}$, considered in the previous problems. These contributions are expressed via coefficients of conductivity σ, Seebeck S, Peltier Π and thermal conductivity $\varkappa_E$ which have been introduced in the previous problems. So, we get

$$\boldsymbol{j} = \sigma(\boldsymbol{E}' - S\nabla T) - [\boldsymbol{H}\times\boldsymbol{E}']\frac{1}{3}\int\frac{e^3v^2\tau^2}{mc}(-\frac{\partial f_0}{\partial\varepsilon})\,d\Gamma_p$$
$$+[\boldsymbol{H}\times\nabla T]\frac{1}{3}\int\frac{e^2v^2\tau^2}{mc}\frac{(\varepsilon-\mu)}{T}(-\frac{\partial f_0}{\partial\varepsilon})\,d\Gamma_p,$$
$$\boldsymbol{q} = \sigma\Pi\boldsymbol{E}' - \varkappa_E\nabla T - [\boldsymbol{H}\times\boldsymbol{E}']\frac{1}{3}\int\frac{e^2v^2\tau^2(\varepsilon-\mu)}{mc}(-\frac{\partial f_0}{\partial\varepsilon})\,d\Gamma_p$$
$$+[\boldsymbol{H}\times\nabla T]\frac{1}{3}\int\frac{ev^2\tau^2}{mc}\frac{(\varepsilon-\mu)^2}{T}(-\frac{\partial f_0}{\partial\varepsilon})\,d\Gamma_p\,.$$

With the aid of transport coefficients these relations demonstrate the linear interconnection between dissipative flows $\boldsymbol{j}, \boldsymbol{q}$ and small perturbations $\boldsymbol{E}'$, ∇T of equilibrium state in the system

$$j = \sigma(E' - S\nabla T) - \sigma^2 R[H \times E'] - \sigma N_E[H \times \nabla T],$$
$$q = \sigma\Pi E' - \varkappa_E \nabla T + \sigma M_E[H \times E'] + L_E[H \times \nabla T].$$

Restricting ourselves only with the linear terms in H, we can rewrite these expressions in the equivalent form as

$$E' = \sigma^{-1} j + S\nabla T + R[H \times j] + N[H \times \nabla T], \quad N = N_E + \sigma S R,$$
$$q = \Pi j - \varkappa \nabla T + M[H \times j] + L[H \times \nabla T], \quad M = M_E + \sigma \Pi R,$$
$$L = L_E + \sigma \Pi N_E + \sigma S M_E + \sigma^2 \Pi S R, \qquad \varkappa = \varkappa_E - \sigma \Pi S.$$

It is readily to check that the cross-coefficients N_E and M_E are related with the formula

$$M_E = N_E T.$$

Reminding the second Thomson relation $\Pi = ST$, we see that the similar property

$$M = NT,$$

as an example of the Onsager reciprocal principle for the transport coefficients, holds for the coefficients N and M. A set of the Hall R, Nernst N, and Righi-Leduc L coefficients helps to describe the thermomagnetic phenomena and, correspondingly, *Hall*, *Nernst-Ettingshausen* and *Righi-Leduc effects*.

(b) Let us turn to calculating the transport coefficients. The integrals determining the coefficients R, N_E, and L_E are similar to the integrals that occur in the thermoelectric phenomena. Here it is also convenient to introduce the density of states $\nu(\varepsilon)$ and cross over to the integration over energy

$$\nu(\varepsilon) = \frac{m\sqrt{2m\varepsilon}}{\pi^2\hbar^3}, \quad d\Gamma_p = \nu(\varepsilon)\, d\varepsilon.$$

The Hall coefficient is given by the expression

$$\sigma^2 R = \frac{1}{3}\int \frac{e^3 v^2(\varepsilon)\tau^2(\varepsilon)}{mc}\Big(-\frac{\partial f_0}{\partial \varepsilon}\Big)\nu(\varepsilon)\, d\varepsilon = \frac{e^3}{3mc}(v^2\tau^2\nu)_\mu .$$

Here $(...)_\mu$ implies that the expression is taken at the energy ε equal to chemical potential μ. Taking into account that $\sigma = e^2(v^2\tau\nu)_\mu/3 = ne^2\tau_\mu/m$ where n is the electron density, we find the Hall coefficient

$$R = \frac{1}{nec}.$$

The Nernst coefficient N_E, determined under the lack of electrical field, is given by the following formula:

$$\sigma N_E = -\frac{1}{3}\int \frac{e^2 v^2\tau^2}{mc}\frac{\varepsilon-\mu}{T}\Big(-\frac{\partial f_0}{\partial \varepsilon}\Big)\,\nu(\varepsilon)\,d\varepsilon = -\frac{\pi^2}{9}\frac{e^2T}{mc}(v^2\tau^2\nu)'_\mu\,.$$

Hence we obtain

$$N_E = -\frac{\pi^2 T}{3mc}\frac{(v^2\tau^2\nu)'_\mu}{(v^2\tau\nu)_\mu}\,.$$

For $\tau(\varepsilon) = \text{const}$, one has

$$N_E = -\frac{\pi^2\tau T}{3mc}\frac{(v^2\nu)'_\mu}{(v^2\nu)_\mu} = -\frac{\pi^2\tau}{2mc}\frac{T}{\mu} = -\frac{\tau}{mcn}s(T),$$

$$s(T) = \frac{\pi^2}{3}T\nu(\mu) = \frac{\pi^2}{2}\frac{nT}{\mu}\,,$$

where $s(T)$ is the density of entropy in the degenerate electron gas.

Next, we consider the Nernst coefficient N according to $N = N_E + \sigma SR$. Substituting the following relations:

$$\sigma S = \frac{\pi^2}{9}eT(v^2\tau\nu)'_\mu\,,\quad \sigma^2 R = \frac{e^3}{3mc}(v^2\tau^2\nu)_\mu\,,\quad \sigma = \frac{1}{3}e^2(v^2\tau\nu)_\mu$$

into the formula for N, we find that

$$\begin{aligned} N = N_E + \sigma S\frac{\sigma^2 R}{\sigma^2} &= -\frac{\pi^2 T}{3mc}\frac{(v^2\tau^2\nu)'_\mu}{(v^2\tau\nu)_\mu} + \frac{\pi^2 T}{3mc}(v^2\tau\nu)'_\mu\frac{(v^2\tau^2\nu)_\mu}{(v^2\tau\nu)^2_\mu} \\ &= -\frac{\pi^2 T}{3mc}\frac{\tau_\mu\tau'_\mu(v^2\nu)_\mu}{(v^2\tau\nu)_\mu} = -\frac{\pi^2 T}{3mc}\frac{\partial\tau(\mu)}{\partial\mu}\,. \end{aligned}$$

The last expression represents the *Sondheimer formula*. For the constant relaxation time $\tau(\varepsilon) = \text{const}$, the Nernst coefficient proves to be zero.

Let us consider the calculation of the Righi-Leduc coefficient L_E determined under zero electric field

$$\begin{aligned} L_E &= \frac{1}{3}\int \frac{ev^2\tau^2}{mc}\frac{(\varepsilon-\mu)^2}{T}\Big(-\frac{\partial f_0}{\partial\varepsilon}\Big)\nu(\varepsilon)\,d\varepsilon \\ &= \frac{\pi^2}{9}\frac{eT}{mc}(v^2\tau^2\nu)_\mu = \frac{\pi^2 T}{3}\frac{(\sigma\tau)_\mu}{emc}\,. \end{aligned}$$

In contrast to the Nernst coefficients N and N_E in metals where the condition of electron degeneracy $T \ll \mu$ is fulfilled, the difference in the Righi-Leduc coefficients L and L_E can be neglected within the accuracy to $T^2/\mu^2 \ll 1$ or, in other words, $L \approx L_E$. In fact,

$$L - L_E = 2\sigma STN_E + \sigma^2 S^2 TR$$
$$= -\frac{\pi^4}{27}\frac{eT^3}{mc}\left[2\frac{(v^2\tau\nu)'_\mu(v^2\tau^2\nu)'_\mu}{(v^2\tau\nu)_\mu} - (v^2\tau\nu)'^2_\mu\frac{(v^2\tau^2\nu)_\mu}{(v^2\tau\nu)^2_\mu}\right].$$

Then we straightforwardly obtain

$$\frac{L - L_E}{L_E} = -\frac{\pi^2 T^2}{3}\frac{(v^2\tau\nu)'_\mu}{(v^2\tau\nu)_\mu}\left[\frac{2(v^2\tau^2\nu)'_\mu - (v^2\tau\nu)'_\mu\tau_\mu}{(v^2\tau^2\nu)_\mu}\right] \sim \left(\frac{T}{\mu}\right)^2 \ll 1.$$

8. Let us write the equation for the distribution function of electrons experiencing the Lorentz force

$$e\left(\boldsymbol{E} + \frac{1}{c}[\boldsymbol{v} \times \boldsymbol{H}]\right)\frac{\partial n}{\partial \boldsymbol{p}} = -\frac{n - n_0}{\tau(\varepsilon_{\boldsymbol{p}})}.$$

Here $n_0 = n_0(\varepsilon_{\boldsymbol{p}})$ is the equilibrium distribution function. In order to find electrical conductivity, it is sufficient in the equation to keep the linear field $\boldsymbol{E}$ terms alone. In addition, concerning the magnetic field term, we will subtract the similar term but with the equilibrium distribution function n_0. We use here that $[\boldsymbol{v} \times \boldsymbol{H}]\partial n_0/\partial \boldsymbol{p} \equiv 0$. Then we have

$$e(\boldsymbol{vE})\frac{\partial n_0}{\partial \varepsilon_{\boldsymbol{p}}} + \frac{e}{c}[\boldsymbol{v} \times \boldsymbol{H}]\frac{\partial(n - n_0)}{\partial \boldsymbol{p}} = -\frac{n - n_0}{\tau(\varepsilon_{\boldsymbol{p}})}.$$

We look for the solution of equation, representing $n - n_0 = \boldsymbol{v}\boldsymbol{\chi}(\varepsilon_{\boldsymbol{p}})\partial n_0/\partial \varepsilon_{\boldsymbol{p}}$. The substitution yields

$$e(\boldsymbol{vE}) + \frac{e}{mc}[\boldsymbol{v} \times \boldsymbol{H}]\boldsymbol{\chi}(\varepsilon_{\boldsymbol{p}}) = -\frac{\boldsymbol{v}\boldsymbol{\chi}(\varepsilon_{\boldsymbol{p}})}{\tau(\varepsilon_{\boldsymbol{p}})}.$$

Here we should keep in mind that $\boldsymbol{v} = \boldsymbol{p}/m$ and the terms with the momentum derivative vanish identically due to scalar triple product since these terms are proportional to velocity.

Emphasizing that $[\boldsymbol{v} \times \boldsymbol{H}]\boldsymbol{\chi} = [\boldsymbol{H} \times \boldsymbol{\chi}]\boldsymbol{v}$, we compare the coefficients in the front of velocity $\boldsymbol{v}$ and then arrive at

$$e\boldsymbol{E} + [\boldsymbol{\Omega} \times \boldsymbol{\chi}] = -\frac{\boldsymbol{\chi}}{\tau}, \quad \boldsymbol{\Omega} = \frac{e\boldsymbol{H}}{mc},$$

where $\Omega = eH/mc$ is the *cyclotron frequency*. In order to find vector $\boldsymbol{\chi}$, let us multiply the both sides of equation by vector $\boldsymbol{\Omega}$ as follows: first, in the scalar manner and, second, in the vector manner

$$e(\boldsymbol{\Omega E}) = -(\boldsymbol{\Omega\chi})/\tau,$$
$$e[\boldsymbol{\Omega} \times \boldsymbol{E}] + \boldsymbol{\Omega}(\boldsymbol{\Omega\chi}) - \Omega^2\boldsymbol{\chi} = -[\boldsymbol{\Omega} \times \boldsymbol{\chi}]/\tau.$$

At first, we find unknown function χ and then distribution function $n(\boldsymbol{p})$

$$\chi = -\frac{e\tau}{1+\Omega^2\tau^2}\big(\boldsymbol{E} + [\boldsymbol{E}\times\boldsymbol{\Omega}]\tau + \boldsymbol{\Omega}(\boldsymbol{E\Omega})\tau^2\big),$$

$$n = n_0 + \frac{e\tau}{1+\Omega^2\tau^2}\big(\boldsymbol{v} + [\boldsymbol{\Omega}\times\boldsymbol{v}]\tau + \boldsymbol{\Omega}(\boldsymbol{\Omega v})\tau^2\big)\cdot\boldsymbol{E}\left(-\frac{\partial n_0}{\partial\varepsilon}\right).$$

The current density $\boldsymbol{j}$, implying two spin projections, is determined by the expression

$$\begin{aligned}\boldsymbol{j} &= \int e\boldsymbol{v}n(\boldsymbol{p})\frac{2d^3p}{(2\pi\hbar)^3}\\ &= \int\frac{e^2\tau}{1+\Omega^2\tau^2}\boldsymbol{v}\big(\boldsymbol{v} + [\boldsymbol{\Omega}\times\boldsymbol{v}]\tau + \boldsymbol{\Omega}(\boldsymbol{\Omega v})\tau^2\big)\cdot\boldsymbol{E}\left(-\frac{\partial n_0}{\partial\varepsilon}\right)\frac{2d^3p}{(2\pi\hbar)^3}.\end{aligned}$$

Then we know that the equilibrium distribution function n_0 does not contribute to the current. Finally, the conductivity tensor $\sigma_{\alpha\beta}$, which we determine according to $j_\alpha = \sigma_{\alpha\beta}E_\beta$, will be equal to

$$\sigma_{\alpha\beta} = \frac{e^2}{3}\int\frac{v^2\tau}{1+\Omega^2\tau^2}\big(\delta_{\alpha\beta} + e_{\alpha\gamma\beta}\Omega_\gamma\tau + \Omega_\alpha\Omega_\beta\tau^2\big)\left(-\frac{\partial n_0}{\partial\varepsilon}\right)\frac{2d^3p}{(2\pi\hbar)^3}.$$

Here we have already employed the spherical symmetry of electron spectrum for averaging over the solid angle of momentum vector $\boldsymbol{p}$

$$\langle v_\alpha v_\beta\rangle = (v^2/3)\delta_{\alpha\beta},$$

$e_{\alpha\beta\gamma}$ being the completely antisymmetric unit tensor. In a metal, as a degenerate Fermi liquid, the integral is gained in the region of energies close to chemical potential $\varepsilon_p = \mu$. Thus, we can put that the derivative of equilibrium density function with respect to energy is equal approximately to $n_0' \approx -\delta(\varepsilon-\mu)$. We have finally

$$\sigma_{\alpha\beta} = \frac{ne^2}{m}\frac{\tau(\mu)}{1+\Omega^2\tau^2(\mu)}\big[\delta_{\alpha\beta} + e_{\alpha\gamma\beta}\Omega_\gamma\tau(\mu) + \Omega_\alpha\Omega_\beta\tau^2(\mu)\big],$$

where n is the electron concentration. Such conductivity tensor corresponds to the following relation between the current and electric field:

$$\boldsymbol{j} = \frac{ne^2}{m}\frac{\tau}{1+\Omega^2\tau^2}\big[\boldsymbol{E} + [\boldsymbol{\Omega}\times\boldsymbol{E}]\tau + \boldsymbol{\Omega}(\boldsymbol{\Omega E})\tau^2\big].$$

Let us write the conductivity matrix $\sigma_{\alpha\beta}$ and the resistivity matrix $\rho_{\alpha\beta} = \sigma_{\alpha\beta}^{-1}$ reciprocal to the conductivity one

$$\sigma_{\alpha\beta} = \begin{pmatrix} \frac{\sigma_0}{1+\Omega^2\tau^2} & -\frac{\sigma_0\Omega\tau}{1+\Omega^2\tau^2} & 0 \\ \frac{\sigma_0\Omega\tau}{1+\Omega^2\tau^2} & \frac{\sigma_0}{1+\Omega^2\tau^2} & 0 \\ 0 & 0 & \sigma_0 \end{pmatrix}, \quad \rho_{\alpha\beta} = \begin{pmatrix} \sigma_0^{-1} & -RH & 0 \\ RH & \sigma_0^{-1} & 0 \\ 0 & 0 & \sigma_0^{-1} \end{pmatrix}.$$

The magnetic field is applied in the z-axis direction. The notations here are as follows: $\sigma_0 = ne^2\tau/m$, $RH = -\Omega\tau/\sigma_0$ and $R = -1/(nec)$ is the *Hall constant*. The dependence of electric field $\boldsymbol{E}$ upon current density $\boldsymbol{j}$ is clearly expressed with the aid of the resistivity matrix. This dependence can be represented in the vector form as

$$\boldsymbol{E} = \sigma_0^{-1}\boldsymbol{j} + R[\boldsymbol{H} \times \boldsymbol{j}].$$

In conclusion, we underline that in two-dimensional metal (metal film), the resistance is commonly measured in units of the *von Klitzing constant* $R_K = h/e^2 \approx 25.8\ \text{k}\Omega$ $(h = 2\pi\hbar)$. The Hall resistance is represented as $\rho_{xy} = -\text{sgn}\,(e)\,R_K/\nu$. The dimensionless parameter $\nu = 2\pi n l_H^2$ is called the *filling factor* of Landau levels and is determined by a product of surface charge density n and *magnetic length* $l_H = \sqrt{\hbar c/|e|H}$.

3.4 Kinetic Phenomena in a Metal and Normal Fermi Liquid

9. The collision integral is commonly represented as a difference of two terms. The first term is responsible for the number of events with injecting the particles to the state of interest (*in*) and the second one does for the number of events with ejecting the particles from the same state of interest (*out*). We can write the number of all particles coming to the state with momentum $\boldsymbol{p}$ from the state with some momentum $\boldsymbol{p}'$, i.e. $\boldsymbol{p}' \to \boldsymbol{p}$, as follows:

$$\text{St}[n_{\boldsymbol{p}}]_{in} = \sum_{\boldsymbol{p}'} W_{\boldsymbol{p},\boldsymbol{p}'}\, n_{\boldsymbol{p}'}(1 - n_{\boldsymbol{p}}).$$

Here $W_{\boldsymbol{p},\boldsymbol{p}'}$ is the particle transition probability from the state with momentum $\boldsymbol{p}'$ to the state with the momentum $\boldsymbol{p}$ and $n_{\boldsymbol{p}}$ is the distribution function of particles. The origination of an additional factor $(1 - n_{\boldsymbol{p}})$ in comparison with the factor $n_{\boldsymbol{p}'}$ reflects simply the number of free states. This point is directly associated with the Fermi statistics since the transition of a particle after collision is only possible to the unoccupied state.

The term responsible for ejecting the particles from the state with momentum $\boldsymbol{p}$ to a certain state with momentum $\boldsymbol{p}'$ can be represented in the analogous form $(\boldsymbol{p} \to \boldsymbol{p}')$ as

$$\text{St}[n_{\boldsymbol{p}}]_{out} = \sum_{\boldsymbol{p}'} W_{\boldsymbol{p}',\boldsymbol{p}}\, n_{\boldsymbol{p}}(1 - n_{\boldsymbol{p}'}),$$

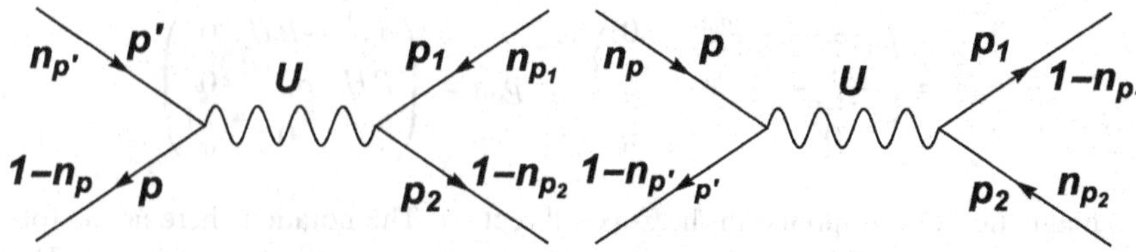

Fig. 3.2 Two possible diagrams for the fermion-fermion scattering. The solid line labels a fermion and the wavy one does the interaction. The factor n should be put on the input line and factor $(1-n)$ should be put on the output line

$W_{\boldsymbol{p}',\boldsymbol{p}}$ being the particle transition probability from the state with momentum $\boldsymbol{p}$ to the state with the momentum $\boldsymbol{p}'$. An additional factor $(1-n_{\boldsymbol{p}'})$, resulted from the Fermi statistics, characterizes the number of final unoccupied states.

We determine the probabilities of transitions $\boldsymbol{p}' \to \boldsymbol{p}$ and $\boldsymbol{p} \to \boldsymbol{p}'$ using the Born approximation and Fermi statistics

$$W_{\boldsymbol{p},\boldsymbol{p}'} = \frac{2\pi}{\hbar} \sum_{\boldsymbol{p}_1,\boldsymbol{p}_2} |U_{\boldsymbol{p},\boldsymbol{p}'}|^2$$
$$\times\, n_{\boldsymbol{p}_1}(1-n_{\boldsymbol{p}_2})\delta(\varepsilon_{\boldsymbol{p}}+\varepsilon_{\boldsymbol{p}_2}-\varepsilon_{\boldsymbol{p}'}-\varepsilon_{\boldsymbol{p}_1})\delta(\boldsymbol{p}+\boldsymbol{p}_2-\boldsymbol{p}'-\boldsymbol{p}_1),$$

$$W_{\boldsymbol{p}',\boldsymbol{p}} = \frac{2\pi}{\hbar} \sum_{\boldsymbol{p}_1,\boldsymbol{p}_2} |U_{\boldsymbol{p}',\boldsymbol{p}}|^2$$
$$\times\, n_{\boldsymbol{p}_2}(1-n_{\boldsymbol{p}_1})\delta(\varepsilon_{\boldsymbol{p}'}+\varepsilon_{\boldsymbol{p}_1}-\varepsilon_{\boldsymbol{p}}-\varepsilon_{\boldsymbol{p}_2})\delta(\boldsymbol{p}'+\boldsymbol{p}_1-\boldsymbol{p}-\boldsymbol{p}_2).$$

Here $U_{\boldsymbol{p}',\boldsymbol{p}} = U_{\boldsymbol{p}'-\boldsymbol{p}} = \int d^3r\, U(\boldsymbol{r})e^{-i(\boldsymbol{p}'-\boldsymbol{p})\boldsymbol{r}}$ is the Fourier transform of two-particle coupling potential $U(\boldsymbol{r})$. The factors like $n_{\boldsymbol{p}}$ and $1-n_{\boldsymbol{p}}$ correspond to the initial occupied and to the final unoccupied states. In Fig. 3.2 the diagrams are shown to make clear the scattering processes of two fermions and the formulas for two contributions to the collision integral. Two δ-functions take the conservation laws of energy and momentum for the particle collisions into account.

The collision integral is given by a difference of two contributions: the gain ("incomings") (*in*) and the loss (*out*).

$$\mathrm{St}[n_{\boldsymbol{p}}] = \mathrm{St}[n_{\boldsymbol{p}}]_{in} - \mathrm{St}[n_{\boldsymbol{p}}]_{out}.$$

For the further calculations it is useful to introduce the following notations for momenta and energies:

$$\boldsymbol{p}=\boldsymbol{k}, \quad \boldsymbol{p}_1=\boldsymbol{k}', \quad \boldsymbol{p}_2=\boldsymbol{k}'+\boldsymbol{q}, \quad \boldsymbol{p}'=\boldsymbol{p}+\boldsymbol{q}=\boldsymbol{k}+\boldsymbol{q}, \quad \omega=\varepsilon_{\boldsymbol{k}}-\varepsilon_{\boldsymbol{k}+\boldsymbol{q}}.$$

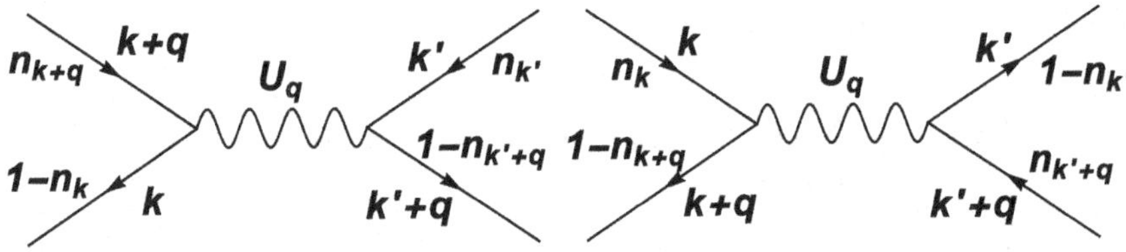

Fig. 3.3 Two possible diagrams of fermion-fermion scattering. The solid line labels a fermion and the wavy one does the interaction. The momentum transfer in the scattering process is equal to q. The factor n should be put on the input line and factor$(1-n)$ should be put on the output line

Next, involving integration over momentum $\boldsymbol{p}_2$ which takes one δ-function out, we arrive at the following expression with two momentum integration alone:

$$
\begin{aligned}
\mathrm{St}[n_{\boldsymbol{k}}] &= \frac{2\pi}{\hbar}\sum_{\boldsymbol{k}',\boldsymbol{q}}|U_{\boldsymbol{q}}|^2 \\
&\times\left[n_{\boldsymbol{k}+\boldsymbol{q}}(1-n_{\boldsymbol{k}})n_{\boldsymbol{k}'}(1-n_{\boldsymbol{k}'+\boldsymbol{q}})-n_{\boldsymbol{k}}(1-n_{\boldsymbol{k}+\boldsymbol{q}})n_{\boldsymbol{k}'+\boldsymbol{q}}(1-n_{\boldsymbol{k}'})\right] \\
&\times\delta(\varepsilon_{\boldsymbol{k}+\boldsymbol{q}}+\varepsilon_{\boldsymbol{k}'}-\varepsilon_{\boldsymbol{k}}-\varepsilon_{\boldsymbol{k}'+\boldsymbol{q}}).
\end{aligned}
$$

In the course of derivation we have paid attention that $|U_{\boldsymbol{q}}|^2 = U_{\boldsymbol{q}}U_{-\boldsymbol{q}} = |U_{-\boldsymbol{q}}|^2$ since the interaction potential $U(\boldsymbol{r})$ is a real-valued quantity. The diagrams in Fig. 3.3 are shown to elucidate the fermion-fermion scattering processes and the writing of formula for the collision integral in new notations.

For the further transformation $\mathrm{St}[n_{\boldsymbol{k}}]$, we use the following trick here. So, let us introduce two additional integration over energy with the aid of two δ-functions, namely, $\delta(\varepsilon_{\boldsymbol{k}+\boldsymbol{q}}-\varepsilon_{\boldsymbol{k}}-\omega)$ and $\delta(\varepsilon'-\varepsilon_{\boldsymbol{k}'})$, each of them will produce identical unity after integrating over ω and ε'. The momentum $\boldsymbol{q}$ and energy ω have the meaning of the momentum and energy variation as a result of scattering one particle with the other. Denoting $\varepsilon = \varepsilon_{\boldsymbol{k}}$ for brevity, we write the collision integral $\mathrm{St}[n_{\boldsymbol{k}}]$ as follows:

$$
\begin{aligned}
\mathrm{St}[n_{\boldsymbol{k}}] &= \int d\omega\int d\varepsilon'\frac{2\pi}{\hbar}\sum_{\boldsymbol{k}',\boldsymbol{q}}|U_{\boldsymbol{q}}|^2 \\
&\times\left[n_{\boldsymbol{k}+\boldsymbol{q}}(1-n_{\boldsymbol{k}})n_{\boldsymbol{k}'}(1-n_{\boldsymbol{k}'+\boldsymbol{q}})-n_{\boldsymbol{k}}(1-n_{\boldsymbol{k}+\boldsymbol{q}})n_{\boldsymbol{k}'+\boldsymbol{q}}(1-n_{\boldsymbol{k}'})\right] \\
&\times\delta(\varepsilon_{\boldsymbol{k}+\boldsymbol{q}}+\varepsilon_{\boldsymbol{k}'}-\varepsilon_{\boldsymbol{k}}-\varepsilon_{\boldsymbol{k}'+\boldsymbol{q}})\,\delta(\varepsilon'-\varepsilon_{\boldsymbol{k}'})\,\delta(\varepsilon_{\boldsymbol{k}+\boldsymbol{q}}-\varepsilon_{\boldsymbol{k}}-\omega)
\end{aligned}
$$

or with regard to δ-functions

$$
\begin{aligned}
\mathrm{St}[n_{\varepsilon}] &= \int d\omega\int d\varepsilon'\frac{2\pi}{\hbar}\sum_{\boldsymbol{k}',\boldsymbol{q}}|U_{\boldsymbol{q}}|^2 \\
&\times\left[n_{\varepsilon+\omega}(1-n_{\varepsilon})n_{\varepsilon'}(1-n_{\varepsilon'+\omega})-n_{\varepsilon}(1-n_{\varepsilon+\omega})n_{\varepsilon'+\omega}(1-n_{\varepsilon'})\right] \\
&\times\delta(\varepsilon_{\boldsymbol{k}'+\boldsymbol{q}}-\varepsilon_{\boldsymbol{k}'}-\omega)\delta(\varepsilon'-\varepsilon_{\boldsymbol{k}'})\,\delta(\varepsilon_{\boldsymbol{k}+\boldsymbol{q}}-\varepsilon_{\boldsymbol{k}}-\omega).
\end{aligned}
$$

Introducing the spectral density ($\varepsilon = \varepsilon_{\boldsymbol{k}}$)

$$K(\varepsilon, \varepsilon', \omega) = \frac{2\pi}{\hbar} \sum_{\boldsymbol{k}', \boldsymbol{q}} |U_{\boldsymbol{q}}|^2 \delta(\varepsilon_{\boldsymbol{k}+\boldsymbol{q}} - \varepsilon_{\boldsymbol{k}} - \omega)\, \delta(\varepsilon' - \varepsilon_{\boldsymbol{k}'})\, \delta(\varepsilon_{\boldsymbol{k}'+\boldsymbol{q}} - \varepsilon_{\boldsymbol{k}'} - \omega),$$

we write down the collision integral in the following evident form:

$$\mathrm{St}[n_\varepsilon] = \int d\omega \int d\varepsilon'\, K(\varepsilon, \varepsilon', \omega)$$
$$\times \left[n_{\varepsilon+\omega}(1 - n_\varepsilon) n_{\varepsilon'}(1 - n_{\varepsilon'+\omega}) - n_\varepsilon(1 - n_{\varepsilon+\omega}) n_{\varepsilon'+\omega}(1 - n_{\varepsilon'})\right].$$

As a first step, we estimate the order of magnitude for the spectral density $K(\varepsilon, \varepsilon', \omega)$ from the dimensional speculations. In fact, each of three δ-functions has dimension $1/\varepsilon$ but two integrations over momenta $\boldsymbol{k}'$ and $\boldsymbol{q}$ entail the dimensional contribution of the order of $(\nu(\varepsilon)\, \varepsilon)^2$ where $\nu(\varepsilon)$ is the density of states. As a result, we obtain

$$K \sim \frac{(\nu(\varepsilon)\, \varepsilon)^2}{\varepsilon^3} \frac{|U|^2}{\hbar} \sim \frac{g^2}{\hbar \varepsilon}, \qquad g = \nu |U|.$$

Here the so-called dimensionless quantity g represents the fermion-fermion coupling constant. Since in degenerate Fermi liquid, all the typical energies are of the order of Fermi energy ε_F, one may put $\varepsilon \sim \varepsilon_F$ for the estimate and, correspondingly, $\nu \sim \nu(\varepsilon_F) \sim \varepsilon_F^{-1} a^{-3}$ and $|U_{\boldsymbol{q}}| \sim \varepsilon_F a^3$ where a is about the mean spacing between particles. We obtain readily the following estimate:

$$K \sim \frac{g^2}{\hbar \varepsilon_F} \quad \text{and} \quad g \sim 1.$$

Let us discuss calculation $K(\varepsilon, \varepsilon', \omega)$ more detailed. We are interested in the energies $\varepsilon, \varepsilon'$ close to the Fermi energy ε_F and in the small energy transfer $|\omega| \ll \varepsilon_F$. In the integrand the region of small momentum transfer $q \ll p_F$ is of most interest. The point is that the integrand expression diverges most drastically as a result of coinciding the arguments in the δ-functions. For small q, we have the following expansion:

$$\varepsilon_{\boldsymbol{k}+\boldsymbol{q}} \approx \varepsilon_{\boldsymbol{k}} + q v_{\boldsymbol{k}} \boldsymbol{e}_{\boldsymbol{q}} \boldsymbol{e}_{\boldsymbol{k}} = \varepsilon_{\boldsymbol{k}} + q v_F \cos\vartheta, \quad v_{\boldsymbol{k}} \sim v_F\,,$$
$$\varepsilon_{\boldsymbol{k}'+\boldsymbol{q}} \approx \varepsilon_{\boldsymbol{k}'} + q v_{\boldsymbol{k}'} \boldsymbol{e}_{\boldsymbol{q}} \boldsymbol{e}_{\boldsymbol{k}'} = \varepsilon_{\boldsymbol{k}'} + q v_F \cos\vartheta', \quad v_{\boldsymbol{k}'} \sim v_F,$$

where $\boldsymbol{e}_{\boldsymbol{q}}, \boldsymbol{e}_{\boldsymbol{k}}, \boldsymbol{e}_{\boldsymbol{k}'}$ are the unit vectors aligned in the directions of vectors $\boldsymbol{q}$, $\boldsymbol{k}$, and $\boldsymbol{k}'$, respectively. So, it is required to calculate the integral

$$K = \frac{2\pi}{\hbar} \int_0^\infty \frac{4\pi q^2\, dq}{(2\pi\hbar)^3} |U_{\boldsymbol{q}}|^2 \int \delta(\varepsilon' - \varepsilon_{\boldsymbol{k}'}) \nu(\varepsilon_{\boldsymbol{k}'})\, d\varepsilon_{\boldsymbol{k}'}$$
$$\times \int \frac{d\Omega_{\boldsymbol{q}}}{4\pi} \int \frac{d\Omega_{\boldsymbol{k}'}}{4\pi} \delta(q v_F \cos\vartheta - \omega)\, \delta(q v_F \cos\vartheta' - \omega).$$

After integration over two solid angles the integral reduces to the form

$$K = \frac{2\pi}{\hbar}\int_0^{\infty}\frac{q^2\,dq}{2\pi^2\hbar^3}|U_q|^2\nu(\varepsilon')\frac{\theta(qv_F-|\omega|)}{(2qv_F)^2} = \frac{2\pi}{\hbar}\nu(\varepsilon')\int_{|\omega|/v_F}^{\infty}\frac{dq}{8\pi^2\hbar^3}\frac{|U_q|^2}{v_F^2}\,.$$

We see that singularity $1/q^2$ at small momenta q is completely compensated in the three-dimensional case and, therefore, the integrand function proves to be non-singular at small transfer of energy $\omega\to 0$ and momentum $q\to 0$. Thus, dependence K upon ω becomes insignificant and, for the approximate estimate, we can put the lower integration limit equal to zero. It is worthwhile to emphasize that this is wrong for two-dimensional Fermi liquid since singularity $1/q^2$ cannot be compensated with the nominator proportional only to q in first power in this case. Thus the spectral function K will essentially be ω-dependent.

The upper limit of integrating over momentum q_{max} can approximately be restricted by the momentum about $\sim 2p_F$. In fact, let us consider argument $\varepsilon_{\boldsymbol{k}'+\boldsymbol{q}}-\varepsilon_{\boldsymbol{k}'}-\omega$ in the δ-function. As we have seen above, it is possible to put $\omega=0$ and nonzero contribution from the δ-function takes place only in the momentum region q obeying the equality $\varepsilon_{\boldsymbol{k}'+\boldsymbol{q}}=\varepsilon_{\boldsymbol{k}'}$. For the region of large momenta q, it is obvious that this equality can be satisfied at $\boldsymbol{q}=-2\boldsymbol{k}'$, i.e. $q=2k'$. For spectrum $\varepsilon_{\boldsymbol{k}}=k^2/2m$, the exact result would be simply $q\leqslant q_{max}=2k'$. Since we are interested in the region of values $\varepsilon'\sim\varepsilon_F$ and $k'\sim p_F$, we can approximately take that $q_{max}\sim 2p_F$. So,

$$K \approx \frac{2\pi}{\hbar}\nu(\varepsilon_F)\int_0^{2p_F}\frac{dq}{2\pi^2\hbar^3}\frac{|U_q|^2}{(2v_F)^2}\sim\frac{\pi\nu^2(\varepsilon_F)}{\hbar}\frac{|U_0|^2}{p_Fv_F}\sim\frac{g^2}{\hbar\varepsilon_F}\,,$$

$$\nu(\varepsilon_F)=\frac{p_F^2}{2\pi^2\hbar^3v_F}\,,\qquad g=\nu(\varepsilon_F)|U_0|,$$

where $\nu(\varepsilon_F)$ is the density of states with the fixed spin projection at the Fermi surface and g is the dimensionless coupling constant. We assume also that the radius of interaction potential $U(\boldsymbol{r})$ does not exceed the average spacing between particles $\sim\hbar/p_F$. Accordingly, we can use $U(q\lesssim 2p_F)\approx U_0$ for the estimate.

Let us return to the collision integral and rewrite it in the following form:

$$\mathrm{St}[n_\varepsilon] = K\int d\omega\,d\varepsilon'\big[n_{\varepsilon+\omega}(1-n_\varepsilon)n_{\varepsilon'}(1-n_{\varepsilon'+\omega}) - n_\varepsilon(1-n_{\varepsilon+\omega})n_{\varepsilon'+\omega}(1-n_{\varepsilon'})\big].$$

As it should be, the collision integral vanishes at the equilibrium Fermi distribution function $n=n_0(\varepsilon)$. This is readily seen from the equality

$$\frac{n_{\varepsilon+\omega}}{1-n_{\varepsilon+\omega}}\,\frac{n_{\varepsilon'}}{1-n_{\varepsilon'}}=\frac{n_{\varepsilon'+\omega}}{1-n_{\varepsilon'+\omega}}\,\frac{n_\varepsilon}{1-n_\varepsilon}$$

which satisfies identically at $n = n_0(\varepsilon)$ due to

$$\frac{n_0}{1-n_0} = \exp\left(-\frac{\varepsilon-\mu}{T}\right).$$

Let us determine relaxation time or collision time τ_ε according to the relation

$$\mathrm{St}[n_0+\delta n_\varepsilon] - \mathrm{St}[n_0] = -\frac{\delta n_\varepsilon}{\tau_\varepsilon} \quad \text{or} \quad \frac{1}{\tau_\varepsilon} = -\frac{\delta}{\delta n_\varepsilon}(\mathrm{St}[n_\varepsilon])\bigg|_{n_\varepsilon=n_0},$$

where δn_ε is the small deviation from equilibrium. Then we find

$$\frac{1}{\tau_\varepsilon} = K\int d\omega\, d\varepsilon'\big[n_{\varepsilon+\omega}n_{\varepsilon'}(1-n_{\varepsilon'+\omega}) + (1-n_{\varepsilon+\omega})(1-n_{\varepsilon'})n_{\varepsilon'+\omega}\big]$$
$$= K\int d\omega\, d\varepsilon'\big[n_{\varepsilon+\omega}(n_{\varepsilon'} - n_{\varepsilon'+\omega}) + n_{\varepsilon'+\omega}(1-n_{\varepsilon'})\big],$$

and take the equilibrium distribution functions $n = n_0$ as functions n. Using the values of the following integrals:

$$\int_{-\infty}^{\infty} d\varepsilon'\big[n_0(\varepsilon') - n_0(\varepsilon'+\omega)\big] = -T\ln\frac{1+e^{-(\varepsilon'-\mu)/T}}{1+e^{-(\varepsilon'-\mu+\omega)/T}}\bigg|_{-\infty}^{\infty} = \omega,$$

$$\int_{-\infty}^{\infty} d\varepsilon' n_0(\varepsilon'+\omega)\big[1-n_0(\varepsilon')\big]$$
$$= \frac{T}{e^{\omega/T}-1}\ln\frac{1+e^{-(\varepsilon'-\mu+\omega)/T}}{1+e^{-(\varepsilon'-\mu)/T}}\bigg|_{-\infty}^{\infty} = \frac{\omega}{e^{\omega/T}-1},$$

we obtain

$$\frac{1}{\tau_\varepsilon} = K\int_{-\infty}^{\infty} d\omega\left(\frac{\omega}{e^{(\varepsilon+\omega-\mu)/T}+1} + \frac{\omega}{e^{\omega/T}-1}\right)$$
$$= \frac{K}{2}\int_{-\infty}^{\infty} d\omega\,\omega\left(\coth\frac{\omega}{2T} - \tanh\frac{\varepsilon+\omega-\mu}{2T}\right).$$

Calculating the integral with the aid of formula

$$\int_{-\infty}^{\infty} dx\, x\big(\coth x - \tanh(x+a)\big) = \int_{-\infty}^{\infty} dx\, \frac{x^2}{2}\left(\frac{1}{\sinh^2 x} + \frac{1}{\cosh^2(x+a)}\right) = \frac{\pi^2}{4} + a^2,$$

we arrive at the final answer for the relaxation time

$$\frac{1}{\tau_\varepsilon(T)} = \frac{K}{2}\left[(\pi T)^2 + (\varepsilon-\mu)^2\right] \sim g^2 \frac{(\pi T)^2 + (\varepsilon-\mu)^2}{\hbar \varepsilon_F}.$$

As is seen, the magnitude of damping for elementary excitations or quasiparticles $\hbar/\tau$ will be small as compared with their typical excitation energy about $|\varepsilon-\mu|$ or T while the inequalities T and $|\varepsilon-\mu| \ll \mu \sim \varepsilon_F$ are valid. In usual metals this condition assumes the temperature $T \ll T_F \sim 10^4 \div 10^5$ K and $T < 0.3$ K for neutral Fermi liquid of helium isotope ^{3}He.

10. Let us employ the results of the previous problem for the three-dimensional case and write down the collision integral in the same form

$$\mathrm{St}[n_k] = \frac{2\pi}{\hbar}\sum_{k',q}|U_q|^2$$
$$\times \left[n_{k+q}(1-n_k)n_{k'}(1-n_{k'+q}) - n_k(1-n_{k+q})n_{k'+q}(1-n_{k'})\right]$$
$$\times\, \delta(\varepsilon_{k+q} + \varepsilon_{k'} - \varepsilon_k - \varepsilon_{k'+q}).$$

Now all the momenta are two-dimensional vectors. We define the relaxation time or collision time τ_k according to the previous relation

$$\mathrm{St}[n_0+\delta n_k] - \mathrm{St}[n_0] = -\frac{\delta n_k}{\tau_k} \quad \text{or} \quad \frac{1}{\tau_k} = -\frac{\delta}{\delta n_k}\left(\mathrm{St}[n_k]\right)\Big|_{n_k=n_0},$$

δn_k being the small deviation from equilibrium. Then we find

$$\frac{1}{\tau_k} = \frac{2\pi}{\hbar}\sum_{k',q}|U_q|^2\left[n_{k+q}n_{k'}(1-n_{k'+q}) + (1-n_{k+q})(1-n_{k'})n_{k'+q}\right]$$
$$\times\delta(\varepsilon_{k+q} + \varepsilon_{k'} - \varepsilon_k - \varepsilon_{k'+q}).$$

Here we should insert the equilibrium Fermi functions $n = n_0(\varepsilon_k) = (\exp[(\varepsilon_k - \mu)/T] + 1)^{-1}$ instead of all functions n. For the next transformation $1/\tau_k$, we apply the following two formulas for the equilibrium distribution Fermi functions $n = n_0$:

$$n_{k'}\left(1-n_{k'+q}\right) = \frac{n_{k'} - n_{k'+q}}{1 - e^{-(\varepsilon_{k'+q}-\varepsilon_{k'})/T}},$$

$$\left(1-n_{k'}\right)n_{k'+q} = \frac{n_{k'} - n_{k'+q}}{e^{(\varepsilon_{k'+q}-\varepsilon_{k'})/T} - 1},$$

and introduce an additional integration with the aid of δ-function. As a result, we obtain

$$\frac{1}{\tau_k} = \frac{2\pi}{\hbar}\int d\omega \sum_q |U_q|^2\left[\frac{n_{k+q}}{1-e^{-\omega/T}} + \frac{1-n_{k+q}}{e^{\omega/T}-1}\right]\delta(\varepsilon_{k+q}-\varepsilon_k-\omega)$$
$$\times\sum_{k'}(n_{k'}-n_{k'+q})\delta(\varepsilon_{k'+q}-\varepsilon_{k'}-\omega)$$

and rewrite it with notation $\varepsilon = \varepsilon_k$

$$\frac{1}{\tau_\varepsilon} = \frac{2\pi}{\hbar}\int d\omega \sum_q |U_q|^2\left[\frac{n_{\varepsilon+\omega}}{1-e^{-\omega/T}} + \frac{1-n_{\varepsilon+\omega}}{e^{\omega/T}-1}\right]\delta(\varepsilon_{k+q}-\varepsilon_k-\omega)$$
$$\times\sum_{k'}(n_{\varepsilon'}-n_{\varepsilon'+\omega})\delta(\varepsilon_{k'+q}-\varepsilon_{k'}-\omega).$$

Then we use the relation valid for the equilibrium functions $n_\varepsilon = n_0(\varepsilon)$

$$\frac{n_{\varepsilon+\omega}}{1-e^{-\omega/T}} + \frac{1-n_{\varepsilon+\omega}}{e^{\omega/T}-1} = \frac{1}{2}\left[\coth\frac{\omega}{2T} - \tanh\frac{\varepsilon-\mu+\omega}{2T}\right],$$

and introduce also function $\chi(\boldsymbol{q},\omega)$ according to definition

$$\chi(\boldsymbol{q},\omega) = \pi\sum_{k'}(n_{\varepsilon'}-n_{\varepsilon'+\omega})\delta(\varepsilon_{k'+q}-\varepsilon_{k'}-\omega), \quad \varepsilon' = \varepsilon_{k'}.$$

After that the relaxation time can be written in the following transparent form:

$$\frac{1}{\tau_\varepsilon} = \frac{1}{\hbar}\int d\omega \sum_q |U_q|^2\left[\coth\frac{\omega}{2T} - \tanh\frac{\varepsilon-\mu+\omega}{2T}\right]\delta(\varepsilon_{k+q}-\varepsilon_k-\omega)\chi(\boldsymbol{q},\omega).$$

In what follows, we will be interested in the small excitation energies $|\varepsilon-\mu| \ll \mu$ and low temperatures $T \ll \mu$. Accordingly, we consider $\varepsilon_k \sim \mu$ and the magnitude of momenta k about Fermi momentum p_F. As it will be seen below, the regions for the values of variables ω, q and k' which are essential and give the main contribution with integration are determined by the following inequalities $\omega \ll \mu$, $\omega/v_F \lesssim q \lesssim p_F$ and $k' \sim p_F$.

Let us analyze behavior $\chi(\boldsymbol{q},\omega)$ at small momenta $q \ll p_F$ and energies $\omega \lesssim qv_F \ll \mu$. For small values $q \ll k' \sim p_F$, putting approximately

$$\varepsilon_{k'+q} - \varepsilon_{k'} \approx qv_{k'}\cos\varphi, \quad v_{k'} \sim v_F,$$

we will have

$$\chi(q,\omega)=\pi\int_0^{\infty}\frac{k'dk'}{(2\pi\hbar)^2}(n_{\varepsilon'}-n_{\varepsilon'+\omega})\int_0^{2\pi}d\varphi'\,\delta(qv_F\cos\varphi'-\omega).$$

Integrating over angle φ' between vectors $\boldsymbol{k}'$ and $\boldsymbol{q}$ yields

$$\frac{2\vartheta(qv_F-|\omega|)}{\sqrt{q^2v_F^2-\omega^2}}\approx\frac{2}{qv_F}\vartheta(qv_F-|\omega|)\quad\text{at}\quad qv_F\gtrsim|\omega|.$$

While integrating over k', it is profitable to go over to variable $\xi'=\varepsilon'-\mu$. The principle region where the integral is gained corresponds to inequality $|\xi'|\ll\mu$. Also, due to rapid convergence one can extend the integration limits from $-\infty$ to $+\infty$ and replace $k'\,dk'$ with $(p_F/v_F)d\xi'$. As a result, we obtain

$$\begin{aligned}&\chi(q\gtrsim|\omega|v_F^{-1},\omega)\\&=\frac{\pi}{(2\pi\hbar)^2}\int_{-\infty}^{+\infty}\frac{d\xi'}{2}\left[\tanh\frac{\xi'+\omega}{2T}-\tanh\frac{\xi'}{2T}\right]\frac{2}{qv_F}\vartheta(qv_F-|\omega|)\\&\qquad=\frac{p_F}{2\pi\hbar^2v_F}\frac{\omega}{qv_F}\vartheta(qv_F-|\omega|)=\nu(\varepsilon_F)\frac{\omega}{qv_F}\vartheta(qv_F-|\omega|),\end{aligned}$$

where $\nu(\varepsilon_F)=p_F/(2\pi\hbar^2v_F)$ is the density of states at the Fermi surface with the fixed spin projections.

Let us turn to integrating over momentum $\boldsymbol{q}$. We will assume the particle-particle interaction $U(r)$ to be isotropic, i.e. $U(\boldsymbol{q})=U(q)$. Integration over angle φ between vectors $\boldsymbol{k}$ and $\boldsymbol{q}$ can be performed likewise the integration over angle φ'. Then we obtain the following estimate:

$$\begin{aligned}\frac{\hbar}{\tau_\varepsilon}&\approx\int d\omega\left[\coth\frac{\omega}{2T}-\tanh\frac{\varepsilon-\mu+\omega}{2T}\right]\int_0^{\infty}\frac{q\,dq}{(2\pi\hbar)^2}|U_q|^2\frac{2\vartheta(qv_F-|\omega|)}{qv_F}\chi(q,\omega)\\&=\frac{2\nu(\varepsilon_F)}{(2\pi\hbar)^2v_F^2}\int\left[\coth\frac{\omega}{2T}-\tanh\frac{\varepsilon-\mu+\omega}{2T}\right]\omega\,d\omega\int_0^{\infty}|U_q|^2\frac{dq}{q}\vartheta(qv_F-|\omega|).\end{aligned}$$

In the similar way as in the previous problem, we assume that the radius of interaction potential $U(\boldsymbol{r})$ does not exceed the average interparticle spacing $\sim\hbar/p_F$ and we put $U(q\lesssim p_F)\approx U_0$ for the estimate. First of all, we emphasize the following. In the integral over q the estimate of integrand is valid, strictly speaking, in the momentum region $q\gtrsim|\omega|/v_F$ alone. On the other hand, the integral has a logarithmic character and, thus, is not sensitive to the exact integration limits within the logarithmic accuracy. Cutting the integral at the lower limit $q=|\omega|/v_F$ and at the upper one $q\sim p_F$, we obtain

$$\frac{\hbar}{\tau_\varepsilon} \approx \frac{\nu(\varepsilon_F)|U_0|^2 p_F}{2\pi\hbar^2 v_F}\frac{1}{\pi p_F v_F}\int d\omega\left[\coth\frac{\omega}{2T} - \tanh\frac{\varepsilon-\mu+\omega}{2T}\right]\omega\ln\frac{p_F v_F}{|\omega|}.$$

For the simple estimate, we put $p_F v_F \sim \varepsilon_F$ and determine also the dimensionless coupling constant g according to $g = |U_0|\nu(\varepsilon_F)$. Then the expression for estimating the relaxation time takes the form

$$\frac{1}{\tau_\varepsilon} \approx \frac{g^2}{\hbar\varepsilon_F}\int\limits_{-\infty}^{\infty} d\omega\left[\coth\frac{\omega}{2T} - \tanh\frac{\varepsilon-\mu+\omega}{2T}\right]\omega\ln\frac{\varepsilon_F}{|\omega|}.$$

As compared with the three-dimensional case, we see the logarithmic amplification in the integrand at small ω. Since the logarithmic function is sufficiently slow, its contribution can be estimated with taking ω equal to $\omega_0 \sim T$ or $\varepsilon - \mu$. The choice depends on the size of the region where the rest part of the integrand differs noticeably from zero. For $T = 0$, we find with the logarithmic accuracy

$$\frac{1}{\tau_\varepsilon} \sim \frac{g^2}{\hbar\varepsilon_F}|\varepsilon-\mu|^2\ln\frac{\varepsilon_F}{|\varepsilon-\mu|}, \qquad |\varepsilon-\mu| \ll \mu.$$

For the finite temperatures and $|\varepsilon - \mu| \ll T$, it is readily to see that

$$\frac{1}{\tau_\varepsilon} \sim \frac{g^2}{\hbar\varepsilon_F}(\pi T)^2\ln\frac{\varepsilon_F}{T}, \qquad |\varepsilon-\mu| \ll T \ll \mu.$$

Thus, as compared with the three-dimensional case, in the two-dimensional Fermi liquid the magnitude of the damping $\hbar/\tau_\varepsilon$ of elementary excitations or quasiparticles is multiplied by the large logarithmic factor about

$$\ln\frac{\varepsilon_F}{max\{T,\ |\varepsilon-\mu|\}}.$$

In the one-dimensional Fermi liquid, the damping magnitude $\hbar/\tau_\varepsilon$ even for the slightly excited states becomes comparable with their typical energy $|\varepsilon - \mu|$ or T. Therefore, the general representation about an existence of weakly damped elementary excitations in the case of one-dimensional Fermi liquid becomes inadequate.

11. Let us study small deviations of distribution function $\delta n = n - n_0$ from equilibrium one n_0. Accordingly, the quasiparticle energy $H = H(\boldsymbol{p}, \boldsymbol{r})$ varies as well

$$n = n_0(\boldsymbol{p}) + \delta n(\boldsymbol{r},\ \boldsymbol{p},\ t) \quad\text{and}\quad H = \varepsilon_0(\boldsymbol{p}) + \delta\varepsilon[n(\boldsymbol{r},\ \boldsymbol{p},\ t)]\,.$$

In the linear approximation in δn and $\delta\varepsilon$, we find using the Liouville equation

$$\frac{\partial\delta n}{\partial t} + \frac{\partial\delta n}{\partial\boldsymbol{r}}\frac{\partial\varepsilon_0}{\partial\boldsymbol{p}} - \frac{\partial n_0}{\partial\boldsymbol{p}}\frac{\partial\delta\varepsilon}{\partial\boldsymbol{r}} = 0\,.$$

The quasiparticle velocity equals $\boldsymbol{v} = \partial\epsilon_0/\partial\boldsymbol{p}$ and the derivative of distribution function has the form $\partial n_0/\partial\boldsymbol{p} = \boldsymbol{v}\partial n_0/\partial\varepsilon$. The solution is attempted as a traveling wave

$$\delta n(\boldsymbol{r},\ \boldsymbol{p},\ t) = \delta n(\boldsymbol{p})e^{i\boldsymbol{kr}-i\omega t},$$

where $\boldsymbol{k}$ is the wave vector and ω is the oscillation frequency. The variation of quasiparticle energy in terms of the interaction function $f(\boldsymbol{p},\ \boldsymbol{p}')$ is given by

$$\delta\varepsilon = \int f(\boldsymbol{p},\ \boldsymbol{p}')\delta n(\boldsymbol{p}')e^{i\boldsymbol{kr}-i\omega t}d\tau_{p'}\,,\quad d\tau_{p'} = 2\frac{d^3p'}{(2\pi\hbar)^3}\,.$$

Thus

$$(\omega - \boldsymbol{kv})\,\delta n(\boldsymbol{p}) = -(\boldsymbol{kv})\frac{\partial n_0}{\partial\varepsilon}\int f(\boldsymbol{p},\ \boldsymbol{p}')\,\delta n(\boldsymbol{p}')\,d\tau_{p'}\,.$$

To simplify the analysis, we put $f(\boldsymbol{p},\ \boldsymbol{p}') = f_0 = \text{const}$ and temperature $T = 0$. Then

$$(\omega - \boldsymbol{kv})\,\delta n(\boldsymbol{p}) = -(\boldsymbol{kv})\frac{\partial n_0}{\partial\varepsilon}\,f_0\int \delta n(\boldsymbol{p}')\,d\tau_{p'}\,.$$

Looking at the structure of the equation, we disclose that the solution should have the form like $\delta n(\boldsymbol{p}) = n_0'(\varepsilon)\nu(\theta)$ where θ is the angle between vectors $\boldsymbol{v}$ and $\boldsymbol{k}$. Since $n_0'(\varepsilon) = -\delta(p - p_F)/v_F$, the correction $\delta n(\boldsymbol{p})$ is nonzero at the Fermi surface alone, i.e. at $|\boldsymbol{p}| = p_F$ and $|\boldsymbol{v}| = v_F$. Substituting $\delta n(\boldsymbol{p})$, we find

$$(\omega - \boldsymbol{kv})\nu(\theta) = \boldsymbol{kv}F_0\int \nu(\theta')\frac{d\Omega_{\theta'}}{4\pi}\,,\quad F_0 = \frac{p_F^2}{\pi^2\hbar^3 v_F}\,f_0 = \frac{m^* p_F}{\pi^2\hbar^3}\,f_0.$$

Here F_0 is the dimensionless *Landau parameter*, $m^* p_F/\pi^2\hbar^3$ is the density of states, and $d\Omega = 2\pi\sin\theta\,d\theta$ is the element of a solid angle. Then we obtain

$$\frac{\nu(\theta)}{F_0} = \frac{\cos\theta}{s - \cos\theta}\int \nu(\theta')\frac{d\Omega_{\theta'}}{4\pi}\quad\text{and}\quad s = \frac{\omega}{v_F k}\,.$$

From the condition of self-consistency

$$\frac{1}{F_0}\int \nu(\theta)\frac{d\Omega_\theta}{4\pi} = \int \frac{\cos\theta}{s - \cos\theta}\frac{d\Omega_\theta}{4\pi}\left(\int \nu(\theta')\frac{d\Omega_{\theta'}}{4\pi}\right)$$

we find

$$\frac{1}{F_0} = \int_0^\pi \frac{\cos\theta}{s - \cos\theta}\,\frac{2\pi\sin\theta\,d\theta}{4\pi} = \frac{1}{2}\int_{-1}^{1}\frac{x\,dx}{s - x}\,.$$

The dispersion equation

$$\frac{1}{F_0} = \frac{s}{2}\ln\frac{s+1}{s-1} - 1$$

has the real solutions for $F_0 > 0$ alone since the right-hand side varies within the limits $(+\infty, 0)$ as s runs from 1 to $+\infty$. Thus, for $F_0 > 0$ that means the repulsion of quasiparticles at the Fermi surface, the dispersion equation has a single real solution $s = s(F_0)$ which corresponds to undamped wave with the sound dispersion spectrum

$$\omega(k) = s(F_0) v_F k.$$

Such sound-like waves in the Fermi liquid are commonly called the *collisionless sound* or *zero-sound*. If $F_0 \to +0$, one sees $s \to 1$. For $F_0 \gg 1$, $s \approx \sqrt{F_0/3}$. Therefore, the undamped wave processes are only possible, provided that the quasiparticle velocity v_F does not exceed the phase velocity of wave ω/k. In liquid ^{3}He the magnitude of the Landau parameter is $F_0 = 10.8$ and zero-sound velocity is approximately 197 m/s. In the process of zero-sound propagation, the Fermi surface is stretched in the direction of wave propagation and flattened in the opposite direction.

At $F_0 < 0$ there is no real solutions of the dispersion equation and the modes of the Fermi-surface oscillations are damped.

12. Let us write down the transport equation for the distribution function $n_{\boldsymbol{p}}(\boldsymbol{r}, t)$

$$\frac{\partial n_{\boldsymbol{p}}}{\partial t} + \frac{\partial H}{\partial \boldsymbol{p}}\frac{\partial n_{\boldsymbol{p}}}{\partial \boldsymbol{r}} - \frac{\partial H}{\partial \boldsymbol{r}}\frac{\partial n_{\boldsymbol{p}}}{\partial \boldsymbol{p}} = -\frac{1}{\tau}\big[\delta n_{\boldsymbol{p}} - \langle \delta n_{\boldsymbol{p}}\rangle - 3\langle \delta n_{\boldsymbol{p}} \cos\theta\rangle \cos\theta\big].$$

Varying the distribution function $\delta n = n - n_0$ results in changing the quasiparticle energy H

$$H = \varepsilon_0(\boldsymbol{p}) + \delta\varepsilon[n_{\boldsymbol{p}}(\boldsymbol{r}, t)].$$

The change of quasiparticle energy, expressed via the interaction function $f(\boldsymbol{p}, \boldsymbol{p}')$, equals

$$\delta\varepsilon = \int f(\boldsymbol{p}, \boldsymbol{p}')\delta n_{\boldsymbol{p}'}(\boldsymbol{r}, t)\frac{2d^3p'}{(2\pi\hbar)^3} = f_0 \int \delta n_{\boldsymbol{p}'}(\boldsymbol{r}, t)\frac{2d^3p'}{(2\pi\hbar)^3}.$$

In the linear approximation in $\delta n = n - n_0$ one has

$$\frac{\partial\,\delta n_{\boldsymbol{p}}}{\partial t} + \frac{\partial \varepsilon_0}{\partial \boldsymbol{p}}\frac{\partial\,\delta n_{\boldsymbol{p}}}{\partial \boldsymbol{r}} - \frac{\partial n_0}{\partial \boldsymbol{p}}\frac{\partial\,\delta\varepsilon}{\partial \boldsymbol{r}} = -\frac{1}{\tau}\big[\delta n_{\boldsymbol{p}} - \langle \delta n_{\boldsymbol{p}}\rangle - 3\langle \delta n_{\boldsymbol{p}} \cos\theta\rangle \cos\theta\big].$$

The solution of equation is sought as a traveling wave

$$\delta n_{\boldsymbol{p}}(\boldsymbol{r}, t) = \delta n_{\boldsymbol{p}} e^{i\boldsymbol{k}\boldsymbol{r} - i\omega t}.$$

Here $\boldsymbol{k}$ is the wave vector and ω is the oscillation frequency. Taking into account that $\boldsymbol{v} = \partial\varepsilon_0/\partial\boldsymbol{p}$ and $\partial n_0/\partial\boldsymbol{p} = \boldsymbol{v}\partial n_0/\partial\varepsilon$, we arrive at

$$(\omega - \boldsymbol{k}\boldsymbol{v})\delta n_{\boldsymbol{p}} + \boldsymbol{k}\boldsymbol{v}\frac{\partial n_0}{\partial\varepsilon} f_0 \int \delta n_{\boldsymbol{p}'}\frac{2d^3p'}{(2\pi\hbar)^3} = -\frac{i}{\tau}\big[\delta n_{\boldsymbol{p}} - \langle \delta n_{\boldsymbol{p}}\rangle - 3\langle \delta n_{\boldsymbol{p}} \cos\theta\rangle \cos\theta\big].$$

Here $\boldsymbol{kv} = kv\cos\theta$ and θ is the angle between vectors $\boldsymbol{p}$ and $\boldsymbol{k}$. We present an unknown function $\delta n_{\boldsymbol{p}}$ as

$$\delta n_{\boldsymbol{p}} = \frac{\partial n_0}{\partial \varepsilon}\nu(\theta)$$

and then we find readily

$$(\omega - kv\cos\theta)\nu(\theta) + f_0 kv\cos\theta \int \frac{\partial n_0}{\partial \varepsilon'}\frac{p'^2 dp'}{\pi^2\hbar^3}\nu(\theta')\frac{d\Omega'}{4\pi} = -\frac{i}{\tau}\big[\nu(\theta) - \langle\nu\rangle - 3\langle\nu\cos\theta\rangle\cos\theta\big].$$

For integration over the modulus of momentum p', only the region of the magnitudes close to the Fermi momentum p_F is essential. In other words, at low temperature we have approximately $\partial n_0/\partial\varepsilon \approx -\delta(p - p_F)/v_F$ where v_F is the Fermi velocity depending on both the Fermi momentum and the effective mass m^* of quasiparticle excitations according to $v_F = p_F/m^*$.

Introduce the following notations for convenience:

$$\nu_0 = \langle\nu(\theta)\rangle, \quad \nu_1 = \langle\nu(\theta)\cos\theta\rangle \quad \text{and} \quad F_0 = \frac{p_F^2}{\pi^2\hbar^3}f_0 = \frac{m^* p_F}{\pi^2\hbar^3}f_0,$$

where the last quantity F_0 is the dimensionless Landau parameter and $m^* p_F/(\pi^2\hbar^3)$ is the density of states of quasiparticles at the Fermi surface. Then we obtain

$$(\omega - kv_F\cos\theta)\nu(\theta) - kv_F F_0\nu_0\cos\theta = -\frac{i}{\tau}\big(\nu(\theta) - \nu_0 - 3\nu_1\cos\theta\big).$$

Denoting

$$s = \frac{\omega}{kv_F}, \quad \sigma = \frac{1}{kv_F\tau} \quad \text{and} \quad \xi = s + i\sigma,$$

we find that

$$\nu(\theta) = \frac{(F_0\cos\theta + i\sigma)\nu_0 + 3i\sigma\nu_1\cos\theta}{\xi - \cos\theta}.$$

To solve the equation obtained, we calculate

$$\frac{1}{2}\int_0^\pi \nu(\theta)\sin\theta\,d\theta = \Big[F_0 W(\xi) + i\sigma\frac{1 + W(\xi)}{\xi}\Big]\nu_0 + 3i\sigma W(\xi)\nu_1 = \nu_0\,,$$

$$\frac{1}{2}\int_0^\pi \nu(\theta)\cos\theta\sin\theta\,d\theta = [F_0\xi W(\xi) + i\sigma W(\xi)]\nu_0 + 3i\sigma\xi W(\xi)\nu_1 = \nu_1,$$

where $W(\xi) = \frac{\xi}{2}\ln\frac{\xi+1}{\xi-1} - 1$.

The condition of compatibility of the last two equations will give us the equation for determining the dispersion behavior $\omega = \omega(k)$ for the proper oscillations of distribution function

$$\frac{s}{s+i\sigma} - W(s+i\sigma)\left[F_0 + \frac{i\sigma}{s+i\sigma} + 3i\sigma s\right] = 0.$$

Let us first analyze two limiting cases as low- and high-frequency ones. We start by studying the low-frequency $\omega\tau \ll 1$ part of oscillation spectrum. This limit corresponds to $\sigma \to \infty$. Expanding $W(\xi)$ in powers $1/\xi$

$$W(\xi \to \infty) \approx \frac{1}{3\xi^2} + \frac{1}{5\xi^4} + \ldots$$

and taking into account that $\sigma \gg s$, we only keep the terms of the order of $1/\sigma$ in expanding the dispersion equation. As a result, we obtain

$$s^2 = \frac{1+F_0}{3} - \frac{4}{15}\frac{is}{\sigma} \quad \text{or} \quad \left(\frac{\omega}{kv_F}\right)^2 = \frac{1+F_0}{3} - \frac{4}{15}i\omega\tau.$$

The real term yields the velocity of ordinary low-frequency sound

$$u_0 = v_F\sqrt{\frac{1+F_0}{3}}$$

and the imaginary term results in damping and absorbing the sound. (In liquid ^{3}He the sound velocity u_0 is about 187 m/s.) The sound absorption coefficient α_0 is determined as an imaginary part of wave vector

$$\alpha_0(\omega) = \operatorname{Im} k = \frac{2\omega^2 v_F^2 \tau}{15u^3}.$$

The sound attenuation is small, i.e. $\alpha_0 \ll \omega/u$, since $\omega\tau \ll 1$.

Here we would like to remind that the attenuation of normal sound in the Fermi liquid at low $T \ll \varepsilon_F$ temperatures is mainly connected with the coefficient of shear viscosity η and is given by the hydrodynamic formula in this case

$$\alpha_0 = \frac{4}{3}\eta\frac{\omega^2}{2\rho u^3}.$$

Here ρ is the liquid density and we can estimate the coefficient of shear viscosity η as

$$\eta = \frac{1}{5}\rho v_F^2 \tau.$$

It is obvious that our estimate for viscosity correlates with the dimensional one given by a well-known elementary formula in the kinetic theory of gases. In a Fermi liquid, as we have seen in the previous problems, the collision time of quasiparticle excitations increases rapidly according to the law $\tau(T) \sim \varepsilon_F/T^2$ as the temperature lowers. Thus the coefficient of viscosity grows also in accordance with the same law $\eta(T) \sim 1/T^2$.

Since the collision time increases as the temperature lowers, the condition $\omega\tau \sim 1$ can be achieved, implying that the mean free path of excitations $l = v_F\tau$ is also comparable with the wavelength $\lambda = 2\pi/k$. In this situation the sound oscillations become overdamped and the sound ceases to propagate in the Fermi liquid. However, as one continues further to lower the temperature and reaches the high-frequency limit $\omega\tau \gg 1$, the propagation of weakly damped wave processes, called commonly the *zero sound*, becomes possible again.

So, let us analyze dispersion equation in the high-frequency limit $\omega\tau \gg 1$ and, correspondingly, $\sigma \to 0$. For $\sigma = 0$, we obtain the following dispersion equation:

$$1 - W(s)F_0 = 0$$

studied in the previous problem. Its real solution, corresponding to the undamped oscillations, exists at $F_0 > 0$ alone and entails the sound spectrum $\omega = s(F_0)v_F k$. In what follows, we find an imaginary correction to frequency ω, solving the dispersion equation

$$y_\sigma(s) = \frac{s}{s+i\sigma} - W(s+i\sigma)\left[F_0 + \frac{i\sigma}{s+i\sigma} + 3i\sigma s\right] = 0$$

in the linear approximation $\sigma \ll 1$ and using the following formula:

$$s \approx s_0 - \sigma\frac{\left.\frac{\partial y_\sigma(s)}{\partial\sigma}\right|_{\substack{\sigma=0\\ s=s_0}}}{\left.\frac{\partial y_0(s)}{\partial s}\right|_{s=s_0}} = -i\sigma\left[1 + \frac{1+(1+3s_0^2)W(s_0)}{s_0F_0W'(s_0)}\right],$$

where s_0 is the root of equation $W(s_0) = 1/F_0$. Let us return to the usual variables and introduce the notation $u_\infty = s_0v_F$ for the zero-sound velocity. Then from the dispersion equation

$$\omega = u_\infty k - \frac{i}{\tau}\left[1 + \frac{1+(1+3s_0^2)/F_0}{s_0F_0W'(s_0)}\right]$$

we can find the zero-sound attenuation coefficient α_∞

$$\alpha_\infty(\omega) = \operatorname{Im} k = \frac{1}{u_\infty\tau}\left[1 + \frac{1+(1+3s_0^2)/F_0}{s_0F_0W'(s_0)}\right]$$
$$= \frac{1}{u_\infty\tau}\begin{cases} 1 - 16F_0^{-2}e^{-2/F_0}, & F_0 \ll 1, \\ 1/2, & F_0 \gg 1. \end{cases}$$

So, in the high-frequency $\omega\tau \gg 1$ limit or when the mean free path of excitations l is much larger than wavelength λ, there also appears a possibility for the propagation of weakly damping oscillations in the Fermi liquid. In liquid ^{3}He the zero-sound velocity is about 197 m/s. The coefficient of zero-sound attenuation enhances with increasing the temperature as $1/\tau$ and, therefore, is proportional to T^2.

A ratio of high-frequency sound velocity u_∞ to low-frequency one u_0 is always larger than unity for any value of Landau parameter $F_0 > 0$

$$\frac{u_\infty}{u_0} = \begin{cases} \sqrt{3}, & F_0 \ll 1, \\ 1 + \frac{2}{5F_0}, & F_0 \gg 1. \end{cases}$$

The region $\omega\tau \sim 1$ between low-frequency and high-frequency sounds requires a numerical treatment. However, for describing the sound attenuation coefficient $\alpha(\omega)$ in the Fermi liquid in the whole frequency range, one can apply a simple interpolation formula

$$\alpha(\omega) = \frac{B_0\omega^2\tau}{1 + (B_0/B_\infty)(\omega\tau)^2},$$

where B_0 and B_∞ are the fitting parameters independent of frequency and temperature. Because of behavior $\tau \sim T^{-2}$ the temperature T_m corresponding to the sound attenuation maximum depends on the frequency as $T_m \sim \sqrt{\omega}$.

When selecting the effective relaxation time approximation for the collision integral, we have completely neglected a possible dependence for the collision frequency of quasiparticle excitations τ^{-1} upon their energy. Strictly speaking, this is not so. As we have seen in the previous problems, the quasiparticle collision frequency in the Fermi liquid depends on the energy as well

$$\tau_\varepsilon^{-1}(T) \sim \frac{(\pi T)^2 + (\varepsilon - \mu)^2}{\varepsilon_F}.$$

Here μ is the chemical potential indiscernible practically from the Fermi energy ε_F at $T \ll \varepsilon_F$.

In the collision of quasiparticles in the sound wave field of frequency ω their total energy, in general, does not conserve and changes by the magnitude of the order of $\hbar\omega$. Since for the finite temperature the typical excitation energy is already about the same temperature, such inexact conservation of quasiparticle energy is inessential while $T \gg \hbar\omega$. In the opposite $T \ll \hbar\omega$, which concerns mainly the region of zero-sound existence, we may expect that the quantity $\hbar\omega$ will play a role of temperature dispersion in the energy of excitations and the typical collision frequency of quasiparticles will behave as

$$\tau_\omega^{-1}(T) \sim \frac{(\hbar\omega)^2}{\varepsilon_F}.$$

This should entail that the zero-sound attenuation at ultralow $T \ll \hbar\omega$ temperatures becomes proportional to the square of frequency but the temperature

$$\alpha_\infty(\omega) \sim \omega^2 \quad \text{at} \quad T \ll \hbar\omega.$$

3.5 Electron-Phonon Scattering in a Metal: The Transport Coefficients

13. (a) Variation over $\Phi(\boldsymbol{p})$ and condition $\delta F[\Phi(\boldsymbol{p})]/\delta\Phi(\boldsymbol{p}) = 0$ result in the equation required.

(b) For brevity, let us introduce the operator symbol

$$\hat{W}\Phi = \int d^3p'\, W(\boldsymbol{p}, \boldsymbol{p}')[\Phi(\boldsymbol{p}) - \Phi(\boldsymbol{p}')]$$

and determine the scalar product as

$$(\Phi, \chi) = \int d^3p\, \Phi(\boldsymbol{p})\chi(\boldsymbol{p}).$$

Operator $\hat{W}$ is symmetrical, i.e. equality $(\Phi, \hat{W}\chi) = (\chi, \hat{W}\Phi)$ is valid for any functions $\Phi(\boldsymbol{p})$ and $\chi(\boldsymbol{p})$. Using property $W(\boldsymbol{p}, \boldsymbol{p}') = W(\boldsymbol{p}', \boldsymbol{p})$, it is readily to check that

$$\begin{aligned}(\Phi, \hat{W}\chi) &= (\chi, \hat{W}\Phi) \\ &= \frac{1}{2}\iint W(\boldsymbol{p}, \boldsymbol{p}')[\Phi(\boldsymbol{p}) - \Phi(\boldsymbol{p}')]\,[\chi(\boldsymbol{p}) - \chi(\boldsymbol{p}')]\,d^3p\,d^3p'.\end{aligned}$$

Putting $\chi = \Phi$ and presuming $W(\boldsymbol{p}, \boldsymbol{p}') \geqslant 0$, we see that operator $\hat{W}$ is positively definite. For all functions $\Phi(\boldsymbol{p})$, the following inequality is satisfied:

$$(\Phi, \hat{W}\Phi) \geqslant 0.$$

Let $\chi(\boldsymbol{p})$ be an arbitrary function. Next, we introduce the normalization factor C equal to the ratio

$$C = \frac{(\chi, f)}{(\chi, \hat{W}f)}$$

and go over to function $\tilde{\chi}(\boldsymbol{p}) = C\chi(\boldsymbol{p})$ satisfying the relation

$$(\tilde{\chi}, \hat{W}\tilde{\chi}) = (\tilde{\chi}, f).$$

Then, let function $\Phi(\boldsymbol{p})$ be solution of integral equation, i.e. $\hat{W}\Phi = f$, and consider the scalar product which is obviously larger or equal to zero.

$$\begin{aligned}(\Phi - \tilde{\chi}, \hat{W}(\Phi - \tilde{\chi})) &= (\Phi, \hat{W}\Phi) + (\tilde{\chi}, \hat{W}\tilde{\chi}) - (\tilde{\chi}, \hat{W}\Phi) - (\Phi, \hat{W}\tilde{\chi}) \\ = (\Phi, \hat{W}\Phi) + (\tilde{\chi}, \hat{W}\tilde{\chi}) - 2(\tilde{\chi}, \hat{W}\Phi) &= (\Phi, \hat{W}\Phi) + (\tilde{\chi}, \hat{W}\tilde{\chi}) - 2(\tilde{\chi}, f) \\ = (\Phi, \hat{W}\Phi) - (\tilde{\chi}, \hat{W}\tilde{\chi}) \geqslant 0 \quad &\text{or} \quad (\Phi, \hat{W}\Phi) \geqslant (\tilde{\chi}, \hat{W}\tilde{\chi}).\end{aligned}$$

Accordingly, one has for the inverse quantities

$$\frac{1}{(\Phi, \hat{W}\Phi)} \leqslant \frac{1}{(\tilde{\chi}, \hat{W}\tilde{\chi})} = \frac{1}{C(\chi, f)} = \frac{(\chi, \hat{W}\chi)}{(\chi, f)^2} = R[\chi(\boldsymbol{p})],$$

that is to be demonstrated.

With the aid of reasonable choice of trial function $\chi(\boldsymbol{p})$ with a few variational parameters, we can estimate the genuine solution of transport equation within satisfactory accuracy, minimizing a ratio $R[\chi(\boldsymbol{p})]$.

14. Let us start from analyzing the coupling between an electron and the oscillations of positive ion lattice. Displacing the ions from equilibrium site disturbs electrical neutrality and results in the electrical polarization. The origination of polarization vector $\boldsymbol{P}$, being the dipole moment per the unit volume, is equivalent to generating the polarization (bound) charges with the bulk density $\rho_{\text{pol}} = -\text{div}\,\boldsymbol{P}$. The energy of coupling between the electron at the point $\boldsymbol{r}$ and the polarization charges can be written as $U_{e-ph}(\boldsymbol{r}) = -e\varphi(\boldsymbol{r})$ where $\varphi(\boldsymbol{r})$ is the potential induced by the polarization charges. While determining $\varphi(\boldsymbol{r})$, we take into account that any excess charge in a metal is screened. So, in lieu of the Coulomb potential there arises a screened potential written here as

$$\varphi(\boldsymbol{r}) = \int K(\boldsymbol{r} - \boldsymbol{r}')\rho_{\text{pol}}(\boldsymbol{r}')\, d^3r'.$$

The kernel $K(\boldsymbol{r} - \boldsymbol{r}')$ determines the behavior of screening an excess charge as a function of distance.

The relatively good approximation is usually the Debye screening

$$K(\boldsymbol{r} - \boldsymbol{r}') = \frac{e^{-|(\boldsymbol{r}-\boldsymbol{r}'|/r_D}}{|(\boldsymbol{r} - \boldsymbol{r}'|}.$$

Here r_D is the Debye screening radius which is usually about interatomic distance a in a metal. The short-range screening allows us approximately to replace function K with the δ-function according to $K(\boldsymbol{r} - \boldsymbol{r}') = K(0)\delta(\boldsymbol{r} - \boldsymbol{r}')$, where $K(0) = \int K(\boldsymbol{r})\, d^3r \sim r_D^2 \sim a^2$. Then $\varphi(\boldsymbol{r}) = a^2\rho_{\text{pol}}(\boldsymbol{r})$ and the coupling potential equals

$$U_{\text{e-ph}} = ea^2\text{div}\,\boldsymbol{P} \sim e^2a^2n\,\text{div}\,\boldsymbol{u} = g\,\text{div}\,\boldsymbol{u}.$$

Here we estimate approximately the polarization or dipole moment per unit volume as $\boldsymbol{P} = en\boldsymbol{u}$ where $\boldsymbol{u} = \boldsymbol{u}(\boldsymbol{r}, t)$ is the *displacement vector* for an ion of mass M and charge e from the equilibrium site at point $\boldsymbol{r}$ and time moment t, and n is the ion density. The electron-phonon coupling of such form is commonly referred to as the *deformation potential*.

Keeping in mind that $n \sim 1/a^3$, we find the following estimate: $g \sim e^2/a$ for the coupling constant g. The latter has a scale of the Coulomb interaction for the charges at the interatomic distance. As it should be, the uniform deformation $\boldsymbol{u}(\boldsymbol{r}) = \text{const}$, corresponding to the displacement of the lattice as whole, does not result in the electron-phonon coupling. It is known from the theory of elasticity that div $\boldsymbol{u}$ equals a relative change of the volume or the density of body but with a minus sign, i.e. $-\delta\rho/\rho$.

In what follows, it is convenient to decompose the displacement vector $\boldsymbol{u}(\boldsymbol{r}, t)$ into a Fourier series in wave vectors $\boldsymbol{k}$, representing it as a sum over all normal harmonic oscillations. Since $\boldsymbol{u}$ is a real quantity, we have

$$\boldsymbol{u}(\boldsymbol{r}, t) = \sum_{\boldsymbol{k}} \left(\boldsymbol{u}_{\boldsymbol{k}} e^{i\boldsymbol{k}\boldsymbol{r} - i\omega_{\boldsymbol{k}} t} + \boldsymbol{u}^*_{\boldsymbol{k}} e^{-i\boldsymbol{k}\boldsymbol{r} + i\omega_{\boldsymbol{k}} t}\right), \qquad \omega_{\boldsymbol{k}} = sk.$$

Accordingly, the electron-phonon coupling can also be given as a sum of independent harmonics

$$U_{\text{e-ph}}(\boldsymbol{r}, t) = ig \sum_{\boldsymbol{k}} \left(\boldsymbol{k}\boldsymbol{u}_{\boldsymbol{k}} e^{i\boldsymbol{k}\boldsymbol{r} - i\omega_{\boldsymbol{k}} t} - \boldsymbol{k}\boldsymbol{u}^*_{\boldsymbol{k}} e^{-i\boldsymbol{k}\boldsymbol{r} + i\omega_{\boldsymbol{k}} t}\right).$$

Let us turn now to determining the connection between the displacement vector amplitude $\boldsymbol{u}_{\boldsymbol{k}}$ playing a role of the complex normal coordinate of oscillator and the number of phonons $N_{\boldsymbol{k}}$ relating to wave vector $\boldsymbol{k}$. Substituting the expansion for $\boldsymbol{u}(\boldsymbol{r}, t)$ into the kinetic energy of oscillations

$$E_{\text{kin}} = \frac{1}{2} \int \rho \dot{\boldsymbol{u}}^2 d^3r, \qquad \rho = Mn$$

where ρ is the density of a metal, M is the ion mass, n is the ion concentration, and $\dot{\boldsymbol{u}}$ is the velocity of ions, we find after averaging over time that

$$\langle E_{\text{kin}} \rangle = \frac{\rho}{2} \sum_{\boldsymbol{k}} \omega_{\boldsymbol{k}}^2 (\boldsymbol{u}^*_{\boldsymbol{k}} \boldsymbol{u}_{\boldsymbol{k}} + \boldsymbol{u}_{\boldsymbol{k}} \boldsymbol{u}^*_{\boldsymbol{k}}).$$

For harmonic oscillations, the average kinetic energy of oscillations equals their average potential energy. Thus, the total energy of oscillations equals the double kinetic energy $E_{\rm tot} = 2\langle E_{\rm kin}\rangle$. On the other hand, the energy of the system consisting of a set of oscillators with frequencies ω_k is given by a simple sum if the zero-oscillation energy is subtracted

$$E_{\rm tot} = \sum_k \hbar\omega_k N_k.$$

Here N_k is the energy level number of oscillator, which we will interpret as the *number of phonons* with wave vector $\boldsymbol{k}$. The transitions between the oscillator levels will be described as *emission* or *absorption* of phonons with momentum $\boldsymbol{k}$.

Now we can relate the Fourier transform of displacement vector $\boldsymbol{u}_k$ representing the oscillator coordinate with the phonon number N_k and its variation. Using the results of matrix theory for the linear oscillator and comparing two energies $2\langle E_{\rm kin}\rangle$ and $E_{\rm tot}$, we write down the following nonzero matrix elements of oscillator coordinate:

$$\langle N_k|\boldsymbol{u}_k|N_k+1\rangle = \langle N_k+1|\boldsymbol{u}_k^*|N_k\rangle = \sqrt{1+N_k}\left(\frac{\hbar}{2\rho\omega_k}\right)^{1/2},$$
$$\langle N_k|\boldsymbol{u}_k^*|N_k-1\rangle = \langle N_k-1|\boldsymbol{u}_k|N_k\rangle = \sqrt{N_k}\left(\frac{\hbar}{2\rho\omega_k}\right)^{1/2},$$

and the square of displacement $\boldsymbol{u}_k$ proportional to the number N_k:

$$\langle N_k|\boldsymbol{u}_k^*\boldsymbol{u}_k|N_k\rangle = \frac{N_k\hbar}{2\rho\omega_k}.$$

As it is known, the other matrix elements for the harmonic oscillator coordinate vanish. This means that the transitions changing the oscillator state or phonon number by unity are only important in our consideration, i.e. $N_k \to N_k \pm 1$.

The Born approximation is proportional to the interaction modulus-squared. Accordingly, the elementary act of scattering an electron with momentum $\boldsymbol{p}$ on the thermal oscillations of ions reduces to emitting or absorbing a single phonon with some wave vector $\boldsymbol{k}$. From the above equalities we obtain the matrix elements of electron-phonon coupling and the squares of their moduli with emitting a phonon of wave vector $\boldsymbol{k}$

$$\left(U_{\text{e-ph}}\right)_{\boldsymbol{p}-\hbar\boldsymbol{k},\boldsymbol{p}} = -igk\,e^{-i\boldsymbol{kr}+i\omega_k t}\sqrt{1+N_k}\left(\frac{\hbar}{2\rho\omega_k}\right)^{1/2},$$
$$|\left(U_{\text{e-ph}}\right)_{\boldsymbol{p}-\hbar\boldsymbol{k},\boldsymbol{p}}|^2 = \frac{g^2\hbar k^2}{2\rho\omega_k}(1+N_k) = w_k(1+N_k), \quad w_k = \frac{g^2\hbar k^2}{2\rho\omega_k} \sim k,$$

and absorbing a phonon of wave vector $\boldsymbol{k}$

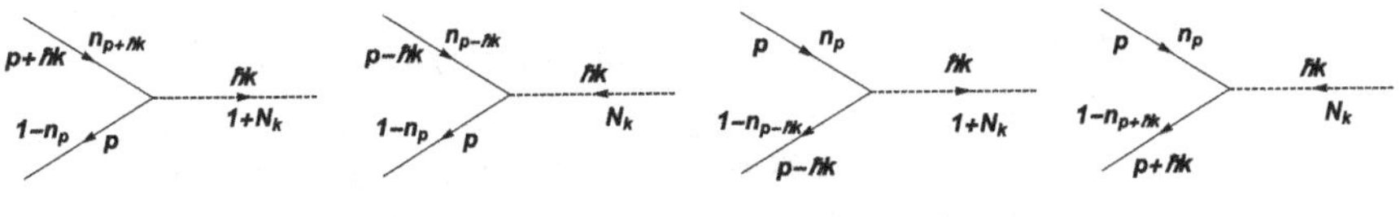

Fig. 3.4 The processes of electron scattering with emitting or absorbing a phonon. The solid line corresponds to an electron and the dashed one means a phonon

$$\left(U_{\text{e-ph}}\right)_{p+\hbar k,p}=igk\,e^{ikr-i\omega_k t}\sqrt{N_k}\left(\frac{\hbar}{2\rho\omega_k}\right)^{1/2},$$

$$|\left(U_{\text{e-ph}}\right)_{p+\hbar k,p}|^2=\frac{g^2\hbar k^2}{2\rho\omega_k}N_k=w_kN_k,\qquad w_k=\frac{g^2\hbar k^2}{2\rho\omega_k}\sim k.$$

Similarly to emitting the photons, the factor $(1+N_k)$ involves also a combination of spontaneous and induced emissions of phonons. The equilibrium phonon distribution

$$N_k=\left(e^{\hbar\omega_k/T}-1\right)^{-1}$$

is the Planck one.

Now we can go over to determining the collision integral for the electron distribution function. The collision integral is a difference between the number of gain (injecting) acts and the number of loss (ejecting) acts

$$\text{St}[n_p]=\sum_{p'}W_{p,p'}\,n_{p'}(1-n_p)-\sum_{p'}W_{p',p}\,n_p(1-n_{p'}).$$

Here $W_{p,q}$ is the probability of electron scattering from the state with momentum $\boldsymbol{q}$ to the state with momentum $\boldsymbol{p}$. The term with the gain acts consists of two contributions. The first is due to electrons with momentum $\boldsymbol{p}'=\boldsymbol{p}+\hbar\boldsymbol{k}$ and energy $\varepsilon_{p+\hbar k}$, which emit a phonon with momentum $\hbar\boldsymbol{k}$ and acquire momentum $\boldsymbol{p}$ and energy ε_p. The second results from the electrons with momentum $\boldsymbol{p}'=\boldsymbol{p}-\hbar\boldsymbol{k}$ and energy $\varepsilon_{p-\hbar k}$, which absorb a phonon with momentum $\hbar\boldsymbol{k}$ and acquire momentum $\boldsymbol{p}$ and energy ε_p.

The term with the loss acts consists of two contributions as well. The first is due to emitting the phonons with momentum $\hbar\boldsymbol{k}$ when the electron momentum becomes equal to $\boldsymbol{p}'=\boldsymbol{p}-\hbar\boldsymbol{k}$ and the energy equals to $\varepsilon_{p-\hbar k}$. The second is absorption of a phonon with momentum $\hbar\boldsymbol{k}$ when the electron momentum becomes equal to $\boldsymbol{p}'=\boldsymbol{p}+\hbar\boldsymbol{k}$ and energy does to $\varepsilon_{p+\hbar k}$. Figure 3.4 shows these possible elementary one-phonon processes.

As a result, the collision integral, written in the Born approximation or *Bloch collision integral*, will consist of four terms

$$\begin{aligned}\text{St}\,[n_p]=\frac{2\pi}{\hbar}\sum_k\big[&w_k(1+N_k)n_{p+\hbar k}(1-n_p)\delta(\varepsilon_p-\varepsilon_{p+\hbar k}+\hbar\omega_k)\\&+w_kN_kn_{p-\hbar k}(1-n_p)\delta(\varepsilon_p-\varepsilon_{p-\hbar k}-\hbar\omega_k)\big]\end{aligned}$$

$$-\frac{2\pi}{\hbar}\sum_{k}\big[w_k(1+N_k)n_p(1-n_{p-\hbar k})\delta(\varepsilon_p-\varepsilon_{p-\hbar k}-\hbar\omega_k)$$
$$+w_kN_kn_p(1-n_{p+\hbar k})\delta(\varepsilon_p-\varepsilon_{p+\hbar k}+\hbar\omega_k)\big].$$

Here we have involved that quantity w_k depends on the wave vector modulus $\boldsymbol{k}$. The energy conservation law in the course of scattering is engaged with the aid of δ-functions. Replacing $-\boldsymbol{k}$ with $\boldsymbol{k}$ in two sums over wave vector, we arrive at the following expression for the electron-phonon collision integral:

$$\begin{aligned}\mathrm{St}\,[n_p]=\frac{2\pi}{\hbar}\sum_{k}w_k\big[&(1+N_k)n_{p+\hbar k}(1-n_p)\\&-N_kn_p(1-n_{p+\hbar k})\big]\delta(\varepsilon_p-\varepsilon_{p+\hbar k}+\hbar\omega_k)\\&+\frac{2\pi}{\hbar}\sum_{k}w_k\big[N_kn_{p+\hbar k}(1-n_p)\\&-(1+N_k)n_p(1-n_{p+\hbar k})\big]\delta(\varepsilon_p-\varepsilon_{p+\hbar k}-\hbar\omega_k).\end{aligned}$$

Below we will suppose that the equilibrium in the phonon subsystem is set so rapidly that its deviation from equilibrium can be neglected. Because of this we take the phonon distribution function same as the equilibrium one $N_k=N(\omega_k)$. (It is implied here that the necessary time to establish the equilibrium in the phonon subsystem is significantly shorter as compared with the typical times of electron-phonon scattering. In general, this is not always valid.) As it should be, the collision integral $\mathrm{St}\,[n_p]$ vanishes identically on substituting the equilibrium distribution functions of electrons $n_0(\varepsilon_p)$ and phonons $N(\omega_k)$. This property is straightforwardly verified, for example, with the aid of the following relation for equilibrium functions:

$$\frac{1+N(\varepsilon'-\varepsilon)}{N(\varepsilon'-\varepsilon)}=\frac{n_0(\varepsilon)}{1-n_0(\varepsilon)}\,\frac{1-n_0(\varepsilon')}{n_0(\varepsilon')}\,.$$

Let us consider the collision integral in the linear approximation for the deviation from equilibrium in the following form:

$$\begin{aligned}n_p\approx n_0(\varepsilon_p)+\delta n_p&=n_0(\varepsilon_p)-\frac{\partial n_0}{\partial\varepsilon}\,\Phi_p\\&=n_0(\varepsilon_p)+\frac{n_0(\varepsilon_p)\big(1-n_0(\varepsilon_p)\big)}{T}\,\Phi_p.\end{aligned}$$

In order to linearize the collision integral, we rewrite first the expression in the square brackets from the first term as

$$\begin{aligned}&(1+N_k)n_{p+\hbar k}(1-n_p)-N_kn_p(1-n_{p+\hbar k})\\&=(1+N_{\varepsilon'-\varepsilon})(1-n_{p'})(1-n_p)\left[\frac{n_{p'}}{1-n_{p'}}-\frac{N_{\varepsilon'-\varepsilon}}{1+N_{\varepsilon'-\varepsilon}}\,\frac{n_p}{1-n_p}\right],\quad p'=p+\hbar k.\end{aligned}$$

Hence, denoting for brevity $n_0 = n_0(\varepsilon)$ and $n'_0 = n_0(\varepsilon')$ where $\varepsilon = \varepsilon_{\boldsymbol{p}}$ and $\varepsilon' = \varepsilon_{\boldsymbol{p}'}$, we find in the linear approximation for the deviation from equilibrium

$$(1+N_{\varepsilon'-\varepsilon})(1-n_0)(1-n'_0)\left[\frac{\delta n_{\boldsymbol{p}'}}{(1-n'_0)^2} - \frac{N_{\varepsilon'-\varepsilon}}{1+N_{\varepsilon'-\varepsilon}}\frac{\delta n_{\boldsymbol{p}}}{(1-n_0)^2}\right]$$
$$= \frac{(1+N_{\varepsilon'-\varepsilon})(1-n_0)(1-n'_0)}{T}\left[\frac{n_0\Phi_{\boldsymbol{p}'}}{(1-n'_0)} - \frac{N_{\varepsilon'-\varepsilon}}{1+N_{\varepsilon'-\varepsilon}}\frac{n_0\Phi_{\boldsymbol{p}}}{(1-n_0)}\right]$$
$$= \frac{(1+N_{\varepsilon'-\varepsilon})n'_0(1-n_0)}{T}\left[\Phi_{\boldsymbol{p}'} - \Phi_{\boldsymbol{p}}\right]$$
$$= -\frac{(1+N_{\varepsilon'-\varepsilon})N_{\varepsilon'-\varepsilon}}{T}(n'_0 - n_0)\left[\Phi_{\boldsymbol{p}'} - \Phi_{\boldsymbol{p}}\right].$$

We will do the same with the second term. Transforming it, we use equality $(1+N_{\varepsilon-\varepsilon'})N_{\varepsilon-\varepsilon'} = (1+N_{\varepsilon'-\varepsilon})N_{\varepsilon'-\varepsilon}$. As a result, we obtain the following linearized collision integral:

$$\mathrm{St}\,[\Phi_{\boldsymbol{p}}] = -\frac{2\pi}{\hbar}\sum_{\boldsymbol{k}} w_k \frac{N(\omega_k)\big(1+N(\omega_k)\big)}{T}\left[n_0(\varepsilon_{\boldsymbol{p}+\hbar\boldsymbol{k}}) - n_0(\varepsilon_{\boldsymbol{p}})\right]$$
$$\times\left[\delta(\varepsilon_{\boldsymbol{p}} - \varepsilon_{\boldsymbol{p}+\hbar\boldsymbol{k}} + \hbar\omega_{\boldsymbol{k}}) - \delta(\varepsilon_{\boldsymbol{p}} - \varepsilon_{\boldsymbol{p}+\hbar\boldsymbol{k}} - \hbar\omega_{\boldsymbol{k}})\right](\Phi_{\boldsymbol{p}+\hbar\boldsymbol{k}} - \Phi_{\boldsymbol{p}}).$$

or

$$\mathrm{St}\,[\Phi_{\boldsymbol{p}}] = -\frac{2\pi}{\hbar}\sum_{\boldsymbol{k}} w_k \frac{N_k(1+N_k)}{T}$$
$$\times\Big\{\left[n_0(\varepsilon_{\boldsymbol{p}} + \hbar\omega_k) - n_0(\varepsilon_{\boldsymbol{p}})\right]\delta(\varepsilon_{\boldsymbol{p}} - \varepsilon_{\boldsymbol{p}+\hbar\boldsymbol{k}} + \hbar\omega_{\boldsymbol{k}})$$
$$-\left[n_0(\varepsilon_{\boldsymbol{p}} - \hbar\omega_k) - n_0(\varepsilon_{\boldsymbol{p}})\right]\delta(\varepsilon_{\boldsymbol{p}} - \varepsilon_{\boldsymbol{p}+\hbar\boldsymbol{k}} - \hbar\omega_{\boldsymbol{k}})\Big\}(\Phi_{\boldsymbol{p}+\hbar\boldsymbol{k}} - \Phi_{\boldsymbol{p}}).$$

In the temperature $T \ll \varepsilon_F$ region the physical properties of a metal or degenerate Fermi liquid are associated with the momenta close to the Fermi momentum $|p - p_F| \ll p_F$. This region corresponds to the electron energy about the Fermi energy or chemical potential $\varepsilon_{\boldsymbol{p}} \sim \varepsilon_F \sim \mu$. Since the phonon energies are small as compared with the Fermi energy $\hbar\omega_k \leqslant \hbar\omega_D \ll \varepsilon_F$, we expand the differences in the square brackets in powers ω_k

$$n_0(\varepsilon_{\boldsymbol{p}} \pm \hbar\omega_k) - n_0(\varepsilon_{\boldsymbol{p}}) \approx \pm\frac{\partial n_0(\varepsilon_{\boldsymbol{p}})}{\partial\varepsilon_{\boldsymbol{p}}}\hbar\omega_k.$$

After that one can neglect the phonon energy $\pm\hbar\omega_k$ in the arguments of δ-functions. In fact, we have for the electron dispersion law $\varepsilon_{\boldsymbol{p}} = \boldsymbol{p}^2/2m$

$$\varepsilon_{p+\hbar k} - \varepsilon_p \pm \hbar\omega_k = \frac{\hbar k p}{m}\left(\cos\vartheta + \frac{\hbar k}{2p} \pm \frac{ms}{p}\right) \approx \frac{\hbar k p}{m}\left(\cos\vartheta + \frac{\hbar k}{2p}\right).$$

Here ϑ is the angle between vectors $\boldsymbol{p}$ and $\boldsymbol{k}$ and we have used inequality $ms/p \sim ms/p_F \sim s/v_F \ll 1$. As a result, we obtain

$$\mathrm{St}\,[\Phi_p] \approx -\frac{\partial n_0(\varepsilon_p)}{\partial \varepsilon_p}\frac{2\pi}{\hbar}\sum_{\boldsymbol{k}} w_k \frac{N_k(1+N_k)}{T} \times 2\hbar\omega_k\,\delta(\varepsilon_{p+\hbar k} - \varepsilon_p)(\Phi_{p+\hbar k} - \Phi_p).$$

So, we can turn now to analyzing the electrical conductivity and determining the transport time τ_{tr} for scattering of electrons with phonons. The electron transport equation has a usual form for the linear approximation in electric field $\boldsymbol{E}$

$$e\boldsymbol{E}\frac{\partial n_0(\varepsilon_p)}{\partial \boldsymbol{p}} = \mathrm{St}\,[\Phi_p].$$

Let us rewrite this equation using that the electron velocity equals $\boldsymbol{v} = \partial\varepsilon/\partial\boldsymbol{p} = \boldsymbol{p}/m$. Substituting the latter in $\mathrm{St}\,[\Phi_p]$, we have an expression

$$e\boldsymbol{v}\boldsymbol{E}\,\frac{\partial n_0(\varepsilon_p)}{\partial \varepsilon_p} = -\frac{\partial n_0(\varepsilon_p)}{\partial \varepsilon_p}\frac{4\pi}{T}\sum_{\boldsymbol{k}} w_k N_k(1+N_k)\omega_k\,\delta(\varepsilon_{p+\hbar k} - \varepsilon_p)(\Phi_{p+\hbar k} - \Phi_p).$$

We seek for the solution of equation as

$$\Phi_p = e(\boldsymbol{v}\boldsymbol{E})\tau_{tr}(\varepsilon_p) = e\frac{\boldsymbol{p}\boldsymbol{E}}{m}\tau_{tr}, \quad \text{i.e.} \quad \delta n_p = e(\boldsymbol{v}\boldsymbol{E})\tau_{tr}(\varepsilon_p)\left(-\frac{\partial n_0}{\partial \varepsilon_p}\right).$$

The unknown function $\tau_{tr}(\varepsilon_p)$, energy dependent alone, will enter the formula for conductivity, and play a role of the electron transport for scattering with the thermal lattice vibrations. Then we have

$$\boldsymbol{v}\boldsymbol{E} = -\frac{4\pi}{T}\sum_{\boldsymbol{k}} w_k N_k(1+N_k)\omega_k\,\delta(\varepsilon_{p+\hbar k} - \varepsilon_p)\frac{\hbar\boldsymbol{k}\boldsymbol{E}}{m}\tau_{tr}(\varepsilon_p).$$

In order to integrate over solid angle Ω_k, we direct the polar axis along vector $\boldsymbol{p}$ and introduce polar angle ϑ and azimuthal angle φ which determine the direction of vector $\boldsymbol{k}$. In the similar way we introduce the analogous angles ϑ_E and φ_E indicating the direction of electric field $\boldsymbol{E}$. The angle α between vectors $\boldsymbol{k}$ and $\boldsymbol{E}$ equals

$$\cos\alpha = \cos\vartheta\cos\vartheta_E + \sin\vartheta\sin\vartheta_E\cos(\varphi - \varphi_E).$$

Using this relation in the last equation, we obtain

$$pE\cos\vartheta_E\tau_{tr}^{-1}(\varepsilon) = -\frac{4\pi}{T}\int_0^{k_D}\frac{k^2dk}{(2\pi)^3}\int_0^{\pi}\sin\vartheta\, d\vartheta\int_0^{2\pi}d\varphi\, w_kN_k(1+N_k)\omega_k$$
$$\times\,\delta\left(\frac{\hbar kp}{m}\cos\vartheta+\frac{\hbar^2k^2}{2m}\right)\hbar kE\big[\cos\vartheta\cos\vartheta_E+\sin\vartheta\sin\vartheta_E\cos(\varphi-\varphi_E)\big].$$

Here the summation over $\boldsymbol{k}$ is replaced with integration and the phonon wave vector is limited by the maximum magnitude k_D. The term with $\cos(\varphi-\varphi_E)$ after integrating over angle φ yields zero contribution to the answer. After dividing the left-hand and right-hand sides of the equation by the common factor $E\cos\vartheta_E$ and substituting $x=\cos\vartheta$, we find that

$$p\tau_{tr}^{-1}(\varepsilon) = -\frac{4\pi}{T}\int_0^{k_D}\frac{k^2dk}{4\pi^2}w_kN_k(1+N_k)\omega_k\hbar k\frac{m}{\hbar kp}\int_{-1}^{1}dx\, x\delta\left(x+\frac{\hbar k}{2p}\right).$$

As it is seen, the presence of δ-function, resulting from the energy conservation law, leads to an additional restriction for the magnitude of phonon vector $k\leqslant 2p/\hbar$. Henceforth, integration over k is limited by the minimal magnitude from two wave vectors k_D and $2p/\hbar$

$$k_m = \min\,(k_D,\ 2p/\hbar).$$

Performing the integration over x yields

$$\tau_{tr}^{-1}(\varepsilon) = \frac{\hbar}{2\pi T}\frac{m}{p^3}\int_0^{k_m}dk\, k^2w_kN_k(1+N_k)k\omega_k,\quad w_k=\frac{g^2\hbar k^2}{2\rho\omega_k}\quad\text{and}\quad \omega_k=sk.$$

Since we are interested in the electron momentum close to the Fermi one p_F, we will not discern the magnitudes of vectors k_D and $2p_F/\hbar$ being the same order of magnitude and about inverse interatomic distance a, namely $k_D\sim 2p_F/\hbar\sim 1/a$. For definiteness, we put $k_m=k_D$. The opposite assumption leads only to inessential qualitative distinctions. Putting $z=\hbar sk/T$ and introducing the Debye temperature as $\Theta_D=\hbar sk_D$, we obtain the following formula:

$$\tau_{tr}^{-1}(\varepsilon) = \frac{m}{4\pi p^3}\frac{g^2\hbar}{\rho s}k_D^5\left(\frac{T}{\Theta_D}\right)^5\int_0^{\Theta_D/T}dz\frac{z^5e^z}{(e^z-1)^2}$$
$$=\frac{\pi}{4}\zeta\left(\frac{\hbar k_D}{p}\right)^4\frac{\Theta_D}{4\hbar}\left(\frac{T}{\Theta_D}\right)^5\int_0^{\Theta_D/T}dz\frac{z^5}{\sinh^2 z/2},$$

$$\text{with notation} \quad \zeta = \frac{g^2}{\rho s^2}\frac{mp(\varepsilon)}{\pi^2\hbar^3} = \frac{g^2}{\rho s^2}\nu(\varepsilon).$$

Here $\nu(\varepsilon)$ is the density of electron states and ζ is the dimensionless *electron-phonon coupling constant*. Let us evaluate the constant ζ by the order of magnitude at $\varepsilon = \varepsilon_F$

$$\zeta = \frac{g^2}{\rho s^2}\nu(\varepsilon_F) \sim \frac{(e^2/a)^2}{Mn(v_F^2 m/M)}\frac{mp_F}{\hbar^3} \sim \frac{(e^2/a)^2}{mv_F^2}\frac{mp_F}{(\hbar/a)^3} \sim \left(\frac{e^2/a}{\varepsilon_F}\right)^2.$$

In the typical metals the characteristic energy of the Coulomb interaction e^2/a is comparable with the average electron kinetic energy ε_F and thus $\zeta \sim 1$. In addition, we put $\hbar k_D/p \sim 1$ at $p \sim p_F$. Eventually, we arrive at the following dimensional estimate for the transport scattering time, the numerical factors being neglected

$$\frac{\hbar}{\tau_{tr}(\varepsilon_F)} \sim \zeta\Theta_D\left(\frac{T}{\Theta_D}\right)^5 \int_0^{\Theta_D/T} dz \frac{z^5}{\sinh^2 z/2}$$
$$\sim \zeta\Theta_D\left(\frac{T}{\Theta_D}\right)^5 \begin{cases} \left(\frac{\Theta_D}{T}\right)^4, & T \gg \Theta_D, \\ 480\zeta(5), & T \ll \Theta_D. \end{cases}$$

Here $\zeta(5) \approx 1.037$ is the Riemann zeta function.

In order to find the electrical conductivity, we write down the expression for the electric current in a metal with the perturbed distribution function $n(\boldsymbol{p}) = n_0(\boldsymbol{p}) + \delta n(\boldsymbol{p})$:

$$\boldsymbol{j} = \int e\boldsymbol{v}\big[n_0(\boldsymbol{p}) + \delta n(\boldsymbol{p})\big]\frac{2d^3p}{(2\pi\hbar)^3} = \int e\boldsymbol{v}\,\delta n(\boldsymbol{p})\,\frac{2d^3p}{(2\pi\hbar)^3}\,.$$

The first integral with n_0 vanishes since the electric current is absent in equilibrium at $\boldsymbol{E} = 0$. One finds for the second integral

$$\boldsymbol{j} = e^2\int \boldsymbol{v}(\boldsymbol{v}\boldsymbol{E})\tau_{tr}(\varepsilon)\left(-\frac{\partial n_0}{\partial\varepsilon}\right)\frac{2d^3p}{(2\pi\hbar)^3}$$
$$= \boldsymbol{E}\frac{e^2}{3}\int v^2\tau_{tr}(\varepsilon)\left(-\frac{\partial n_0}{\partial\varepsilon}\right)\frac{2d^3p}{(2\pi\hbar)^3}.$$

Determining conductivity σ from relation $\boldsymbol{j} = \sigma\boldsymbol{E}$, we can approximately take $\partial n_0/\partial\varepsilon \approx -\delta(\varepsilon - \varepsilon_F)$ for the degenerate Fermi system at $T \ll \varepsilon_F$. Then

$$\sigma = \frac{e^2}{3}v_F^2\tau_{tr}(\varepsilon_F)\nu(\varepsilon_F) = \frac{e^2\tau_{tr}(\varepsilon_F)}{3}\frac{p_F^2}{m^2}\frac{mp_F}{\pi^2\hbar^3} = \frac{ne^2\tau_{tr}(\varepsilon_F)}{m},$$

where $n = p_F^3/(3\pi^2\hbar^3)$ is the electron concentration. Accordingly, electrical resistivity $\rho_{\text{e-ph}} = \sigma^{-1}$, due to electron scattering on thermal vibrations of ion lattice, will

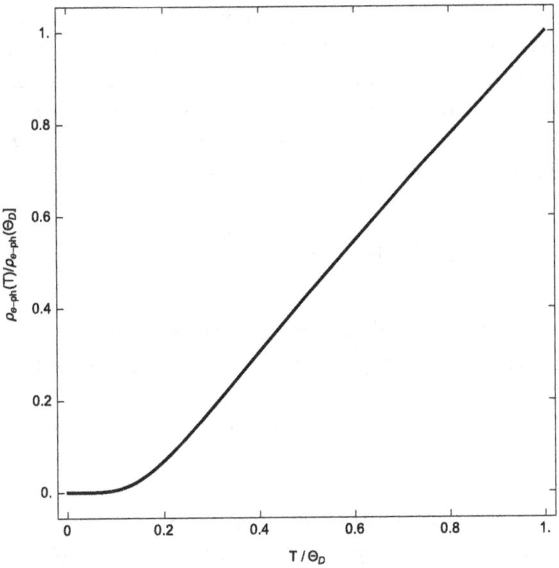

Fig. 3.5 The normalized resistivity as a function of the normalized temperature

be described with the *Bloch-Grüneisen formula*

$$\frac{\rho_{\text{e-ph}}(T)}{\rho_{\text{e-ph}}(\Theta_D)} = \frac{\tau_{tr}(\theta_D)}{\tau_{tr}(T)} = \left(\frac{T}{\Theta_D}\right)^5 \frac{\int\limits_0^{\Theta_D/T} \frac{z^5 dz}{\sinh^2 z/2}}{\int\limits_0^1 \frac{z^5 dz}{\sinh^2 z/2}}$$

$$= \begin{cases} \sim (T/\Theta_D)^5, \; T \ll \Theta_D, \\ \sim T/\Theta_D, \quad T \gg \Theta_D. \end{cases}$$

It is worthwhile to note the following. In simplest model treated above the resistivity of a metal proves to be a universal function of a ratio Θ_D/T. This is similar to the Debye model for the phonon specific heat. In Fig. 3.5 the plot of a ratio $\rho_{\text{e-ph}}(T)/\rho_{\text{e-ph}}(\Theta_D)$ is shown as a function of a ratio T/Θ_D. To conclude, for high $T > \Theta_D$ temperatures the resistivity of a metal grows in the direct proportion to the temperature as $\rho_{\text{e-ph}}(T) \sim T$. As the temperature lowers and $T \lesssim \Theta_D/2\pi$, the resistivity of a metal crosses over to the *Bloch law* when $\rho_{\text{e-ph}}(T) \sim T^5$.

15. The thermal conductivity $\varkappa$ is determined with formula $\boldsymbol{q} = -\varkappa \nabla T$ relating the dissipative heat flow $\boldsymbol{q}$ and temperature gradient ∇T. It is implied here that the average electron current is absent, i.e. $\boldsymbol{j} = 0$, and the energy flux $\boldsymbol{Q}$ is the same as a dissipative heat flow $\boldsymbol{q}$.

Provided that there is no effective field $\boldsymbol{E}' = \boldsymbol{E} - \nabla\mu/e$ acting on electrons, the relation between heat flow and temperature gradient is determined by coefficient $\varkappa_E$

which differs from thermal conductivity $\varkappa$ by quantity $\sigma \Pi S$ in accordance with the previous problems. As we have seen before, the relative difference between $\varkappa$ and $\varkappa_E$ is negligibly small in the low $T \ll \mu$ temperature region. Below, strictly speaking, while calculating $\varkappa_E$, we will not discern $\varkappa_E$ and $\varkappa$.

Henceforth, we assume everywhere that $\boldsymbol{E}' = \boldsymbol{E} - \nabla\mu/e = 0$. Then the linearized transport equation takes the form

$$(\boldsymbol{v}_p \cdot \nabla T)\frac{\varepsilon_p - \mu}{T}\left(-\frac{\partial n_0(\varepsilon_p)}{\partial \varepsilon_p}\right) = \mathrm{St}\,[\delta n_p], \quad \delta n_p = \frac{\partial n_0(\varepsilon_p)}{\partial \varepsilon_p}\Phi_p,$$

where δn_p is the deviation of distribution function from equilibrium one $n_0(\varepsilon_p)$.

The linearized collision integral can be represented in the general form as

$$\mathrm{St}\,[\delta n_p] = \mathrm{St}\,[\Phi_p] = \hat{W}\Phi = \sum_{p'} W(\boldsymbol{p}, \boldsymbol{p}')(\Phi_p - \Phi_{p'}).$$

Here the kernel $W(\boldsymbol{p}, \boldsymbol{p}')$ is positive definite and symmetrical to permuting the momenta as $W(\boldsymbol{p}, \boldsymbol{p}') = W(\boldsymbol{p}', \boldsymbol{p})$. The summation or integration, according to,

$$\sum_p \to \int \frac{2d^3p}{(2\pi\hbar)^3},$$

is performed over all possible momenta $\boldsymbol{p}$ of an electron with two spin projections. The corresponding transport equation takes the form

$$(\boldsymbol{v}_p \cdot \nabla T)\frac{\varepsilon_p - \mu}{T}\left(-\frac{\partial n_0(\varepsilon_p)}{\partial \varepsilon_p}\right) = \sum_{p'} W(\boldsymbol{p}, \boldsymbol{p}')(\Phi_p - \Phi_{p'}).$$

Multiplying the both sides of equation by Φ_p and then summing over all possible momenta $\boldsymbol{p}$, we arrive at

$$\sum_p (\boldsymbol{v}_p \cdot \nabla T)\frac{\varepsilon_p - \mu}{T}\left(-\frac{\partial n_0(\varepsilon_p)}{\partial \varepsilon_p}\right)\Phi_p$$
$$= \sum_{p,p'} W(\boldsymbol{p}, \boldsymbol{p}')(\Phi_p - \Phi_{p'})\Phi_p = \frac{1}{2}\sum_{p,p'} W(\boldsymbol{p}, \boldsymbol{p}')(\Phi_p - \Phi_{p'})^2.$$

Obtaining the last equality, we use the symmetry $W(\boldsymbol{p}, \boldsymbol{p}') = W(\boldsymbol{p}', \boldsymbol{p})$. Choosing the trial function as

$$\Phi_p = \Phi_p^{(0)}\tau(T),$$

we find that

$$\frac{1}{\tau} = \frac{\frac{1}{2}\sum\limits_{p,p'} W(p,p')(\Phi_p^{(0)} - \Phi_{p'}^{(0)})^2}{\sum\limits_p (v_p \cdot \nabla T)\left(\frac{\varepsilon_p - \mu}{T}\right)\left(-\frac{\partial n_0}{\partial \varepsilon_p}\right)\Phi_p^{(0)}}.$$

On the other hand, heat flow q with relation $\delta n_p = n_0'(\varepsilon_p)\Phi_p$ is determined by the equation

$$q = \sum_p v_p(\varepsilon_p - \mu)n_p = \sum_p v_p(\varepsilon_p - \mu)\delta n_p = \sum_p v_p(\varepsilon_p - \mu)\frac{\partial n_0(\varepsilon_p)}{\partial \varepsilon_p}\Phi_p$$
$$= \tau(T)\sum_p v_p(\varepsilon_p - \mu)\frac{\partial n_0(\varepsilon_p)}{\partial \varepsilon_p}\Phi_p^{(0)} = -\varkappa \nabla T.$$

Here it is taken into account that $n_p = n_0(\varepsilon_p) + \delta n_p$ and the term with $n_0(\varepsilon_p)$ gives no contribution to the heat flow. In addition, the relation of heat flow q with the temperature is involved. Multiplying the last two equalities by ∇T results in

$$\frac{1}{\tau(T)} = \frac{\sum\limits_p (v_p \cdot \nabla T)(\varepsilon_p - \mu)\left(-\frac{\partial n_0}{\partial \varepsilon_p}\right)\Phi_p^{(0)}}{\varkappa(\nabla T)^2}.$$

Comparing two expressions for $\tau^{-1}(T)$, it is readily to see that the following relation is valid for the thermal conductivity

$$\frac{1}{\varkappa T} = \frac{\frac{1}{2}(\nabla T)^2 \sum\limits_{p,p'} W(p,p')(\Phi_p^{(0)} - \Phi_{p'}^{(0)})^2}{\left[\sum\limits_p (v_p \cdot \nabla T)(\varepsilon_p - \mu)\left(-\frac{\partial n_0}{\partial \varepsilon_p}\right)\Phi_p^{(0)}\right]^2}.$$

Since $\Phi_p^{(0)}$ is squared both in the nominator and in the denominator, any multiplier introduced in function $\Phi_p^{(0)}$ as $\Phi_p^{(0)} \to \Phi_p^{(0)}\tau = \Phi_p$ will be cancelled. Therefore, the expression for $(\varkappa T)^{-1}$ can be represented in the following general form:

$$\frac{1}{\varkappa T} = \frac{\frac{1}{2}(\nabla T)^2 \sum\limits_{p,p'} W(p,p')(\Phi_p - \Phi_{p'})^2}{\left[\sum\limits_p (v_p \cdot \nabla T)(\varepsilon_p - \mu)\left(-\frac{\partial n_0}{\partial \varepsilon_p}\right)\Phi_p\right]^2} = \frac{\frac{1}{2}(\nabla T)^2(\Phi, \hat{W}\Phi)}{(\Phi, f)^2} = R_\varkappa[\Phi_p].$$

Note that we can give certain physical sense for this relation, using the expression for the energy dissipation function R in the lack of electrical current

$$R = \frac{\varkappa(\nabla T)^2}{T} = -\frac{(q \cdot \nabla T)}{T}\frac{(q \cdot \nabla T)}{-\varkappa(\nabla T)^2} = \frac{(q \cdot \nabla T)^2}{\varkappa T(\nabla T)^2}.$$

Hence we have also

$$\frac{1}{\varkappa T} = \frac{(\nabla T)^2 R}{(\boldsymbol{q} \cdot \nabla T)^2}\,.$$

As we have seen above, denominator $(\Phi,\ f)^2$ in $R_{\varkappa}[\Phi_p]$ is an exact square $(\boldsymbol{q} \cdot \nabla T)^2$ for heat flow $\boldsymbol{q}$ originating from a deviation of distribution function $n_{\boldsymbol{p}}$ from equilibrium $n_0(\varepsilon_p)$ by quantity $\delta n_p = n_0'(\varepsilon_p)\Phi_p$. Correspondingly, we should ascribe the meaning of energy dissipation power to the quantity

$$R = \frac{1}{2}\sum_{p,p'} W(\boldsymbol{p}, \boldsymbol{p}')(\Phi_p^{(0)} - \Phi_{p'}^{(0)})^2$$

and the entropy growth rate related to R if the deviation from equilibrium composes the magnitude $\delta n_p = n_0'(\varepsilon_p)\Phi_p$.

We turn now to calculating the nominator and denominator in the fraction $R_{\varkappa}[\Phi_p]$. For the denominator in $R_{\varkappa}[\Phi_p]$, we will have putting function Φ_p equal to $\Phi_p^{(0)} = (\varepsilon_p - \mu)(\boldsymbol{v}_p \cdot \nabla T)/T$

$$(\Phi,\ f) = \sum_p (\boldsymbol{v}_p \cdot \nabla T)(\varepsilon_p - \mu)\left(-\frac{\partial n_0}{\partial \varepsilon_p}\right)\Phi_p^{(0)}$$

$$= \int \frac{2d^3p}{(2\pi\hbar)^3}\,\frac{(\boldsymbol{p}\cdot\nabla T)^2}{m^2}\,\frac{(\varepsilon_p - \mu)^2}{T}\left(-\frac{\partial n_0}{\partial \varepsilon_p}\right).$$

Using $\boldsymbol{v}_p = \boldsymbol{p}/m$ and averaging, for example, over all possible directions of vector $\boldsymbol{p}$ according to $\langle p_\alpha p_\beta\rangle = p^2\delta_{\alpha\beta}/3$ $(\alpha, \beta = x,\ y,\ z)$, we find

$$(\Phi,\ f) = \frac{2}{3}\frac{(\nabla T)^2}{mT}\int \frac{2d^3p}{(2\pi\hbar)^3}\varepsilon_p(\varepsilon_p - \mu)^2\left(-\frac{\partial n_0}{\partial \varepsilon_p}\right)$$

$$= \frac{2}{3}\frac{(\nabla T)^2}{mT}\int_0^\infty \nu(\varepsilon)\varepsilon(\varepsilon - \mu)^2\left(-\frac{\partial n_0}{\partial \varepsilon_p}\right)d\varepsilon.$$

Here $\nu(\varepsilon) = mp(\varepsilon)/\pi^2\hbar^3$ is the density of electron states corresponding to energy ε.

For the calculation in the low $T \ll \mu$ temperature region, it is useful to employ the following Sommerfeld expansion in temperature:

$$\int_0^\infty F(\varepsilon)\left(-\frac{\partial n_0}{\partial \varepsilon}\right)d\varepsilon = F(\mu) + \frac{\pi^2 T^2}{6}F''(\mu) + \ldots$$

Then we find straightforwardly

$$(\Phi,\ f) = \frac{2}{3}\frac{(\nabla T)^2}{mT}\frac{\pi^2 T^2}{3}\mu\nu(\mu) \approx T\,\frac{2\varepsilon_F p_F}{9\hbar^3}(\nabla T)^2.$$

Due to low $T \ll \mu$ temperatures we make no distinction between the chemical potential μ and the Fermi energy $\varepsilon_F = p_F^2/2m$.

The calculation $(\Phi,\ \hat{W}\Phi)$ in the nominator of fraction $R_{\varkappa}[\Phi_p]$ is more complicated. For convenience of applying the results of the previous problem on the metal conductivity, in the collision integral expression and scalar product $(\Phi,\ \hat{W}\Phi)$ we go over from summing over all possible finite electron momenta $\boldsymbol{p}'$ to all possible scattering vectors $\boldsymbol{k}$, using the relation $\boldsymbol{p}' = \boldsymbol{p} + \hbar\boldsymbol{k}$, i.e.

$$\hat{W}\Phi = \sum_{\boldsymbol{p}'} W(\boldsymbol{p},\boldsymbol{p}')(\Phi_{\boldsymbol{p}} - \Phi_{\boldsymbol{p}'}) = \sum_{\boldsymbol{k}} W(\boldsymbol{p},\boldsymbol{p}+\hbar\boldsymbol{k})(\Phi_{\boldsymbol{p}} - \Phi_{\boldsymbol{p}+\hbar\boldsymbol{k}}).$$

So, we should calculate two sums as follows: (1) over all initial electron momenta $\boldsymbol{p}$ with two spin projections, and (2) over all phonon wave vectors $\boldsymbol{k}$

$$(\Phi,\ \hat{W}\Phi) = \sum_{\boldsymbol{p},\boldsymbol{p}'} W(\boldsymbol{p},\boldsymbol{p}')(\Phi_{\boldsymbol{p}} - \Phi_{\boldsymbol{p}'})^2 = \sum_{\boldsymbol{p},\boldsymbol{k}} W(\boldsymbol{p},\boldsymbol{k})(\Phi_{\boldsymbol{p}} - \Phi_{\boldsymbol{p}+\hbar\boldsymbol{k}})^2.$$

For the linearized collision integral, we can use the expression obtained in the previous problem

$$\mathrm{St}\,[\Phi_{\boldsymbol{p}}] = \hat{W}\Phi = -\frac{2\pi}{\hbar}\sum_{\boldsymbol{k}} w_k \frac{N(\omega_k)\big(1+N(\omega_k)\big)}{T}\big[n_0(\varepsilon_{\boldsymbol{p}+\hbar\boldsymbol{k}}) - n_0(\varepsilon_{\boldsymbol{p}})\big]$$
$$\times\big[\delta(\varepsilon_{\boldsymbol{p}} - \varepsilon_{\boldsymbol{p}+\hbar\boldsymbol{k}} + \hbar\omega_k) - \delta(\varepsilon_{\boldsymbol{p}} - \varepsilon_{\boldsymbol{p}+\hbar\boldsymbol{k}} - \hbar\omega_k)\big](\Phi_{\boldsymbol{p}} - \Phi_{\boldsymbol{p}+\hbar\boldsymbol{k}}),$$

where $w_k = g^2\hbar k^2/2\rho\omega_k$ is a square of matrix element of electron-phonon coupling, ω_k is the phonon frequency, and $N(\omega_k)$ is the equilibrium phonon distribution function. Let us transform collision integral $\mathrm{St}\,[\Phi_{\boldsymbol{p}}]$ to more convenient representation by introducing an additional integration over frequency ω. Then we have

$$\mathrm{St}\,[\Phi_{\boldsymbol{p}}] = -\frac{2\pi}{\hbar}\sum_{\boldsymbol{k}}\int_{-\infty}^{\infty} d\omega\, w_k \frac{N(\omega_k)\big(1+N(\omega_k)\big)}{T}\big[n_0(\varepsilon_{\boldsymbol{p}+\hbar\boldsymbol{k}}) - n_0(\varepsilon_{\boldsymbol{p}})\big]$$
$$\times\,\delta(\varepsilon_{\boldsymbol{p}} - \varepsilon_{\boldsymbol{p}+\hbar\boldsymbol{k}} + \hbar\omega)\big[\delta(\omega-\omega_k) - \delta(\omega+\omega_k)\big](\Phi_{\boldsymbol{p}} - \Phi_{\boldsymbol{p}+\hbar\boldsymbol{k}}).$$

Next, we get used of the following two identities:

$$N_{-\omega}(1+N_{-\omega}) = N_\omega(1+N_\omega),$$
$$\delta(\omega-\omega_k) - \delta(\omega+\omega_k) = 2\omega\,\delta(\omega^2-\omega_k^2),$$

and arrive at the expression

$$\mathrm{St}\,[\Phi_p] = -\frac{2\pi}{\hbar}\sum_k w_k \int_{-\infty}^{\infty} d\omega\, 2\omega \frac{N_\omega(1+N_\omega)}{T}\,\delta(\omega^2-\omega_k^2)$$
$$\times\big[n_0(\varepsilon_p+\hbar\omega)-n_0(\varepsilon_p)\big]\delta(\varepsilon_p-\varepsilon_{p+\hbar k}+\hbar\omega)[\Phi_p-\Phi_{p+\hbar k}].$$

Thus, it is necessary to calculate

$$(\Phi,\,\hat{W}\Phi) = -\frac{2\pi}{\hbar}\int_{-\infty}^{\infty} d\omega \int \frac{d^3k}{(2\pi)^3}\, w_k \int \frac{2d^3p}{(2\pi\hbar)^3}\, 2\omega \frac{N_\omega(1+N_\omega)}{T}$$
$$\times\,\delta(\omega^2-\omega_k^2)\big[n_0(\varepsilon_p+\hbar\omega)-n_0(\varepsilon_p)\big]\delta(\varepsilon_p-\varepsilon_{p+\hbar k}+\hbar\omega)[\Phi_p-\Phi_{p+\hbar k}]^2$$

for $\Phi_p=(\varepsilon_p-\mu)(\boldsymbol{v}_p\cdot\nabla T)/T$.

Let us consider the integrand multiplier $(\Phi_p-\Phi_{p+\hbar k})^2$, keeping in mind the energy conservation law $\varepsilon_{p+\hbar k}=\varepsilon_p+\hbar\omega$ and the simple relation of electron velocity with momentum $\boldsymbol{v}_p=\boldsymbol{p}/m$

$$(\Phi_p-\Phi_{p+\hbar k})^2 = \big\{[(\varepsilon_p-\mu)\boldsymbol{v}_p-(\varepsilon_{p+\hbar k}-\mu)\boldsymbol{v}_{p+\hbar k}]\cdot\nabla T/T\big\}^2$$
$$= \big\{[(\varepsilon_p-\mu)\boldsymbol{p}-(\varepsilon_p-\mu+\hbar\omega)(\boldsymbol{p}+\hbar\boldsymbol{k})]\cdot\nabla T\big\}^2/(mT)^2$$
$$= \big\{[(\varepsilon_p-\mu+\hbar\omega)\hbar\boldsymbol{k}+\hbar\omega\boldsymbol{p}]\cdot\nabla T\big\}^2/(mT)^2.$$

For the further transformation of integrals, it is practical to use the spherical symmetry of electron and phonon spectra. The lack of any selected directions in the isotropic model of a metal allows us to average, for example, over all orientations of vector ∇T according to

$$\langle(\nabla T)_\alpha(\nabla T)_\beta\rangle = (\nabla T)^2\delta_{\alpha\beta}/3,\quad \alpha,\,\beta=(x,\,y,\,z).$$

This simplifies the further calculations to great extent. As a result, we find

$$(\Phi_p-\Phi_{p+\hbar k})^2 = \frac{(\nabla T)^2}{3(mT)^2}$$
$$\times\big[(\varepsilon_p-\mu+\hbar\omega)^2\hbar^2k^2+2\hbar\omega(\varepsilon_p-\mu+\hbar\omega)\hbar\boldsymbol{k}\cdot\boldsymbol{p}+\hbar^2\omega^2p^2\big].$$

The scalar product $\boldsymbol{k}\cdot\boldsymbol{p}$ can be estimated with the aid of the condition of zero argument in δ-function. In fact, for the simple electron dispersion law $\varepsilon_p=\boldsymbol{p}^2/2m$ we have

$$\varepsilon_{p+\hbar k}-\varepsilon_p-\hbar\omega = \frac{2\hbar\boldsymbol{k}\cdot\boldsymbol{p}+\hbar^2k^2}{2m}-\hbar\omega = 0 \ \text{ or } \ 2\boldsymbol{k}\cdot\boldsymbol{p} = -\hbar k^2+2m\omega.$$

Accordingly, we obtain

$$(\Phi_p - \Phi_{p+\hbar k})^2 = \frac{(\nabla T)^2}{3(mT)^2}\big[(\varepsilon_p - \mu)^2\hbar^2 k^2 + 2\hbar\omega(\varepsilon_p - \mu)(\hbar^2k^2 + 2m\hbar\omega) + \hbar^2\omega^2(p^2 + 2m\hbar\omega)\big].$$

We consider now integration in $\delta(\varepsilon_p - \varepsilon_{p+\hbar k} + \hbar\omega)$ over the angle between vectors $\boldsymbol{k}$ and $\boldsymbol{p}$. The argument in δ-functions equals

$$\varepsilon_{p+\hbar k} - \varepsilon_p - \hbar\omega = \frac{\hbar kp}{m}\left(\cos\theta + \frac{\hbar k}{2p} - \frac{2m\omega}{pk}\right) \approx \frac{\hbar kp}{2m}\left(\cos\theta + \frac{\hbar k}{2p}\right),$$

where θ is the angle between vectors $\boldsymbol{k}$ and $\boldsymbol{p}$. Similarly to the previous problem we are interested in the region of momenta $\boldsymbol{p}$ close to the magnitude of Fermi momentum p_F. We can neglect the term $2m\omega/pk$ as well since

$$\frac{2m\omega}{pk} = \frac{2m\omega_k}{pk} \sim \frac{2ms}{p_F} \sim \frac{s}{v_F} \ll 1$$

and sound velocity s is small as compared with the Fermi velocity v_F. Let us perform a portion of integration over vector $\boldsymbol{k}$, namely, over angle θ between vectors $\boldsymbol{k}$ and $\boldsymbol{p}$. The result is the following:

$$\int \frac{d\Omega}{4\pi}\delta(\varepsilon_{p+\hbar k} - \varepsilon_p - \hbar\omega) \approx -\frac{1}{2}\int_0^\pi d(\cos\theta)\,\delta\left[\frac{\hbar kp}{m}\left(\cos\theta + \frac{\hbar k}{2p}\right)\right] = \frac{m}{2\hbar kp}\theta(2p - \hbar k).$$

Thereby, the integration over the absolute value of vector $\boldsymbol{k}$ is restricted by magnitude $2p/\hbar \sim 2p_F/\hbar$. So, we arrive at the following expression for $(\Phi,\ \hat{W}\Phi)$:

$$(\Phi,\ \hat{W}\Phi) = -\frac{1}{3}\frac{(\nabla T)^2}{m^2T^2}\frac{2\pi}{\hbar}\int_{-\infty}^{\infty} d\omega \int_0^{2p/\hbar}\frac{4\pi k^2 dk}{(2\pi)^3}\int_0^\infty 2\frac{4\pi p^2\,dp}{(2\pi\hbar)^3}w_k$$
$$\times\, 2\omega\frac{N_\omega(1+N_\omega)}{T}\,\delta(\omega^2 - \omega_k^2)\,\frac{m}{2\hbar kp}\big[n_0(\varepsilon_p + \hbar\omega) - n_0(\varepsilon_p)\big]$$
$$\times\,\big[(\varepsilon_p - \mu)^2\hbar^2k^2 + \hbar\omega(\varepsilon_p - \mu)(\hbar^2k^2 + 2m\hbar\omega) + \hbar^2\omega^2(p^2 + 2m\hbar\omega)\big].$$

Let us turn now to the energy variable $\varepsilon = p^2/2m$ in integrating over electron momentum $\boldsymbol{p}$

$$(\Phi,\,\hat{W}\Phi) = -\frac{2\pi}{3}\frac{(\nabla T)^2}{\hbar^2 T^2}\int\limits_{-\infty}^{\infty} d\omega \int\limits_{0}^{\infty} \frac{d\varepsilon}{\pi^2\hbar^3} \int\limits_{0}^{2p(\varepsilon)/\hbar} \frac{k\,dk}{2\pi^2}\, w_k$$
$$\times\, \omega\, \frac{N_\omega(1+N_\omega)}{T}\, \delta(\omega^2-\omega_k^2)\,[n_0(\varepsilon+\hbar\omega)-n_0(\varepsilon)]$$
$$\times\left[(\varepsilon-\mu)^2\hbar^2k^2 + \hbar\omega(\varepsilon-\mu)(\hbar^2k^2+2m\hbar\omega) + \hbar^2\omega^2(p^2(\varepsilon)+2m\hbar\omega)\right].$$

While integrating over ε, we emphasize that the main contribution to the integrals is gained from the energies close to the chemical potential or to the Fermi surface $|\varepsilon - \mu| \ll \mu$. Along with it the phonon energies $\hbar\omega$ do not exceed the Debye temperature Θ_D also small as compared with the Fermi energy $\varepsilon_F \sim \mu$ due to inequality $\Theta_D/\varepsilon_F \sim s/v_F \ll 1$. These features, specific for the integrals with the Fermi distribution at low $T \ll \mu \sim \varepsilon_F$ temperatures, allow one to replace the lower limit of integration over ε with $-\infty$ and put $p(\varepsilon) \approx p(\varepsilon_F) = p_F$ as well.

Using the following values for the integrals with the Fermi function $n_0(\varepsilon)$:

$$\int\limits_{-\infty}^{\infty} d\varepsilon\big[n_0(\varepsilon+\hbar\omega)-n_0(\varepsilon)\big] = -\hbar\omega,$$

$$\int\limits_{-\infty}^{\infty} d\varepsilon\,(\varepsilon-\mu)\big[n_0(\varepsilon+\hbar\omega)-n_0(\varepsilon)\big] = \hbar^2\omega^2/2,$$

$$\int\limits_{-\infty}^{\infty} d\varepsilon\,(\varepsilon-\mu)^2\big[n_0(\varepsilon+\hbar\omega)-n_0(\varepsilon)\big] = -\hbar\omega\big(\hbar^2\omega^2+\pi^2T^2\big)/3,$$

and neglecting the term $m\hbar\omega$ compared with $p^2 \sim p_F^2$, we find that

$$(\Phi,\,\hat{W}\Phi) = \frac{2\pi(\nabla T)^2}{9\hbar^2}\int\limits_{-\infty}^{\infty} d\omega \int\limits_{0}^{2p_F/\hbar} \frac{k\,dk}{2\pi^2}\, w_k\,\omega^2 \frac{N_\omega(1+N_\omega)}{T}$$
$$\times\, \delta(\omega^2-\omega_k^2)\left[k^2 + \left(\frac{\omega}{\pi T}\right)^2\left(3p_F^2 - \frac{\hbar^2k^2}{2}\right)\right].$$

We will perform the further calculation of integrals for the simple phonon spectrum $\omega_k = sk$ with a single acoustic branch. On the full analogy with the previous problem we see that the integration over the absolute value of wave vector $\boldsymbol{k}$ is restricted with value k_m being the minimal magnitude of two wave vectors $2p_F/\hbar$ and k_D. The last wave vector corresponds to the Debye frequency $\omega_D = sk_D$. Below we make no distinction between the magnitudes of vectors k_D and $2p_F/\hbar$ which, in addition, are approximately the same in a metal. For definiteness, we put $k_m = k_D$.

So, we write

$$(\Phi,\ \hat{W}\Phi) = \frac{(\nabla T)^2}{9\pi\hbar^2}\int\limits_0^{k_m=k_D} dk\, k w_k\, \frac{\omega_k}{T}\, \frac{e^{\hbar\omega_k/T}}{\left(e^{\hbar\omega_k/T}-1\right)^2} \times \left[k^2 + \left(\frac{\omega}{\pi T}\right)^2\left(3p_F^2 - \frac{\hbar^2 k^2}{2}\right)\right].$$

Substituting $w_k = g^2\hbar k^2/(2\rho\omega_k)$ where ρ is the metal density and introducing the Debye temperature $\Theta_D = \hbar s k_D$ and new variable $z = \hbar s k/T$, we find finally

$$(\Phi,\ \hat{W}\Phi) = \frac{(\nabla T)^2}{18\pi}\,\frac{g^2 k_D^4}{\rho s^2\hbar^3}\,\frac{T^5}{\Theta_D^4} \times \left[J_5\!\left(\frac{\Theta_D}{T}\right) - \frac{1}{2\pi^2}J_7\!\left(\frac{\Theta_D}{T}\right) + \frac{3}{\pi^2}\left(\frac{p_F}{\hbar k_D}\right)^2\left(\frac{\Theta_D}{T}\right)^2 J_5\!\left(\frac{\Theta_D}{T}\right)\right].$$

The integral Debye functions $J_n(z)$ are defined according to the formula

$$J_n(z) = \int\limits_0^z dz\,\frac{z^n e^z}{(e^z-1)^2} = \frac{1}{4}\int\limits_0^z dz\,\frac{z^n}{\sinh^2 z/2}.$$

At last, we can now determine the thermal conductivity $\varkappa$, taking the results for $(\Phi,\ f)$ and $(\Phi,\ \hat{W}\Phi)$. The substitution of two last quantities into $R_\varkappa[\Phi_p]$ leads to the *Wilson formula* for thermal resistivity $1/\varkappa$

$$\frac{1}{\varkappa} = T\,\frac{\frac{1}{2}(\nabla T)^2(\Phi,\ \hat{W}\Phi)}{(\Phi,\ f)^2} = \frac{9\pi\zeta\hbar^2}{4mp_F v_F^2}\left(\frac{\hbar k_D}{p_F}\right)^4 f\!\left(\frac{\Theta_D}{T}\right)$$
$$= \frac{3}{4\pi}\zeta\frac{m}{n\hbar}\left(\frac{\hbar k_D}{p_F}\right)^4 f\!\left(\frac{\Theta_D}{T}\right),$$
$$f(z) = \frac{1}{z^4}\left[J_5(z) - \frac{1}{2\pi^2}J_7(z) + \frac{3}{\pi^2}\left(\frac{p_F}{\hbar k_D}\right)^2 z^2 J_5(z)\right].$$

In this formula the parameter

$$\zeta = \frac{g^2}{\rho s^2}\nu(\varepsilon_F) = \frac{g^2}{\rho s^2}\,\frac{mp_F}{\pi^2\hbar^2}$$

determines the electron-phonon coupling constant and $n = p_F^3/(3\pi^2\hbar^3)$ is the electron concentration in a metal.

Let us analyze the temperature behavior of thermal conductivity. We start from the high $T \gg \Theta_D$ temperatures. In this case the main contribution results from the first term in the square brackets because $J_5(z \ll 1) \approx z^4/4$ and the other two terms as z^6 are much smaller. Finally, for the temperatures exceeding the Debye temperature Θ_D, the thermal conductivity

$$\varkappa(T) = \frac{16\pi}{3}\frac{1}{\zeta}\frac{n\hbar}{m}\left(\frac{p_F}{\hbar k_D}\right)^4 \sim \frac{n\hbar}{m}\frac{1}{\zeta} \quad (T \gg \Theta_D)$$

is temperature independent.

In the low $T \ll \Theta_D$ temperature region the predominant contribution is associated with the last term in the square brackets. This entails the following temperature behavior:

$$\varkappa(T) = \frac{4\pi^3}{9J_5(\infty)}\frac{1}{\zeta}\frac{n\hbar}{m}\left(\frac{p_F}{\hbar k_D}\right)^2\left(\frac{\Theta_D}{T}\right)^2$$
$$\sim \frac{n\hbar}{m}\frac{1}{\zeta}\frac{\Theta_D^2}{T^2}, \quad (T \ll \Theta_D).$$

Here $J_5(\infty) = 120\zeta(5)$ and $\zeta(5) \approx 1.037$ is the Riemann zeta function. In Fig. 3.6 the example is shown for the behavior of thermal conductivity $\varkappa(T)$ normalized with its limiting high temperature value $\varkappa(\infty)$ if $\hbar k_D = p_F$.

Let us look at the behavior of thermal conductivity $\varkappa$ from the viewpoint of well-known elementary formula in the kinetic theory of gases

$$\varkappa(T) = \frac{1}{3}C(T)v_F^2\tau_\varkappa,$$
$$C(T) = \frac{\pi^2}{3}T\nu(\varepsilon_F) = \frac{\pi^2}{2}n\frac{T}{\varepsilon_F}.$$

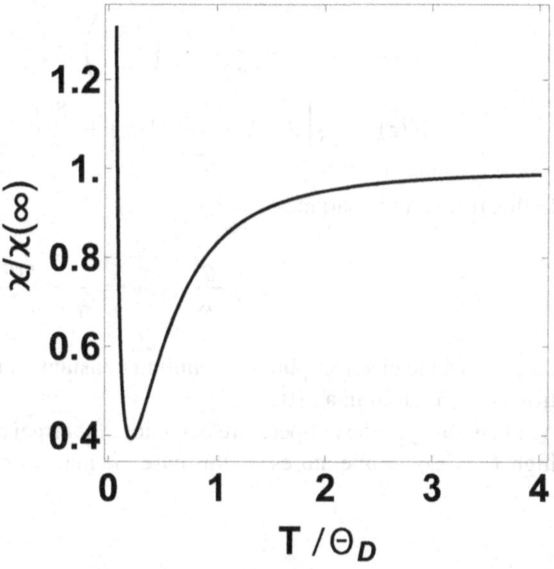

Fig. 3.6 The example of the behavior for the normalized thermal conductivity $\varkappa(T)/\varkappa(\infty)$ as a function of normalized temperature T/Θ_D if $p_F = \hbar k_D$

Here $C(T)$ is the electron specific heat of a metal at temperature $T \ll \varepsilon_F$, v_F is the typical electron velocity equal to the Fermi one, and $\tau_\varkappa$ is the typical electron-phonon collision time. It is more accurate to say about the latter quantity as an effective time of energy exchange between the electron and phonon subsystems. Correspondingly, we find for the electron-phonon collision frequency

$$\nu_\varkappa = \tau_\varkappa^{-1} = \frac{\pi}{4}\left(\frac{\hbar k_D}{p_F}\right)^4 \frac{\zeta T}{\hbar} \left.\frac{J_5(z) - \frac{1}{2\pi^2}J_7(z) + \frac{3}{\pi^2}\left(\frac{p_F}{\hbar k_D}\right)^2 z^2 J_5(z)}{z^4}\right|_{z=\frac{\Theta_D}{T}}.$$

The following formulas below give a clear representation about behavior of effective electron-phonon collision frequency $\nu_\varkappa$ in the limiting cases:

$$\nu_\varkappa = \tau_\varkappa^{-1} = \begin{cases} \frac{\pi}{16}\frac{T}{\hbar}\zeta\left(\frac{\hbar k_D}{p_F}\right)^4 \sim \zeta\frac{T}{\hbar}, & T \gg \Theta_D, \\ \frac{90}{\pi}\zeta(5)\frac{T}{\hbar}\zeta\left(\frac{T}{\Theta_D}\right)^2\left(\frac{\hbar k_D}{p_F}\right)^2 \sim \zeta\frac{T^3}{\hbar\Theta_D^2}, & T \ll \Theta_D. \end{cases}$$

Concerning the Wiedemann-Franz law, we expect that it should exactly hold for the high $T \gg \Theta_D$ temperatures. In this sense the Lorenz number $L = \varkappa/\sigma T$ proves to be temperature independent. This property is a direct consequence of the following circumstance. The effective energy exchange time $\tau_\varkappa$ and the momentum exchange time τ_{tr} between the electron and the phonon subsystems have the same temperature behavior and are the same order of magnitude $\tau_\varkappa = \tau_{tr}$. For low $T \ll \Theta_D$ temperatures, this is wrong since $\tau_\varkappa \sim T^{-3}$ and $\tau_{tr} \sim T^{-5}$. The outcome of the latter is a *violation* of the Wiedemann-Franz law, the Lorenz number being temperature dependent as $L \sim T^2$.

16. Let us write down the transport equation for the electron distribution function $n_{\boldsymbol p}(\boldsymbol r)$

$$\boldsymbol v \cdot \frac{\partial n_{\boldsymbol p}}{\partial \boldsymbol r} + e\boldsymbol E \cdot \frac{\partial n_{\boldsymbol p}}{\partial \boldsymbol p} = \mathrm{St}_e[n, N],$$

where $\boldsymbol E$ is the electron field acting at electrons and the electron velocity equals $\boldsymbol v = \partial\varepsilon_{\boldsymbol p}/\partial\boldsymbol p = \boldsymbol p/m$. The stationary transport equation for phonon distribution function $N_{\boldsymbol k}(\boldsymbol r)$ reads

$$\boldsymbol s \cdot \frac{\partial N_{\boldsymbol k}}{\partial \boldsymbol r} = \mathrm{St}_{ph}[N, n],$$

$\boldsymbol s = \partial\omega_{\boldsymbol k}/\partial\boldsymbol k = s\boldsymbol k/k$ being the phonon velocity.

We multiply the first equation by $\boldsymbol p$ and the second one by $\hbar\boldsymbol k$. Then we integrate them over the electron and phonon variables, respectively. Next, we sum these two equations and involve that the right-hand side vanishes according to conservation of the total momentum for electrons and phonons

$$\int \boldsymbol p\left(\boldsymbol v \cdot \frac{\partial n_{\boldsymbol p}}{\partial \boldsymbol r} + e\boldsymbol E \cdot \frac{\partial n_{\boldsymbol p}}{\partial \boldsymbol p}\right)\frac{2d^3p}{(2\pi\hbar)^3} + \int \hbar\boldsymbol k\left(\boldsymbol s \cdot \frac{\partial N_{\boldsymbol k}}{\partial \boldsymbol r}\right)$$
$$= \int \boldsymbol p\,\mathrm{St}_e[n, N]\frac{2d^3p}{(2\pi\hbar)^3} + \int \hbar\boldsymbol k\,\mathrm{St}_{ph}[N, n]\frac{d^3k}{(2\pi)^3} = 0.$$

We can choose the locally equilibrium electron and phonon distribution functions as a first approximation for the both distribution functions and in the linear approximation in small magnitudes of electric field $\boldsymbol{E}$ and temperature gradient ∇T

$$n_p \approx n_0\left(\frac{\varepsilon_p-\mu}{T}\right) \quad \text{and} \quad N_k \approx N_0\left(\frac{\hbar\omega_k}{T}\right).$$

Here n_0 and N_0 are the Fermi and Planck distributions, respectively. Indicating that $\partial N_0/\partial \boldsymbol{r} = -(\omega\partial N_0/\partial\omega)(\nabla T/T)$, we find

$$\int \boldsymbol{p}(e\boldsymbol{v}\cdot\boldsymbol{E}')\frac{\partial n_0}{\partial\varepsilon_p}\frac{2d^3p}{(2\pi\hbar)^3} - \int \boldsymbol{p}(\boldsymbol{v}\cdot\nabla T)\frac{\varepsilon_p-\mu}{T}\frac{\partial n_0}{\partial\varepsilon_p}\frac{2d^3p}{(2\pi\hbar)^3} - \int \hbar\boldsymbol{k}(\boldsymbol{s}\cdot\nabla T)\frac{\omega_k}{T}\frac{\partial N_0}{\partial\omega_k}\frac{d^3k}{(2\pi)^3} = 0.$$

Here $\boldsymbol{E}' = \boldsymbol{E} - \mu/e$ is an effective field acting on an electron.

Employing the spherical symmetry of electron and phonon spectra, it is straightforwardly to perform averaging over the angles between a series of two vectors as $\boldsymbol{v}$ and $\boldsymbol{E}'$, $\boldsymbol{v}$ and ∇T as well as $\boldsymbol{k}$ and ∇T. This results in

$$e\boldsymbol{E}'\int vp\frac{\partial n_0}{\partial\varepsilon_p}\frac{2d^3p}{(2\pi\hbar)^3} - \nabla T\int vp\frac{\varepsilon_p-\mu}{T}\frac{\partial n_0}{\partial\varepsilon_p}\frac{2d^3p}{(2\pi\hbar)^3} - \nabla T\int\frac{\hbar sk\,\omega_k}{T}\frac{\partial N_0}{\partial\omega_k}\frac{d^3k}{(2\pi)^3} = 0.$$

The integrals containing the Fermi distribution function were formerly calculated in the previous problems. So, we can promptly write down the answers. Integrating by parts yields the last integral

$$\int_0^\infty \frac{\hbar sk\,\omega_k}{T}\frac{\partial N_0}{\partial\omega_k}\frac{k^2dk}{2\pi^2} = \frac{\hbar}{2\pi^2s^3T}\int_0^\infty \omega^4\frac{\partial N_0}{\partial\omega}d\omega$$
$$= -\frac{2\hbar}{\pi^2s^3T}\int_0^\infty \omega^3N_0(\omega)d\omega = -\frac{2\pi^2}{15}\frac{T^3}{\hbar^3s^3}.$$

Hence we have

$$-3en\boldsymbol{E}' + \frac{3\pi^2}{2}\frac{nT}{\varepsilon_F}\nabla T + \frac{2\pi^2}{15}\frac{T^3}{\hbar^3s^3}\nabla T = 0,$$

where n is the electron concentration and ε_F is the Fermi energy.

Because the electron current

$$j = \int v n_0(\varepsilon_p) \frac{2d^3p}{(2\pi\hbar)^3} = 0$$

vanishes for the locally equilibrium distribution function n_0 chosen as zero approximation, we can determine the Seebeck coefficient S as a ratio of $\boldsymbol{E}'$ to temperature gradient ∇T. So, the Seebeck coefficient equals

$$S = \frac{\pi^2}{2e}\left(\frac{T}{\varepsilon_F} + \frac{4\pi^2}{45}\frac{T^3}{\hbar^3 s^3 n}\right) = \frac{\pi^2}{2e}\left[\frac{T}{\varepsilon_F} + \frac{4\pi^4}{15}\left(\frac{\hbar k_D}{p_F}\right)^3\left(\frac{T}{\Theta_D}\right)^3\right],$$

where we introduce the Debye temperature Θ_D and corresponding wave vector $k_D \sim p_F/\hbar$ according to $\Theta_D = \hbar s k_D$. The contribution of the second term to the Seebeck coefficient is wholly due to the presence of phonon subsystem and deviation of its distribution function N_k from equilibrium one. The effect is associated with the slow relaxation in the phonon subsystem, which occurs mainly due to the scattering of electrons with phonons instead of phonon-phonon collisions at low $T \ll \Theta_D$ temperatures.

3.6 The Electron Scattering with Magnetic Impurities: The Kondo Effect

17. Let us write down the collision integral for the distribution function of electrons in the field of randomly located impurities

$$\mathrm{St}\,[n_p] = \int \left[(1-n_p)W(\boldsymbol{p}\,\boldsymbol{p}')n_{p'} - (1-n_{p'})W(\boldsymbol{p}'\boldsymbol{p})n_p\right]\frac{V\,d^3p'}{(2\pi\hbar)^3}\,.$$

Then we formulate the probability $W(\boldsymbol{p}'\boldsymbol{p})$ for the electron elastic transition from the state with momentum $\boldsymbol{p}$ to the state with momentum $\boldsymbol{p}'$ per unit time and per unit volume

$$W(\boldsymbol{p}'\boldsymbol{p}) = \frac{2\pi}{\hbar}|t(\boldsymbol{p}'\boldsymbol{p})|^2\delta(\varepsilon_p - \varepsilon_{p'}).$$

Here the matrix element $t(\boldsymbol{p}'\boldsymbol{p})$ is connected with the scattering amplitude $a(\boldsymbol{p}'\boldsymbol{p})$ as

$$t(\boldsymbol{p}'\boldsymbol{p}) = -\frac{2\pi\hbar^2}{m}a(\boldsymbol{p}'\boldsymbol{p})$$

and, correspondingly, with the differential cross section of scattering as

$$\frac{d\sigma}{d\Omega} = |a(\boldsymbol{p}'\boldsymbol{p})|^2 = \left(\frac{2\pi\hbar^2}{m}\right)^2|t(\boldsymbol{p}'\boldsymbol{p})|^2.$$

We have in the first Born approximation

$$t(\boldsymbol{p}'\boldsymbol{p}) \approx t^{(1)}(\boldsymbol{p}'\boldsymbol{p}) = U(\boldsymbol{p}'\boldsymbol{p}),$$

where $U(\boldsymbol{p}'\boldsymbol{p})$ is the matrix element of electron-impurity coupling potential $U(\boldsymbol{r})$. For the wave function of free electron motion

$$\psi_{\boldsymbol{p}}(\boldsymbol{r}) = \frac{e^{i\boldsymbol{pr}/\hbar}}{\sqrt{V}}$$

with momentum $\boldsymbol{p}$ where the factor $V^{-1/2}$ implies the normalization per one electron in volume V, the matrix element $U(\boldsymbol{p}'\boldsymbol{p})$ is given by

$$U_{\boldsymbol{p}'\boldsymbol{p}} = \int \psi^*_{\boldsymbol{p}'}(\boldsymbol{r})U(\boldsymbol{r})\psi_{\boldsymbol{p}}(\boldsymbol{r})\,d^3r = \frac{u_{\boldsymbol{p}'\boldsymbol{p}}}{V}\sum_a e^{-i(\boldsymbol{p}'-\boldsymbol{p})\boldsymbol{R}_a/\hbar}.$$

Thus, matrix element $u_{\boldsymbol{p}'\boldsymbol{p}}$ acquires an additional factor $e^{-i(\boldsymbol{p}'-\boldsymbol{p})\boldsymbol{R}_a/\hbar}$ from each impurity located at site $\boldsymbol{r} = \boldsymbol{R}_a$. After averaging the modulus-squared $|U_{\boldsymbol{p}'\boldsymbol{p}}|^2$ over random impurity sites, in the double sum there remains the only part corresponding to the same impurity sites $\boldsymbol{R}_a = \boldsymbol{R}_b$, i.e.

$$|U_{\boldsymbol{p}'\boldsymbol{p}}|^2_{\text{av}} = \frac{N_i}{V^2}|u_{\boldsymbol{p}'\boldsymbol{p}}|^2.$$

Here $N_i = n_i V$ is the total number of impurities in volume V.

So, the transition probability takes the following form after averaging over the impurity sites:

$$W^{(\text{av})}_{\boldsymbol{p}'\boldsymbol{p}} = \frac{2\pi}{\hbar}n_i\frac{|u_{\boldsymbol{p}'\boldsymbol{p}}|^2}{V}\delta(\varepsilon_{\boldsymbol{p}} - \varepsilon'_{\boldsymbol{p}}).$$

If the transition probability is symmetrical, i.e. $W^{(\text{av})}_{\boldsymbol{p}'\boldsymbol{p}} = W^{(\text{av})}_{\boldsymbol{p}\boldsymbol{p}'}$, the collision integral simplifies immediately

$$\mathrm{St}\,[n_{\boldsymbol{p}}] = -\int W^{(\text{av})}_{\boldsymbol{p}\boldsymbol{p}'}\left(n_{\boldsymbol{p}} - n_{\boldsymbol{p}'}\right)\frac{V\,d^3p'}{(2\pi\hbar)^3}\,.$$

For the electric field $\boldsymbol{E}$-linear approximation, the solution of transport equation for the electron distribution function $n_{\boldsymbol{p}}$ allows us to represent the collision integral in the relaxation form with some effective transport collision time

$$\mathrm{St}\,[n_{\boldsymbol{p}}] = -\frac{n(\boldsymbol{p}) - n_0(\boldsymbol{p})}{\tau_{tr}(p)}\,.$$

Here $n_0(\boldsymbol{p})$ is the equilibrium Fermi distribution function. The transport collision time is given by the formula

$$\frac{1}{\tau_{tr}(p)} = \int W^{(\mathrm{av})}_{p'p}(1-\cos\theta')\frac{V\,d^3p'}{(2\pi\hbar)^3}$$

in the assumption that $W^{(\mathrm{av})}_{p'p}$ depends only upon the angle θ' between vectors $\boldsymbol{p}$ and $\boldsymbol{p}'$.

Beyond the framework of first Born approximation the transition probability $W^{(\mathrm{av})}_{p'p}$ is now expressed in terms of the total matrix element $t(\boldsymbol{p}'\boldsymbol{p})$ differing by factor $(-2\pi\hbar^2/m)$ alone from the exact electron scattering amplitude $a(\boldsymbol{p}'\boldsymbol{p})$ at single separate impurity, i.e.

$$W^{(\mathrm{av})}_{p'p} = \frac{2\pi}{\hbar}n_i\frac{|t(\boldsymbol{p}'\boldsymbol{p})|^2}{V}\delta(\varepsilon_p-\varepsilon_{p'})$$

with the obvious first approximation $t(\boldsymbol{p}'\boldsymbol{p}) \approx t^{(1)}_{p'p} = u_{p'p}$.

We start from analyzing the transition probability $W^{(\mathrm{av})}_{p'p}$ for the electron scattering with the exchange *s-d* potential $u(\boldsymbol{r}) = J(\boldsymbol{\sigma}\boldsymbol{S})\delta(\boldsymbol{r})$ of localized spin $\boldsymbol{S}$ in first Born approximation. The Fourier transform of this potential is momentum independent and equal to

$$u_{p'p} = J\boldsymbol{\sigma}\boldsymbol{S}.$$

The matrix elements for various electron spin orientations can be written as follows:

$$\begin{aligned}
t^{(1)}_{\uparrow\uparrow}(\boldsymbol{p}'\boldsymbol{p}) &= \langle\uparrow|u_{p'p}|\uparrow\rangle = J\langle\uparrow|\boldsymbol{\sigma}\boldsymbol{S}|\uparrow\rangle = JS_z,\\
t^{(1)}_{\uparrow\downarrow}(\boldsymbol{p}'\boldsymbol{p}) &= \langle\uparrow|u_{p'p}|\downarrow\rangle = J\langle\uparrow|\boldsymbol{\sigma}\boldsymbol{S}|\downarrow\rangle = JS^-,\\
t^{(1)}_{\downarrow\uparrow}(\boldsymbol{p}'\boldsymbol{p}) &= \langle\downarrow|u_{p'p}|\uparrow\rangle = J\langle\downarrow|\boldsymbol{\sigma}\boldsymbol{S}|\uparrow\rangle = JS^+,\\
t^{(1)}_{\downarrow\downarrow}(\boldsymbol{p}'\boldsymbol{p}) &= \langle\downarrow|u_{p'p}|\downarrow\rangle = J\langle\downarrow|\boldsymbol{\sigma}\boldsymbol{S}|\downarrow\rangle = -JS_z.
\end{aligned}$$

Here $S^{\pm} = S_x \pm iS_y$, S_z are the circular and z projections of impurity spin operator $\boldsymbol{S}$. There are two ways for calculating the collision time.

For our case of non-polarized electron Fermi liquid when all impurity spin orientations are equiprobable, the scattering time of electron is spin independent and identical for the up and down spin directions. In this case we can sum up the transition probability over all final electron spin orientations and average over the initial spin states or, in other words, divide by 2. So, we obtain

$$\begin{aligned}
(1/2)\sum_{\sigma,\sigma'}|t^{(1)}(\boldsymbol{p}'\boldsymbol{p})|^2 &= \left(t^{(1)}_{\uparrow\uparrow}t^{(1)}_{\uparrow\uparrow} + t^{(1)}_{\uparrow\downarrow}t^{(1)}_{\downarrow\uparrow} + t^{(1)}_{\downarrow\uparrow}t^{(1)}_{\uparrow\downarrow} + t^{(1)}_{\downarrow\downarrow}t^{(1)}_{\downarrow\downarrow}\right)/2\\
&= J^2\left(S_z^2 + S^-S^+ + S^+S^- + (-S_z)^2\right)/2 = J^2S(S+1).
\end{aligned}$$

As a result, we find for the collision time

$$\frac{1}{\tau_{tr}^{(1)}(p)} = \frac{2\pi}{\hbar} n_i J^2 S(S+1) \int \delta(\varepsilon_{\boldsymbol{p}} - \varepsilon_{\boldsymbol{p}'}) \frac{d^3 p'}{(2\pi\hbar)^3}$$
$$= \frac{\pi n_i}{\hbar} \nu(\varepsilon_p) J^2 S(S+1),$$

$\nu(\varepsilon) = m(2m\varepsilon)^{1/2}/(\pi^2\hbar^3)$ being the density of states.

On the other hand, the scattering time of an electron, e.g. with spin up direction, can be written as

$$\frac{1}{\tau_{tr\uparrow}^{(1)}(p)} = \frac{2\pi}{\hbar} n_i \int \langle t_{\uparrow\uparrow}^{(1)} t_{\uparrow\uparrow}^{(1)} + t_{\uparrow\downarrow}^{(1)} t_{\downarrow\uparrow}^{(1)} \rangle \, \delta(\varepsilon_{\boldsymbol{p}} - \varepsilon_{\boldsymbol{p}'}) \frac{d^3 p'}{(2\pi\hbar)^3}$$
$$= \frac{\pi n_i}{\hbar} \nu(\varepsilon_p) J^2 \langle S_z^2 + S^- S^+ \rangle,$$

where the angle brackets mean averaging over all impurity spin orientations. Taking into attention that

$$\langle S_z^2 + S^- S^+ \rangle = \langle S_z^2 + S_x^2 + S_y^2 \rangle - \langle S_z \rangle = S(S+1)$$

and that the average magnitude of projection $\langle S_z \rangle$ vanishes, we arrive at the same result for the scattering time. The consideration of scattering the spin down electron is analogous.

In first Born approximation the electrical resistivity ρ is determined with the magnitude of scattering time $\tau_{tr}^{(1)}(p)$ at momentum $p = p_F$ or at energy $\varepsilon = \varepsilon_F$ and is equal to

$$\rho^{(1)} = \frac{m}{ne^2} \frac{1}{\tau_{tr}^{(1)}(p_F)} = \frac{3\pi m}{2e^2} n_i \frac{J^2 S(S+1)}{\hbar \varepsilon_F}.$$

The resistivity is proportional to the impurity concentration n_i and to a square of exchange constant J. In this approximation the resistivity is temperature independent.

Let us turn now to the scattering amplitude in second Born approximation and find the contribution to resistivity proportional to a cube of exchange constant J. Within the second-order approximation in the interaction the matrix element of electron transition from state $\boldsymbol{p}$ to state $\boldsymbol{p}'$ is given by the formula

$$t(\boldsymbol{p}', \boldsymbol{p}) = t^{(1)}(\boldsymbol{p}', \boldsymbol{p}) + t^{(2)}(\boldsymbol{p}', \boldsymbol{p}) + \ldots = u_{\boldsymbol{p}'\boldsymbol{p}} + \sum_k \frac{u_{\boldsymbol{p}'\boldsymbol{k}} u_{\boldsymbol{k}\boldsymbol{p}}}{\varepsilon_{\boldsymbol{p}} - \varepsilon_{\boldsymbol{k}} + i\delta} + \ldots$$

The summation here is performed over all *intermediate* states $\boldsymbol{k}$ for transition $\boldsymbol{p} \to \boldsymbol{p}'$. Such transition occurs as if it is twofold scattering on impurity and as two possible sequences, namely, $\boldsymbol{p} \to \boldsymbol{k}$ and then $\boldsymbol{k} \to \boldsymbol{p}'$ or $\boldsymbol{k} \to \boldsymbol{p}'$ and then $\boldsymbol{p} \to \boldsymbol{k}$.

In accordance with the Pauli principle when we treat the transitions involving an intermediate state, we must take into account whether the intermediate state is free or occupied by other electrons. The presence of electrons in the intermediate state is

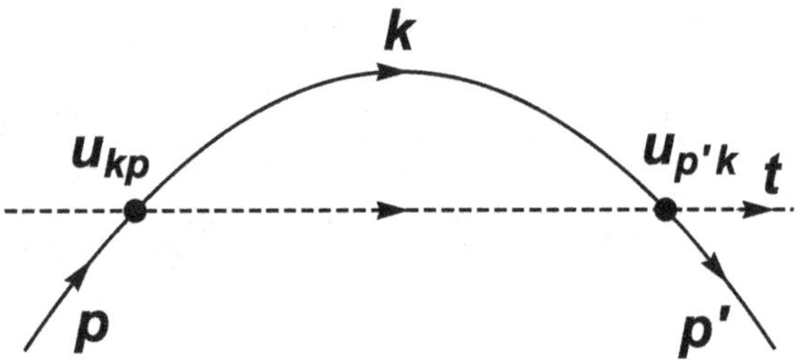

Fig. 3.7 The first option of twofold scattering on an impurity

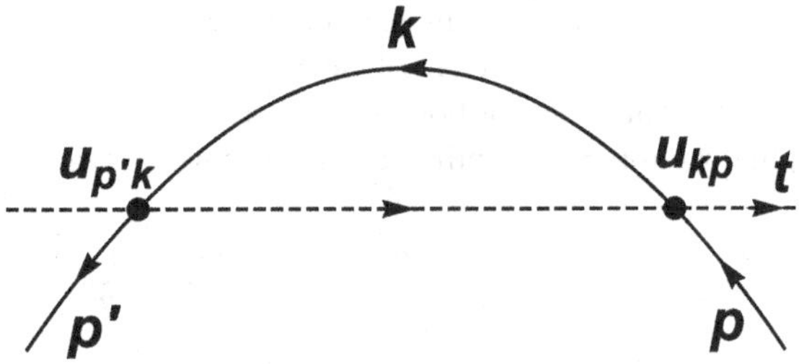

Fig. 3.8 The second option of twofold scattering on an impurity

described by the distribution function n_p. In the first case, in order for the electron to go over from the state $\boldsymbol{p}$ to the state $\boldsymbol{k}$, the latter must be free. This condition can be satisfied by introducing an additional factor $(1 - n_k)$

$$t_1^{(2)}(\boldsymbol{p}', \boldsymbol{p}) = \sum_k u_{p'k} \frac{1 - n_k}{\varepsilon_p - \varepsilon_k + i\delta} u_{kp} \,.$$

The drawing of this scattering option is shown in Fig. 3.7.

The second variant of scattering resembles the first one but occurs in the reverse time order. The electron in state $\boldsymbol{k}$, scattering on impurity, goes over to state $\boldsymbol{p}'$ and later the electron in state $\boldsymbol{p}$, scattering on impurity, arrives at state $\boldsymbol{k}$. Thus, the electron in state $\boldsymbol{p}'$ appears earlier in time as compared with the electron in state $\boldsymbol{p}$. Here, on the contrary, state $\boldsymbol{k}$ should be occupied. To take this aspect into account, we introduce an additional factor n_k. In addition, the reverse time ordering important in quantum mechanics can be performed by the permutation of matrix elements in order to put the time sequence of events from left to right. This second variant of scattering is shown in Fig. 3.8.

Thus we arrive at the relation

$$t_2^{(2)}(\boldsymbol{p}', \boldsymbol{p}) = \sum_k u_{kp} \frac{n_k}{\varepsilon_{p'} - \varepsilon_k + i\delta} u_{p'k} \,.$$

The energy conservation law for scattering between the initial $\boldsymbol{p}$ and final $\boldsymbol{p}'$ states allows us to put $\varepsilon_{p'} = \varepsilon_p$.

Summing these two terms yields

$$t^{(2)}_{\alpha'\alpha}(\boldsymbol{p}',\boldsymbol{p}) = \sum_{k}\left(\frac{u_{p'k}(1-n_k)u_{kp}}{\varepsilon_p-\varepsilon_k+i\delta} + \frac{u_{kp}n_k u_{p'k}}{\varepsilon_p-\varepsilon_k+i\delta}\right)$$

$$= J^2\sum_{\beta}\int\frac{d^3k}{(2\pi\hbar)^3}\left[\frac{(\boldsymbol{\sigma S})_{\alpha'\beta}(\boldsymbol{\sigma S})_{\beta\alpha}}{\varepsilon_p-\varepsilon_k+i\delta}(1-n_k) + \frac{(\boldsymbol{\sigma S})_{\beta\alpha}(\boldsymbol{\sigma S})_{\alpha'\beta}}{\varepsilon_p-\varepsilon_k+i\delta}n_k\right].$$

In the last equality we keep in mind that the state with indices $\boldsymbol{p}, \boldsymbol{k}, \boldsymbol{p}'$ implies the spin projection α, β, α' in addition to electron momentum. Besides, the distribution function n_k is independent of electron spin direction.

In the numerator of both terms there is a product of Pauli matrices and impurity spin operators representing also the following sums over spatial components $i, k = x, y, z$:

$$\sum_{ik}\sigma^i\sigma^k S_i S_k \quad \text{and} \quad \sum_{ik}\sigma^i\sigma^k S_k S_i\,,$$

σ^i and σ^k being the Pauli matrices. Since S_i and S_k are operators, their ordering is essential and the answer will be different for those two expressions:

$$\sum_{ik}\sigma^i\sigma^k S_i S_k = S(S+1) - \sum_{l}\sigma^l S_l = S(S+1) - (\boldsymbol{\sigma S}),$$

$$\sum_{ik}\sigma^i\sigma^k S_k S_i = S(S+1) + \sum_{l}\sigma^l S_l = S(S+1) + (\boldsymbol{\sigma S}).$$

As a result, we obtain the following expression for $t^{(2)}(\boldsymbol{p},\boldsymbol{p}')$:

$$t^{(2)}(\boldsymbol{p}',\boldsymbol{p}) = J^2 S(S+1)\int\frac{1}{\varepsilon_p-\varepsilon_k+i\delta}\frac{d^3k}{(2\pi\hbar)^3} - J^2(\boldsymbol{\sigma S})\int\frac{1-2n_k}{\varepsilon_p-\varepsilon_k+i\delta}\frac{d^3k}{(2\pi\hbar)^3}\,.$$

As we see, one term similar to usual potential scattering does not depend on the electron and impurity spins. The second term, depending on the mutual orientation of electron and impurity spins, has the same exchange structure as first correction $t^{(1)} = u_{pp'} = J(\boldsymbol{\sigma S})$. Emphasize that the second term, unlike the first, depends on temperature T by means of distribution function $n(\varepsilon_k) = [e^{(\varepsilon_k-\mu)/T}+1]^{-1}$ where μ is the chemical potential. Below we do not differ the chemical potential from the Fermi energy ε_F due to inequality $T \ll \varepsilon_F$.

Let us write the total matrix element $t(\boldsymbol{p}'\boldsymbol{p})$ in second Born approximation

$$t(\boldsymbol{p}',\boldsymbol{p}) = J(\boldsymbol{\sigma S})\big(1 - Jg(\varepsilon_p)\big) + J^2 S(S+1) f(\varepsilon_p),$$

where functions $g(\varepsilon_{\boldsymbol{p}})$ and $f(\varepsilon_{\boldsymbol{p}})$ are given by the integrals[2] written below

$$g(\varepsilon_{\boldsymbol{p}}) = \frac{1}{2}\int\limits_{-\mu}^{\infty} \frac{\nu(\xi+\mu)\operatorname{th}\frac{\xi}{2T}\,d\xi}{(\varepsilon_{\boldsymbol{p}}-\mu)-\xi+i\delta} \quad \text{and} \quad f(\varepsilon_{\boldsymbol{p}}) = \frac{1}{2}\int\limits_{-\mu}^{\infty} \frac{\nu(\xi+\mu)\,d\xi}{(\varepsilon_{\boldsymbol{p}}-\mu)-\xi+i\delta}.$$

While treating the integrals, we have crossed over from the momentum integration to integrating over energy $\varepsilon_k = k^2/2m$ and introduced new variable $\xi = \varepsilon_k - \mu$.

Next, in order to estimate the collision time $\tau_{tr}^{(2)}(p)$ required, we should find the average over the electron spins for $(1/2)\sum |t(\boldsymbol{p}'\boldsymbol{p})|^2$ within accuracy to the terms in J not higher than J^3. Since, according to the statement of problem, all the spin orientations are equiprobable, the part $t(\boldsymbol{p}'\boldsymbol{p})$, representing the usual potential scattering and associated with function $f(\varepsilon_p)$, will provide a contribution proportional J^4 alone. The magnitude $\operatorname{Re} f(\varepsilon_p)$ is also small and can be estimated as $\nu(\mu)(\varepsilon_p-\mu)/\mu$. Finally, we find within the accuracy to terms J^3 that

$$(1/2)\sum_{\sigma,\sigma'} |t(\boldsymbol{p}'\boldsymbol{p})|^2 = J^2 S(S+1)\big(1 - 2J\operatorname{Re} g(\varepsilon_p)\big).$$

The collision time is determined with the formula

$$\frac{1}{\tau_{tr}^{(2)}(p)} = \frac{\pi n_i}{\hbar}\nu(\varepsilon_p) J^2 S(S+1)\big(1 - 2J\operatorname{Re} g(\varepsilon_p)\big).$$

As we have seen above, for determining the electrical resistivity ρ, it is essential the collision time of electrons at the Fermi surface, i.e. at $\varepsilon_p = \mu$ or

$$\rho^{(2)} = \frac{m}{ne^2}\frac{1}{\tau_{tr}^{(2)}(\mu)}.$$

Let us estimate function $\operatorname{Re} g(\varepsilon_p)$ at $\varepsilon_p = \mu$

$$\operatorname{Re} g(\mu) = -\frac{1}{2}\int\limits_{-\mu}^{\infty} \nu(\xi+\mu)\frac{\tanh\frac{\xi}{2T}}{\xi}d\xi \sim -\frac{\nu(\mu)}{2}\int\limits_{-\mu}^{\mu\sim\varepsilon_F} \frac{\tanh\frac{\xi}{2T}}{\xi}d\xi\,.$$

At $\mu \sim \varepsilon_F \gg T$ the integral becomes logarithmical and for the large logarithm it is sufficient to know the integration limits only by the order of the magnitude. Hence, $\operatorname{Re} g(\varepsilon = \mu) \approx \nu(\varepsilon_F)\ln(\varepsilon_F/T)$ and we have the following for the collision time and resistivity:

---

[2] The formally divergent integrals are wholly resulted from neglecting the dependence of exchange constant J on the momentum difference $\boldsymbol{p}-\boldsymbol{p}'$. In fact, the exchange interaction decreases rapidly at the distances exceeding the interatomic ones or at the momenta larger than the Fermi momentum. Thus the upper integration limit is confined by a few Fermi momenta p_F.

$$\frac{1}{\tau_{tr}^{(2)}(p_F)} = \frac{\pi n_i}{\hbar}\nu(\varepsilon_F)J^2S(S+1)\left(1+2J\nu(\varepsilon_F)\ln\frac{\varepsilon_F}{T}\right),$$

$$\rho^{(2)} = \frac{m}{ne^2}\frac{1}{\tau_{tr}^{(2)}(p_F)} = \frac{3\pi m}{2e^2}n_i\frac{J^2S(S+1)}{\hbar\varepsilon_F}\left(1+3\frac{Jn}{\varepsilon_F}\ln\frac{\varepsilon_F}{T}\right).$$

In the last relation we have taken that $\nu(\varepsilon_F) = (3/2)n/\varepsilon_F$ into account.

From the last formula one can see that the sign of the temperature correction, resulted from magnetic impurities, depends on the sign of exchange constant J. Provided that sign J is ferromagnetic, i.e. $J < 0$, the resistivity $\rho^{(2)}$ decreases as the temperature lowers. For the antiferromagnetic sign of exchange constant $J > 0$, on the contrary, the magnetic part of resistivity increases with lowering the temperature. The latter gives rise to the minimum in the temperature behavior of resistivity. Thus, if the nonmagnetic contribution to resistivity is due to the presence of nonmagnetic impurities and phonon scattering in accordance with the Bloch law T^5, the resistivity on the whole will be equal to

$$\rho = \rho_0 + n_i a\ln(\varepsilon_F/T) + bT^5, \quad (a \sim J > 0).$$

The position of resistivity minimum should be found by differentiating with respect to temperature and equals

$$T_{\min} = [n_i a/(5b)]^{1/5} \sim n_i^{1/5}.$$

The temperature T_K

$$T_K \sim \varepsilon_F e^{-\frac{1}{|J|\nu(\varepsilon_F)}},$$

for which the logarithmic correction is comparable with unity, is called the *Kondo temperature*. Here it becomes necessary and unavoidable to study higher terms in expanding the scattering amplitude.

3.7 Weak Localization Effects in a Metal with Impurities

18. Let us write the Boltzmann equation in the uniform electric field, taking the homogeneity of distribution function $n(\boldsymbol{r},\boldsymbol{p},t) = n(\boldsymbol{p},t)$ into account

$$\frac{\partial n}{\partial t} + e\boldsymbol{E}\frac{\partial n}{\partial \boldsymbol{p}} = -\frac{n(\boldsymbol{p},t)-n_0(\varepsilon_{\boldsymbol{p}})}{\tau} + \int_{-\infty}^{\infty} dt'\,\alpha_R(t-t')\big[n(-\boldsymbol{p},t)-n_0(\varepsilon_{\boldsymbol{p}})\big].$$

Here we have introduced the retarded response function according to $\alpha_R(t) = \alpha(t)\vartheta(t)$. In the linear approximation in electric field we have for the deviation $\delta n = n - n_0$ of distribution function from equilibrium one

$$\frac{\partial}{\partial t}\delta n + \frac{e}{m}(\boldsymbol{pE})\frac{\partial n_0}{\partial \varepsilon_{\boldsymbol{p}}} = -\frac{\delta n(\boldsymbol{p}, t)}{\tau} + \int_{-\infty}^{\infty} dt'\, \alpha_R(t - t')\delta n(-\boldsymbol{p}, t).$$

Here we use that $\partial n_0/\partial \boldsymbol{p} = \boldsymbol{v}\partial n_0/\partial \varepsilon$ and velocity $\boldsymbol{v} = \boldsymbol{p}/m$.

For the electric field with a single harmonic $\boldsymbol{E}(t) = \boldsymbol{E}_\omega \exp(-i\omega t)$, we also expect $\delta n(\boldsymbol{p}, t) = \delta n_\omega(\boldsymbol{p}) \exp(-i\omega t)$ since the equation is linear. As a result, we obtain

$$-i\omega\delta n_\omega + \frac{e}{m}(\boldsymbol{pE}_\omega)\frac{\partial n_0}{\partial \varepsilon} = -\frac{\delta n_\omega}{\tau} + \alpha_R(\omega)\delta n_\omega(-\boldsymbol{p})$$

where $\alpha_R(\omega)$ represents the Fourier transform of response

$$\alpha_R(\omega) = \int_{-\infty}^{\infty} \alpha_R(t)e^{i\omega t}dt = \int_{0}^{\infty} \alpha(t)e^{i\omega t}dt = \frac{2n_i|V|^2}{\hbar^2}\int \frac{d^3q}{(2\pi)^3}\frac{1}{-i\omega + D_0q^2 + \tau_\varphi^{-1}}.$$

We seek for the solution of equation as

$$\delta n_\omega(\boldsymbol{p}) = \frac{e}{m}(\boldsymbol{pE}_\omega)\frac{\partial n_0}{\partial \varepsilon_{\boldsymbol{p}}}\chi_\omega(\varepsilon)$$

and then find

$$\chi_\omega(\varepsilon) = \frac{1}{-i\omega + \tau^{-1} + \alpha_R(\omega)} \quad \text{and} \quad \delta n_\omega = \frac{\frac{e}{m}\tau(\boldsymbol{pE}_\omega)}{1 - i\omega\tau + \tau\alpha_R(\omega)}\frac{\partial n_0}{\partial \varepsilon_{\boldsymbol{p}}}.$$

Substituting the distribution function $n_\omega = n_0 + \delta n_\omega$ into the expression which determines the electrical current

$$\boldsymbol{j}_\omega = \int \frac{2d^3p}{(2\pi\hbar)^3}e\boldsymbol{v}_{\boldsymbol{p}}n_\omega(\boldsymbol{p}) = \int \frac{2d^3p}{(2\pi\hbar)^3}\frac{e^2\tau}{m}\frac{\boldsymbol{p}(\boldsymbol{pE}_\omega)}{1 - i\omega\tau + \tau\alpha_R(\omega)}\frac{\partial n_0}{\partial \varepsilon_{\boldsymbol{p}}},$$

and involving that the equilibrium distribution function $n_0(\varepsilon_{\boldsymbol{p}})$ does not contribute to the current, we arrive at the following relation between the current density and the electric field: $\boldsymbol{j}_\omega = \sigma(\omega)\boldsymbol{E}_\omega$. This results in the expression

$$\sigma(\omega) = \frac{\sigma_0(\omega)}{1 + \frac{\tau\alpha_R(\omega)}{1 - i\omega\tau}}, \qquad \sigma_0(\omega) = \frac{ne^2\tau}{m}\frac{1}{1 - i\omega\tau},$$

where n is the electron concentration and $\sigma_0(\omega)$ is the *Drude conductivity*.

Assuming the response $\alpha_R(\omega)$ to be small, we have approximately the following estimate for the weak localization correction to conductivity $\delta\sigma(\omega) = \sigma(\omega) - \sigma_0(\omega)$:

$$\frac{\delta\sigma(\omega)}{\sigma_0(\omega)} \approx -\frac{\tau\alpha_R(\omega)}{1-i\omega\tau} = -\frac{2n_i|V|^2}{\hbar^2}\frac{\tau}{1-i\omega\tau}\int\frac{d^3q}{(2\pi)^3}\frac{1}{-i\omega+D_0q^2+\tau_\varphi^{-1}}.$$

For the quantitative estimate of quantum correction to conductivity, we should pay attention to the region of integrating over wave vector q, relation between impurity concentration n_i and mean free time τ, and diffusion coefficient $D_0 = v_F l/3$

$$\frac{\delta\sigma(\omega)}{\sigma_0(\omega)} \approx -\frac{2}{\pi\hbar\nu(\varepsilon_F)}\frac{1}{1-i\omega\tau}\int\limits_{l_\varphi^{-1}\lesssim q\lesssim l^{-1}}\frac{d^3q}{(2\pi)^3}\frac{1}{-i\omega+D_0q^2+\tau_\varphi^{-1}}$$

$$= -\frac{\pi\lambda^2}{1-i\omega\tau}\int\limits_{\tau/\tau_\varphi}^{1}dx\frac{x^2}{-i\omega\tau+\frac{x^2}{3}+\tau/\tau_\varphi}, \qquad \lambda = \frac{\hbar}{\pi p_F l}.$$

Here $p_F = mv_F$ is the Fermi momentum and the limit $\lambda \ll 1$ is implied.

Let us analyze quantum correction to conductivity at zero temperature. Putting $\omega = 0$, we obtain in the limit $\tau/\tau_\varphi \ll 1$

$$\frac{\delta\sigma}{\sigma_0} \approx -3\pi\lambda^2\left(1-\frac{\pi}{2}\sqrt{\frac{3\tau}{\tau_\varphi}}\right).$$

As it is seen, quantum effects of weak localization lead to decreasing the conductivity of a metal as compared with the classical magnitude. Due to our inequality $\lambda \ll 1$ the quantum interference correction for the conductivity is small. The second term here is of most interest since it is temperature dependent as a result of significant temperature dependence of inelastic mean free time $\tau_\varphi = \tau_\varphi(T)$. Then we have

$$\frac{\sigma(T)-\sigma(0)}{\sigma(0)} = \frac{3\pi^2}{2}\lambda^2\sqrt{\frac{3\tau}{\tau_\varphi(T)}} \sim \left(\frac{\hbar}{p_F l}\right)^2\sqrt{\frac{\tau}{\tau_\varphi(T)}}.$$

For the electron-electron interaction, one has $\tau_\varphi \sim \hbar\varepsilon_F/T^2$. As it concerns the electron-phonon coupling, we recall $\tau_\varphi \sim \hbar\Theta_D^2/T^3$ at the temperatures small compared with the Debye one Θ_D. Correspondingly, we find $\delta\sigma(T) \sim T$ for the first case and $\delta\sigma(T) \sim T^{3/2}$ for the second one.

19. Let us write equation to be analyzed

$$\frac{D(\lambda)}{D_0} = \left(1+\pi\lambda^2\frac{2\pi^2 l^3}{\tau}\int\limits_0^{1/l}\frac{d^3q}{(2\pi)^3}\frac{1}{D(\lambda)q^2}\right)^{-1} = \left(1+3\pi\lambda^2\frac{D_0}{D(\lambda)}\right)^{-1}.$$

The simple analysis leads to the result

$$\frac{D(\lambda)}{D_0} = \begin{cases} 1-(\lambda/\lambda_c)^2, & \lambda < \lambda_c \\ 0, & \lambda \geqslant \lambda_c \end{cases} \quad \left(\lambda_c = \frac{1}{\sqrt{3\pi}}\right).$$

For $\lambda > \lambda_c$, there is no real and positive solution for the diffusion coefficient. Thus, as the mean free path decreases and reaches the critical value $l_c = (3/\pi)^{1/2}\hbar/p_F$, the metal-insulator transition occurs. In the vicinity of threshold value λ_c the diffusion coefficient demonstrates the following behavior:

$$D(\lambda) \propto (\lambda_c - \lambda)^s$$

with the critical exponent $s = 1$.

For the phase of insulator ($\lambda > \lambda_c$), the diffusion coefficient $D(\omega)$ vanishes at zero frequency. At sufficiently small $\omega\tau \ll 1$ frequencies the behavior of the diffusion coefficient can be described as

$$D(\omega \to 0) = -i\omega\xi^2(\lambda)$$

where $\xi(\lambda)$ is referred to as the *localization length*. On approaching to the critical value λ_c the localization length ξ grows unlimitedly according to

$$\xi \propto (\lambda_c - \lambda)^{-\nu}$$

with the critical exponent ν. To determine the exponent ν, we write the initial equation

$$\frac{D(\omega)}{D_0}(1 - i\omega\tau) = 1 - \frac{\lambda^2}{\lambda_c^2}\int_0^1 \frac{x^2dx}{x^2 - i\omega\tau\frac{3D_0}{D(\omega)}}$$

in which we go over to the limit $\omega \to 0$, assuming $D(\omega) \to -i\omega\xi^2$. Then we have

$$0 = 1 - \frac{\lambda^2}{\lambda_c^2}\int_0^1 \frac{x^2\,dx}{x^2 + 3D_0\tau/\xi^2} = 1 - \frac{\lambda^2}{\lambda_c^2}\left(1 - \frac{l^2}{\xi^2}\int_0^1 \frac{dx}{x^2 + l^2/\xi^2}\right),$$

using $D_0 = v_F l/3$. Hence,

$$\frac{l}{\xi}\arctan\frac{l}{\xi} = \frac{\lambda^2 - \lambda_c^2}{\lambda^2}.$$

Near the critical value while $\lambda_c - \lambda \ll \lambda_c$ and $\xi \gg l$ we disclose the following behavior for the localization length:

$$\xi(\lambda) \approx \frac{\pi}{4} l \frac{\lambda_c}{\lambda - \lambda_c} \propto (\lambda - \lambda_c)^{-1}.$$

On the penetration deeper into the insulator phase the localization length decreases and becomes comparable with the mean free path $\xi \sim l = \hbar/(\lambda\pi p_F)$.

3.8 Ballistic Electron Transport in the Mesoscopic Systems

20. The current through a microcontact can be represented as a sum of two-electron flows: one directed from left to right and the other directed from right to left

$$I = \sum_{\lambda\lambda'} i^L_{\lambda\lambda'} n(\varepsilon_\lambda)\big(1 - n(\varepsilon_{\lambda'} + eV)\big) + \sum_{\lambda\lambda'} i^R_{\lambda'\lambda} n(\varepsilon_{\lambda'} + eV)\big(1 - n(\varepsilon_\lambda)\big).$$

Here the summation over λ and λ' means a sum over levels n, n' and integration over momenta p, p'. The factor $i^L_{\lambda\lambda'}$ represents an elementary electron flow corresponding to the transitions from the state $\lambda = (n, p)$ at the left to the state $\lambda' = (n', p')$ at the right. On the analogy the quantity $i^R_{\lambda'\lambda}$ corresponds to the electron flow with the transitions from the state $\lambda' = (n', p')$ at the right to the state $\lambda = (n, p)$ at the left. The multipliers $n(1 - n)$ with the distribution functions $n(\varepsilon)$ are responsible for the presence of free and occupied electron states.

In the semiclassical approximation the elementary flows $i^L_{\lambda\lambda'}$ and $i^R_{\lambda'\lambda}$ can be written with the aid of the Heaviside step function ϑ as follows:

$$i^L_{\lambda\lambda'} = 2ev_\lambda \vartheta(v_\lambda)\delta_{\lambda\lambda'},$$
$$i^R_{\lambda'\lambda} = 2ev_{\lambda'} \vartheta(-v_{\lambda'})\delta_{\lambda'\lambda},$$

where $v_\lambda = \partial\varepsilon_\lambda/\partial p = p/m$ is the electron velocity along the microcontact and e is the electron charge. Factor 2 is associated with the existence of two spin projections and the δ-function implies the conservation of electron state under collisionless ballistic passage through the microcontact. The Heaviside function $\vartheta(v)$ selects the necessary directions of electron motion. The substitution results in the following expression for the current:

$$I = \sum_{\lambda} 2ev_\lambda \vartheta(v_\lambda) n(\varepsilon_\lambda)\big(1 - n(\varepsilon_\lambda + eV)\big)$$
$$+ \sum_{\lambda} 2ev_\lambda \vartheta(-v_\lambda) n(\varepsilon_\lambda + eV)\big(1 - n(\varepsilon_\lambda)\big).$$

Then we take into account that $\varepsilon_n(p) = \varepsilon_n(-p)$ and velocity $v_\lambda = p/m$ is an odd function of momentum p. Reminding that a sum over λ means summing over subbands n as well as integrating over momentum p, we obtain

$$I = 2e \sum_n \int_{-\infty}^{\infty} \frac{dp}{2\pi\hbar} v_\lambda \vartheta(v_\lambda)\big[n(\varepsilon_\lambda) - n(\varepsilon_\lambda + eV)\big].$$

So, we find the current I in the linear approximation in voltage V

$$I = GV,$$

where conductance G is determined with the expression

$$G = 2e^2 \sum_n \int_{-\infty}^{\infty} \frac{dp}{2\pi\hbar} \frac{\partial \varepsilon_\lambda}{\partial p} \vartheta(p) \left(-\frac{\partial n}{\partial \varepsilon_\lambda}\right)$$

$$= \frac{e^2}{\pi\hbar} \sum_n \int_0^{\infty} dp \left(-\frac{\partial n}{\partial p}\right) = \frac{e^2}{\pi\hbar} \sum_n n(\varepsilon_n)$$

and $\varepsilon_n = \varepsilon_n(p = 0)$. For zero temperature, one has $n(\varepsilon_n \leqslant \mu) = 1$ if ε_n does not exceed the chemical potential μ, and $n(\varepsilon_n > \mu) = 0$ if ε_n is greater than chemical potential μ. In this case the conductance of quantum microcontact equals

$$G = \frac{e^2}{\pi\hbar} \sum_n \vartheta(\mu - \varepsilon_n) = \frac{e^2}{\pi\hbar} N = \frac{1}{2} G_0 N_e\,, \quad G_0 = \frac{e^2}{\pi\hbar},$$

where N is the number of transverse levels (subbands) occupied with the electrons and attainable for their ballistic transport. Due to two spin projections the number of electrons N_e in these subbands will be twice as larger and equals $2N$.

The universal constant $G_0 = 2e^2/h \approx 7.75 \cdot 10^{-5}\ \mathrm{Ohm}^{-1}$ ($h = 2\pi\hbar$) is referred to as *conductance quantum*. Thus, the conductance of ballistic quantum microcontact proves to be an integer multiple of constant G_0. One may say that a separate ballistic quantum channel has the conductance equal to G_0 or every quantum state (subband) contributes an amount e^2/h per each spin projection to the total conductance of quantum microcontact.

For the plane two-dimensional $D = 2$ case, the number of levels (channels) N, occupied with electrons in the limit $N \gg 1$, can be estimated by putting the level energy equal to

$$\varepsilon_n = \frac{\pi^2\hbar^2 n^2}{2mW^2} \qquad (n = 1,\ 2,\ 3,\ \ldots).$$

Taking $\varepsilon(N) = \mu = p_F^2/2m$, we find

$$N = \frac{N_e}{2} = \frac{p_F W}{\pi\hbar} \qquad (D = 2).$$

Fig. 3.9 The quantization of conductance G as a function of the microcontact width W

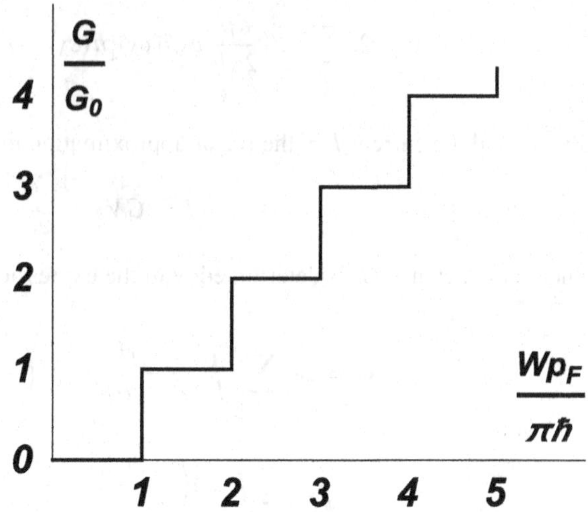

In Fig. 3.9 the behavior of conductance G is schematically shown as a function of microcontact width W.

In the three-dimensional $D = 3$ case we should take the subband (channel) energy equal to

$$\varepsilon_n = \frac{\pi^2\hbar^2(n_1^2 + n_2^2)}{2mW^2} \quad (n_1,\ n_2 = 1,\ 2,\ 3,\ \ldots).$$

So, we have approximately

$$N = \frac{N_e}{2} = \frac{p_F^2 W^2}{4\pi\hbar^2} \quad \text{where} \quad p_F^2 = 2m\mu.$$

If the transmission coefficient $\mathcal{T}$ through the microcontact for an electron with quantum number n or in channel n differs from unity and depends on both the channel number n and the electron energy ε as $\mathcal{T} = \mathcal{T}_n(\varepsilon)$, the generalization of the answer obtained for the conductance leads to the *Landauer formula*

$$G = \frac{e^2}{\pi\hbar}\sum_n \mathcal{T}_n(\varepsilon = \mu).$$

In this case the total conductance is given by a sum of electron transmission coefficients at the Fermi energy.

The quantization of conductance in units of $2e^2/h$ will be well manifested only while the temperature of microcontact does not exceed the spacing between the neighboring levels $\Delta\varepsilon = \varepsilon_{n+1} - \varepsilon_n$.

21. On the analogy with the previous problem let us write energy flux Q from one reservoir to the other provided that the state of electron remains unvaried during its ballistic passage through the microcontact

$$Q = 2\sum_{\lambda} \varepsilon_\lambda v_\lambda \vartheta(v_\lambda) n_L(1 - n_R) + 2\sum_{\lambda} \varepsilon_\lambda v_\lambda \vartheta(-v_\lambda)(1 - n_L) n_R\,.$$

Here factor 2 includes the two possible electron spin projections and the Heaviside step function $\vartheta(v)$ selects the necessary directions of electron motion. The multipliers $n(1-n)$ with the distribution functions $n_L(\varepsilon_\lambda)$ and $n_R(\varepsilon_\lambda)$ for the left-side and right-side reservoirs take the presence of free and occupied states into consideration. Using the following properties of electron spectrum as: $\varepsilon_n(p) = \varepsilon_n(-p)$ and $v(p) = -v(-p)$, we change p to $-p$ in the both second terms and arrive at the expression

$$Q = 2\sum_{\lambda} \varepsilon_\lambda v_\lambda \vartheta(v_\lambda)(n_L - n_R)\,.$$

Let us expand distribution function n_L in small deviations $\Delta\mu$ and ΔT from the quantities μ and T corresponding to distribution function n_R on the right-hand side

$$n_L = n_0 + \frac{\varepsilon_\lambda - \mu}{T}\left(-\frac{\partial n_0}{\partial \varepsilon_\lambda}\right)\Delta T + \left(-\frac{\partial n_0}{\partial \varepsilon_\lambda}\right)\Delta\mu \quad \text{and} \quad n_R = n_0\left(\frac{\varepsilon_\lambda - \mu}{T}\right).$$

The result is the following:

$$Q = 2\sum_{\lambda} \varepsilon_\lambda v_\lambda \vartheta(v_\lambda)\left(-\frac{\partial n_0}{\partial \varepsilon_\lambda}\right)\Delta\mu + 2\sum_{\lambda} \varepsilon_\lambda v_\lambda \vartheta(v_\lambda)\frac{\varepsilon_\lambda - \mu}{T}\left(-\frac{\partial n_0}{\partial \varepsilon_\lambda}\right)\Delta T,$$

where

$$-\frac{\partial n_0}{\partial \varepsilon_\lambda} = \frac{1}{4T\cosh^2(\varepsilon_\lambda - \mu)/2T} \quad \text{and} \quad \varepsilon_\lambda = \varepsilon_n + \frac{p^2}{2m}\,.$$

Here, as in the previous problem, a sum over λ implies summing over subbands n and integrating over momentum p. To integrate with respect to momentum p, it is convenient to go over to variable $x = \xi/T$ where $\xi = \varepsilon_n - \mu + p^2/2m$. Eventually, we obtain the following relation for the energy flux $Q = Q_\mu + Q_T$ as a sum of two terms:

$$Q = \Delta\mu \sum_n \int\limits_{(\varepsilon_n - \mu)/T}^{\infty} \frac{\mu + xT}{4\cosh^2(x/2)}\frac{dx}{\pi\hbar} + \Delta T \sum_n \int\limits_{(\varepsilon_n - \mu)/T}^{\infty} \frac{(\mu + xT)x}{4\cosh^2(x/2)}\frac{dx}{\pi\hbar}\,.$$

Let chemical potential μ be between the energy levels of channels N and $N+1$ and be sufficiently far away from each of them as compared with the temperature, i.e. $E_N < \mu < E_{N+1}$ and $\mu - E_N \gg T$, $E_{N+1} - \mu \gg T$. Due to exponentially fast decay of function $\cosh^{-2}(x/2)$ at $|x| \gtrsim 1$ the magnitudes of integrals are mainly gained within the region of small values $|x| \lesssim 1$. Thus, for $(\varepsilon_n - \mu)/T \ll -1$, the lower integration limit can be taken equal to $x = -\infty$, ensuring only the exponentially small error. In the same approximation at $(\varepsilon_n - \mu)/T \gg 1$ one can put $x = \infty$ as a lower limit, entailing that the values of integrals vanish in the region $\varepsilon_n > \mu$. As a result, the contribution to the energy flux Q from region $|\varepsilon_n - \mu| \gg T$ can be written as follows:

$$Q_N = \Delta\mu \sum_n \frac{\mu}{\pi\hbar}\vartheta(\mu - \varepsilon_n) + \Delta T \sum_n \frac{\pi T}{3\hbar}\vartheta(\mu - \varepsilon_n)$$

$$= \frac{\mu}{\pi\hbar} N\Delta\mu + \frac{\pi T}{3\hbar} N\Delta T = \frac{\mu}{h} N_e \Delta\mu + g_0 N_e \Delta T, \quad g_0 = \frac{\pi^2 T}{3h} \quad (h = 2\pi\hbar).$$

Here N is the total number of channels (subbands) accessible or open for the electron transport. If we take the spin into account, the number N_e of electrons, participating effectively in the energy transfer in these channels, will be twice as larger, i.e. $N_e = 2N$. The universal coefficient g_0 can be called the *thermal conductance quantum* describing the rate at which heat is transported through a single ballistic channel at temperature T.

Let chemical potential μ be kept in the region sufficiently close to the $(N+1)$th channel (subband) and electrons start to occupy this channel, i.e. $|\varepsilon_{N+1} - \mu| \lesssim T$. For estimation of the integrals, one can presume $\varepsilon_n = \mu$ at $n = N+1$ and put the lower integration limit equal to zero. Accordingly, we have approximately for the energy flux

$$Q_{N\to(N+1)} = Q_N + \Delta\mu \frac{\mu + 2T\ln 2}{2\pi\hbar} + \Delta T \frac{\mu \ln 2 + \pi^2 T/6}{\pi\hbar}$$

$$\approx Q_N + \Delta\mu \frac{\mu}{2\pi\hbar} + \Delta T \frac{\mu}{\pi\hbar} \ln 2.$$

As an additional $(N+1)$th channel opens for the electron transport of energy flux fraction connected directly with the temperature difference ΔT, there arises an enhanced contribution as a large factor μ/T compared with the coequal contributions each of already open N channels.

The formulas obtained have a simple physical sense. The energy flux Q, transported with electrons across ballistic contact at sufficiently low temperatures, is quantized in the sense that it is an integer multiple of the N open channels. This statement holds both for the flux component Q_μ associated directly with the chemical potential difference $\Delta\mu$ and for the flux component Q_T associated directly with temperature difference ΔT across the contact. The latter flux component Q_T is similar to dissipative heat flow at the collision electron transport.

In conclusion, let us see how the energy flux Q depends on the microcontact width W. In the previous problem for the plane $(D = 2)$ contact we have estimated the number of open channels or subbands accessible for electrons as

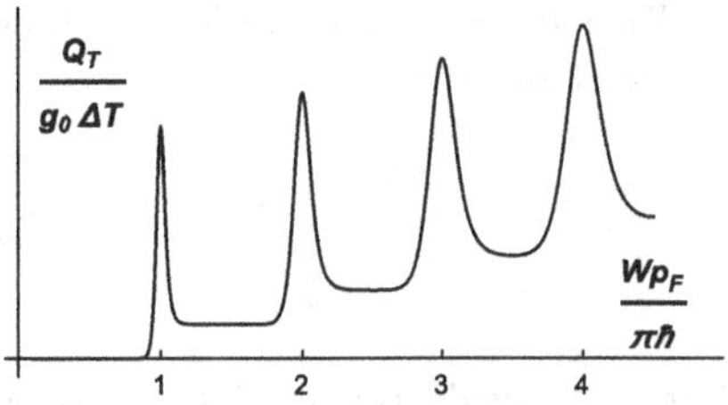

Fig. 3.10 Quantization and oscillations of thermal conductance $Q_T/\Delta T$ in units of g_0 as a function of microcontact width W. A ratio $\mu/T = 30$ is taken as an example

$$N = \frac{N_e}{2} = \frac{p_F W}{\pi\hbar},$$

where the Fermi momentum equals $p_F = \sqrt{2m\mu}$. Therefore, provided that $\Delta\mu$ and ΔT are fixed as the microcontact width increases and opens an additional channel, the energy flux will experience a jump-like singularity diffused in the narrow adjacent region of widths about $\Delta W/W \sim T/\mu \ll 1$. The magnitude of the energy flux remains practically unchanged between the jumps.

In Fig. 3.10 it is shown for example how the thermal conductance in units of g_0 changes with increasing the microcontact width W.

3.9 The Rarified Bose and Fermi Gases

22 In the Bose condensed gas we should distinguish two groups of particles, namely, condensate and noncondensate ones. All condensate particles have zero momentum and, correspondingly, zero velocity. Thus, the wall runs simply on the particles at rest and reflects them away. The noncondensate particles have nonzero momentum and, therefore, finite velocity. The distribution of noncondensate particles over energy is subjected to the Bose-Einstein distribution function with zero chemical potential.

The lack of interparticle interaction allows us to calculate the force exerted on the wall as a sum of separate contributions of condensate particles and noncondensate particles, respectively. The determination of the force exerted on the wall is more convenient to perform in the reference frame in which the wall remains at rest.

We start first from analyzing the condensate contribution. In the wall reference frame the gas of bosons moves as a whole at velocity $-\boldsymbol{V}$. Let us choose axis z in the direction of the normal to the wall. The elastic reflection implies that the z component of particle velocity changes its sign after reflection but the velocity components parallel to the wall plane conserve their magnitudes, i.e.

$$v_{\text{out},z} = -v_{\text{in},z}\,, \quad \boldsymbol{v}_{\text{out},\parallel} = \boldsymbol{v}_{\text{in},\parallel} \quad \text{and} \quad \varepsilon_{\text{in},\boldsymbol{p}} = \varepsilon_{\text{out},\boldsymbol{p}}\,.$$

Prior to reflecting from the wall all the condensate particles have velocity $-\boldsymbol{V}$ and acquire velocity $\boldsymbol{V}$ after elastic reflection. The number of incident condensate particles per the unit time equals the number of reflected condensate particles.

The momentum flux carried by the condensate particles per the unit time and per unit wall area is equal to

$$\Pi_{\text{in}} = \int\limits_{v_z \leqslant 0} p_z v_z n_0(\varepsilon_{\text{in},p,}) d\Gamma_p\,, \quad d\Gamma_p = \frac{d^3p}{(2\pi\hbar)^3}.$$

Here n_0 is the distribution function of condensate particles. The condensate particle density N_0 is determined by integral $N_0 = \int n_0(\varepsilon_p)d\Gamma_p$. Taking $v_z = -V$ and $p_z = mv_z = -mV$ into account, we find

$$\Pi_{\text{in}} = mV^2 \int\limits_{v_z \leqslant 0} n_0(\varepsilon_{\text{in},p,}) d\Gamma_p = mV^2 \frac{N_0}{2}.$$

In the last integral we have kept in mind that the integration should be performed over a half of phase space.

The momentum flux carried away by the condensate particles from the wall is equal to

$$\Pi_{\text{out}} = \int\limits_{v_z \geqslant 0} p_z v_z n_0(\varepsilon_{\text{out}\,,p}) d\Gamma_p = mV^2 \int\limits_{v_z \geqslant 0} n_0(\varepsilon_{\text{out},p}) d\Gamma_p = mV^2 \frac{N_0}{2}.$$

The force exerted to the wall is given by a sum of fluxes Π_{in} and Π_{out}. As a result, the condensate contribution f_0 to the force will be proportional to the square of wall velocity V and the condensate mass density $\rho_0(T) = mN_0(T)$ and equal to

$$f_0 = \rho_0(T)V^2.$$

Let us turn now to calculating the noncondensate particle contribution. The momentum fluxes of incident and reflected noncondensate particles per the unit time and wall area are equal to

$$\Pi_{\text{in}} = \int\limits_{v_z \leqslant 0} p_z v_z n_{\text{ex}}(\varepsilon_{\text{in},p}) d\Gamma_p\,, \quad \varepsilon_{\text{in},p} = \varepsilon_{p-mV} = \varepsilon_p - p_z V + \frac{mV^2}{2};$$

$$\Pi_{\text{out}} = \int\limits_{v_z \geqslant 0} p_z v_z n_{\text{ex}}(\varepsilon_{\text{out},p}) d\Gamma_p\,, \quad \varepsilon_{\text{out},p} = \varepsilon_{p+mV} = \varepsilon_p + p_z V + \frac{mV^2}{2}.$$

Here $\varepsilon_p = \boldsymbol{p}^2/2m$ and $n_{\text{ex}}(\varepsilon_p)$ is the Bose-Einstein distribution with chemical potential $\mu = mV^2/2$ since the gas moves as a whole at velocity $-\boldsymbol{V}$ in the wall reference frame. The total noncondensate particle contribution to the force $F_{\text{ex}} = P + f_{\text{ex}}$ exerted on the wall is given by a sum of momentum fluxes Π_{in} and Π_{out} and consists of two terms. The term independent of wall velocity or at $V = 0$ represents an ordinary pressure resulting from an existence of noncondensate particles

$$P = \int\limits_{v_z \leqslant 0} p_z v_z n_{\text{ex}}(\varepsilon_{\boldsymbol{p}}) d\Gamma_{\boldsymbol{p}} + \int\limits_{v_z \geqslant 0} p_z v_z n_{\text{ex}}(\varepsilon_{\boldsymbol{p}}) d\Gamma_{\boldsymbol{p}}$$

$$= \int p_z v_z n_{\text{ex}}(\varepsilon_{\boldsymbol{p}}) d\Gamma_{\boldsymbol{p}} = \frac{2}{3} \int \varepsilon_{\boldsymbol{p}} n_{\text{ex}}(\varepsilon_{\boldsymbol{p}}) d\Gamma_{\boldsymbol{p}} = \left(\frac{mT}{2\pi\hbar^2}\right)^{3/2} T\zeta(5/2).$$

Here we have used that $\varepsilon_{\text{out},\boldsymbol{p}} = \varepsilon_{\text{in},\boldsymbol{p}} = \varepsilon_{\boldsymbol{p}}$ at $\boldsymbol{V} = 0$. As usual, $\zeta(x)$ denotes the Riemann zeta function.

So, the force $F_{\text{ex}} = P + f_{\text{ex}}(V)$ due to noncondensate particles is determined with the integrals

$$F_{\text{ex}} = \int\limits_{v_z \leqslant 0} p_z v_z n_{\text{ex}}(\varepsilon_{\text{in},\boldsymbol{p}}) d\Gamma_{\boldsymbol{p}} + \int\limits_{v_z \geqslant 0} p_z v_z n_{\text{ex}}(\varepsilon_{\text{out},\boldsymbol{p}}) d\Gamma_{\boldsymbol{p}}$$

$$= \int\limits_{v_z \leqslant 0} p_z v_z n_B(\varepsilon_{\boldsymbol{p}} - p_z V) d\Gamma_{\boldsymbol{p}} + \int\limits_{v_z \geqslant 0} p_z v_z n_B(\varepsilon_{\boldsymbol{p}} + p_z V) d\Gamma_{\boldsymbol{p}},$$

$n_B(\varepsilon) = (e^{\varepsilon/T} - 1)^{-1}$ being the Bose-Einstein distribution with zero chemical potential. Changing the sign of velocity v_z in the first term and expanding $n_B(\varepsilon_{\boldsymbol{p}} + p_z V)$ to second order in V, we obtain

$$F_{\text{ex}} = P + 2 \int\limits_{v_z \geqslant 0} p_z v_z \left(p_z V \frac{\partial n(\varepsilon_{\boldsymbol{p}})}{\partial \varepsilon_{\boldsymbol{p}}} + \frac{1}{2} p_z^2 V^2 \frac{\partial^2 n(\varepsilon_{\boldsymbol{p}})}{\partial \varepsilon_{\boldsymbol{p}}^2} + \ldots \right) d\Gamma_{\boldsymbol{p}}.$$

For calculating the integrals, it is useful first to perform the integration over the polar angle ϑ, keeping in mind for the projections $p_z = p\cos\vartheta$, $v_z = v\cos\vartheta$ and the limits of angle variation between zero and $\pi/2$. Accordingly, the force additional to the pressure is expressed by the integrals

$$f_{\text{ex}} = 2\frac{V}{8} \int p^2 v n_B'(\varepsilon_{\boldsymbol{p}}) d\Gamma_{\boldsymbol{p}} + 2 \cdot \frac{1}{2} \frac{V^2}{10} \int p^3 v n_B''(\varepsilon_{\boldsymbol{p}}) \Big) d\Gamma_{\boldsymbol{p}}.$$

The straightforward calculation of integrals leads to the following result:

$$f_{\text{ex}} = -\frac{(mT)^2}{6\hbar^3} V + \frac{3m}{2} \left(\frac{mT}{2\pi\hbar^2}\right)^{3/2} \zeta(3/2) V^2.$$

For the clear comprehension, we express this result in terms of the mass density of noncondensate particles $\rho_{\text{ex}} = m n_{\text{ex}}(T)$

$$\rho_{\text{ex}}(T) = m\left(\frac{mT}{2\pi\hbar^2}\right)^{3/2}\zeta(3/2)$$

and the mean velocity $\bar{v}_{\text{ex}}$ of noncondensate particles according to formula

$$\bar{v}_{\text{ex}} = \frac{\int v n_{\text{ex}}(\varepsilon_p) d\Gamma_p}{\int n_{\text{ex}}(\varepsilon_p) d\Gamma_p} = \frac{1}{6}\frac{(2\pi)^{3/2}}{\zeta(3/2)}\sqrt{\frac{T}{m}}.$$

Finally, we obtain the following answer for the noncondensate particle contribution to the resistance force:

$$f_{\text{ex}}(T) = -\rho_{\text{ex}}(T)\bar{v}_{\text{ex}}(T)V + \frac{3}{2}\rho_{\text{ex}}(T)V^2.$$

Unlike condensate particles the noncondensate ones represent elementary thermal excitations of the ground state of boson gas when all particles are in the Bose-Einstein condensate. The presence of noncondensate particles entails the drag force proportional to the wall velocity V and, as a result, dissipative character for the wall motion.

The total resistance force f, exerted on the wall, can be described with the formula

$$f = -\rho_{\text{ex}}(T)\bar{v}_{\text{ex}}(T)V + \left(\rho_0(T) + \frac{3}{2}\rho_{\text{ex}}(T)\right)V^2.$$

For the temperatures small compared with the Bose-Einstein condensation temperature, the contribution of condensate particles becomes predominant at the wall velocities $V \gtrsim \bar{v}_{\text{ex}}(\rho_{\text{ex}}/\rho_0)$. This is much smaller than the mean velocity $\bar{v}_{\text{ex}}$ of noncondensate particles.

23. The calculation of force f is readily performed in the reference frame in which the wall is at rest. Accordingly, the gas of fermions moves as a whole at velocity $-\boldsymbol{V}$. The symmetry of the problem implies that the resultant force must be directed along the normal to the wall surface. We choose the z-axis in the direction normal to the wall.

The distribution function of fermions before collisions is the Fermi function $n(\varepsilon'_p)$ with energy $\varepsilon'_p = \varepsilon_{p-mV}$ at temperature T. As it concerns the fermions leaving the wall at rest, they under the full accommodation have the equilibrium Fermi function $An(\varepsilon_p)$ with the same temperature$T$ but at $V = 0$. The latter function should be normalized so that the number of incident fermions per unit time would equal the number of fermions escaping from the wall for the same time. Thus, factor A is determined with equating the incident and reflected fermion fluxes

$$\int\limits_{v_z>0} v_z A n(\varepsilon_p) d\Gamma_p = -\int\limits_{v_z<0} v_z n(\varepsilon_{p-mV}) d\Gamma_p,$$

where $d\Gamma_{\boldsymbol{p}} = 2d^3p/(2\pi\hbar)^3$. Next, we involve that velocity $\boldsymbol{V}$ is directed along z axis and decompose the right-hand side of equality to first order in velocity $\boldsymbol{V}$. Hence we have

$$\int\limits_{v_z>0} v_z A n(\varepsilon_{\boldsymbol{p}}) d\Gamma_{\boldsymbol{p}} = \int\limits_{v_z>0} v_z n(\varepsilon_{\boldsymbol{p}} + p_z V) d\Gamma_{\boldsymbol{p}}$$

$$\approx \int\limits_{v_z>0} v_z n(\varepsilon_{\boldsymbol{p}}) d\Gamma_{\boldsymbol{p}} + V \int\limits_{v_z>0} v_z p_z \left(\frac{\partial n(\varepsilon_{\boldsymbol{p}})}{\partial \varepsilon_{\boldsymbol{p}}}\right) d\Gamma_{\boldsymbol{p}}\,.$$

Integrating over the solid angle yields

$$A - 1 = \frac{2}{3} V \frac{\int vp(-\partial n/\partial \varepsilon) d\Gamma_{\boldsymbol{p}}}{\int vn(\varepsilon) d\Gamma_{\boldsymbol{p}}} \approx \frac{8}{3}\frac{V}{v_F}\,.$$

The momentum flux delivered by the fermions per unit time and per unit area equals

$$\Pi_{\text{in}} = \int\limits_{v_z<0} p_z v_z n(\varepsilon_p - p_z V) d\Gamma_{\boldsymbol{p}}$$

$$\approx \int\limits_{v_z<0} p_z v_z n(\varepsilon_p) d\Gamma_{\boldsymbol{p}} + V \int\limits_{v_z<0} v_z p_z^2 \left(-\frac{\partial n}{\partial \varepsilon_p}\right) d\Gamma_{\boldsymbol{p}}\,.$$

For the momentum flux carried away from the wall per unit time, we have

$$\Pi_{\text{out}} = \int\limits_{v_z>0} p_z v_z A n(\varepsilon_{\boldsymbol{p}}) d\Gamma_{\boldsymbol{p}}\,.$$

The total force $F = P + f$, exerted on the wall, is given by a sum of fluxes Π_{in} and Π_{out} and consists of two terms. The first-term independent of the wall velocity represents an ordinary pressure induced by the fermion gas

$$P = \int\limits_{v_z<0} v_z p_z n(\varepsilon_p) \frac{2d^3p}{(2\pi\hbar)^3} + \int\limits_{v_z>0} v_z p_z n(\varepsilon_p) \frac{2d^3p}{(2\pi\hbar)^3} = \frac{2}{5} n \varepsilon_F\,.$$

The second term is a drag force f proportional to the wall velocity V

$$f = V \int\limits_{v_z<0} v_z p_z^2 \left(-\frac{\partial n}{\partial \varepsilon_p}\right) d\Gamma_{\boldsymbol{p}} + (A-1) \int\limits_{v_z>0} p_z v_z n(\varepsilon_{\boldsymbol{p}}) d\Gamma_{\boldsymbol{p}}\,,$$

which reduces to the following expression after integrating over the solid angle:

$$f=-\frac{V}{8}\int vp^2\left(-\frac{\partial n}{\partial\varepsilon_p}\right)\frac{2d^3p}{(2\pi\hbar)^3}+\frac{A-1}{6}\int pvn(\varepsilon_p)\frac{2d^3p}{(2\pi\hbar)^3}\,.$$

The straightforward treatment results in the drag force desired

$$f=-\frac{V}{8}\frac{p_F^4}{\pi^2\hbar^3}+\frac{A-1}{6}\frac{v_Fp_F^4}{5\pi^2\hbar^3}=-\frac{13}{360}p_FnV=-\frac{13}{360}\rho v_FV.$$

Here $\rho=mn$ is the mass density of fermion gas and v_F is the Fermi velocity. The effect of finite temperature leads to the usual Fermi correction of the order of $(T/\varepsilon_F)^2\ll 1$.

3.10 The Hopping Mott Conductivity and the Coulomb Gap

24. Most important characteristic in the problem is a rate $P_{\lambda'\lambda}$ of electron hops from the impurity site λ to the impurity site λ'. The hopping probability depends both on the distance between sites $r_{\lambda\lambda'}=|\boldsymbol{r}_\lambda-\boldsymbol{r}_{\lambda'}|$ and on the electron energies ε_λ, $\varepsilon_{\lambda'}$ at the sites λ and λ'. In order to find the hopping probability, besides the overlapping of electron wave functions we should also consider the coupling of an electron with phonons. The latter ones must be absorbed or emitted in order to compensate the difference between initial energy ε_λ and final energy $\varepsilon_{\lambda'}$. In addition, we must involve the Pauli principle since the electron hop is only possible to unoccupied state. The necessity to engage the phonon and electron distribution functions $N(\omega)$ and $n(\varepsilon)$ should result in appearing the temperature dependence for the hopping rate $P_{\lambda'\lambda}$.

We write here the transport equation on the analogy with the problem about the electron-phonon resistance

$$\frac{\partial n_\lambda}{\partial t}=\mathrm{St}\,[n_\lambda]=\sum_{\lambda'}(P_{\lambda\lambda'}-P_{\lambda'\lambda})$$
$$=\sum_{\lambda'}\big[W_{\lambda\lambda'}n_{\lambda'}(1-n_\lambda)-W_{\lambda'\lambda}n_\lambda(1-n_{\lambda'})\big].$$

The collision integral $\mathrm{St}\,[n_\lambda]$ determines the rate at which the electron distribution function n_λ varies. Here $W_{\lambda'\lambda}$ and $W_{\lambda\lambda'}$ are the probabilities of hopping from impurity site λ to site λ' and backwards. In what follows, we restrict ourselves with the Born approximation for the hopping probability and take the equilibrium Planck function $N(\omega)$ as a phonon distribution function. So, we have

$$W_{\lambda'\lambda} = \frac{2\pi}{\hbar} \sum_{\boldsymbol{k}} w_{\lambda'\lambda}(k)\big[\big(1 + N(\omega_k)\big)\delta(\varepsilon_\lambda - \varepsilon_{\lambda'} - \hbar\omega_k) + N(\omega_k)\delta(\varepsilon_\lambda - \varepsilon_{\lambda'} + \hbar\omega_k)\big],$$

where $\boldsymbol{k}$ is the transmitted phonon wave vector with the corresponding frequency ω_k and $w_{\lambda'\lambda}(k)$ is the square of the matrix element modulus of electron-phonon coupling represented in simplest deformation potential approximation

$$w_{\lambda'\lambda}(k) = \frac{g^2 \hbar k^2}{2\rho\omega_k} \left| \int \psi^*_{\lambda'}(\boldsymbol{r}) e^{-i\boldsymbol{k}\boldsymbol{r}} \psi_\lambda(\boldsymbol{r})\, d^3r \right|^2 .$$

Here g is the electron-phonon coupling constant, ρ is the density of semiconductor, and $\psi_\lambda(\boldsymbol{r})$ are $\psi_{\lambda'}(\boldsymbol{r})$ are the electron wave functions at impurity sites λ and λ', respectively.

Using the property $N_{-\omega} = -(1 + N_\omega)$ for the Planck function, we represent $W_{\lambda'\lambda}$ as follows:

$$W_{\lambda'\lambda} = \frac{2\pi}{\hbar} \sum_{\boldsymbol{k}} w_{\lambda'\lambda}(k)\big[1 + N(\varepsilon_\lambda - \varepsilon_{\lambda'})\big]\mathrm{sgn}\,(\varepsilon_\lambda - \varepsilon_{\lambda'})\delta(|\varepsilon_\lambda - \varepsilon_{\lambda'}| - \hbar\omega_k)$$

$$= \mathrm{sgn}\,(\varepsilon_\lambda - \varepsilon_{\lambda'})\big[1 + N(\varepsilon_\lambda - \varepsilon_{\lambda'})\big]\widetilde{w}_{\lambda'\lambda} = \frac{\mathrm{sgn}\,(\varepsilon_\lambda - \varepsilon_{\lambda'})}{1 - \exp[-(\varepsilon_\lambda - \varepsilon_{\lambda'})/T]}\, \widetilde{w}_{\lambda'\lambda} .$$

Such representation for probability $W_{\lambda'\lambda}$ as a product of two multipliers, one being temperature dependent and the second

$$\widetilde{w}_{\lambda'\lambda} = \frac{2\pi}{\hbar} \sum_{\boldsymbol{k}} w_{\lambda'\lambda}(k)\delta(|\varepsilon_\lambda - \varepsilon_{\lambda'}| - \hbar\omega_k)$$

being independent, is very useful for the next analysis. For example, this representation allows us simply to check the balance of electron transitions between impurity sites as well as the *principle of detailed balance* for the probabilities of direct and reverse hopping between impurity sites under thermal equilibrium

$$W_{\lambda'\lambda} e^{-\frac{\varepsilon_\lambda}{T}} = \frac{\mathrm{sgn}\,(\varepsilon_\lambda - \varepsilon_{\lambda'})}{e^{\varepsilon_\lambda/T} - e^{\varepsilon_{\lambda'}/T}} \widetilde{w}_{\lambda'\lambda} = \frac{\mathrm{sgn}\,(\varepsilon_{\lambda'} - \varepsilon_\lambda)}{e^{\varepsilon_{\lambda'}/T} - e^{\varepsilon_\lambda/T}} \widetilde{w}_{\lambda\lambda'} = W_{\lambda\lambda'} e^{-\frac{\varepsilon_{\lambda'}}{T}} .$$

Function $\widetilde{w}_{\lambda'\lambda}$ in addition to the electron-phonon coupling constant g contains the square of matrix element with the electron wave functions $\psi_\lambda(\boldsymbol{r})$ and $\psi_{\lambda'}(\boldsymbol{r})$ localized beside the impurity sites at points $\boldsymbol{r}_\lambda$ and $\boldsymbol{r}_{\lambda'}$. Therefore, we can write within an exponential accuracy

$$\widetilde{w}_{\lambda'\lambda} \sim \exp(-2r_{\lambda\lambda'}/a), \quad r_{\lambda\lambda'} = |\boldsymbol{r}_\lambda - \boldsymbol{r}_{\lambda'}|$$

taking into account that the typical spatial separations of impurity sites are large as compared with the localization radius a.

Let us consider the hopping rate $P_{\lambda'\lambda}$ from the selected impurity site λ to site λ' and represent it as

$$P_{\lambda'\lambda} = W_{\lambda'\lambda} n_\lambda (1 - n_{\lambda'}) = \widetilde{w}_{\lambda'\lambda}(\boldsymbol{r}_{\lambda'}, \boldsymbol{r}_\lambda)\gamma(T,\, \varepsilon_\lambda, \varepsilon_{\lambda'}),$$

where

$$\gamma(T,\, \varepsilon_\lambda, \varepsilon_{\lambda'}) = \frac{\mathrm{sgn}\,(\varepsilon_\lambda - \varepsilon_{\lambda'})}{1 - e^{-(\varepsilon_\lambda - \varepsilon_{\lambda'})/T}} \times \frac{1}{e^{(\varepsilon_\lambda - \mu)/T} + 1} \times \frac{1}{1 + e^{-(\varepsilon_{\lambda'} - \mu)/T}}\,.$$

Function $\gamma(T, \varepsilon_\lambda, \varepsilon_{\lambda'})$ incorporates the complete temperature information about the hopping rate. The expression can be simplified

$$\begin{aligned}\gamma(T,\, \varepsilon_\lambda, \varepsilon_{\lambda'}) &= \frac{1}{2}\,\frac{\mathrm{sgn}\,(\varepsilon_\lambda - \varepsilon_{\lambda'})}{\sinh\frac{\varepsilon_\lambda - \varepsilon_{\lambda'}}{T} + \sinh\frac{\varepsilon_\lambda - \mu}{T} - \sinh\frac{\varepsilon_{\lambda'} - \mu}{T}} \\ &= \frac{1}{8}\left(\sinh\frac{|\varepsilon_\lambda - \varepsilon_{\lambda'}|}{2T}\cosh\frac{\varepsilon_\lambda - \mu}{2T}\cosh\frac{\varepsilon_{\lambda'} - \mu}{2T}\right)^{-1}.\end{aligned}$$

If the typical variations of electron energy in the hopping exceed significantly the temperature, the main exponential dependence alone can approximately be retained from the expression for $\gamma(T,\, \varepsilon_\lambda, \varepsilon_{\lambda'})$ at $|\varepsilon_\lambda - \mu| \gg T$, $|\varepsilon_\lambda - \varepsilon_{\lambda'}| \gg T$

$$\gamma \sim \exp\left(-\frac{E_{\lambda\lambda'}}{T}\right) \quad \text{where} \quad E_{\lambda\lambda'} = \frac{|\varepsilon_\lambda - \varepsilon_{\lambda'}| + |\varepsilon_\lambda - \mu| + |\varepsilon_{\lambda'} - \mu|}{2}\,.$$

On the other hand, if the energy level dispersion is inessential, the temperature behavior of hopping rate may be non-exponential. So, for $|\varepsilon_\lambda - \mu| \ll T$, $|\varepsilon_\lambda - \varepsilon_{\lambda'}| \ll T$, we have $\gamma \sim T/|\varepsilon_\lambda - \varepsilon_{\lambda'}|$. Finally, this may result in the linear temperature behavior of hopping conductivity.

Selecting the main exponential dependence, we rewrite the expression for the electron hopping rate as

$$P_{\lambda'\lambda} = P^{(0)}_{\lambda'\lambda}\exp(-\mathcal{R}_{\lambda\lambda'}) \quad \text{where} \quad \mathcal{R}_{\lambda\lambda'} = \frac{2r_{\lambda\lambda'}}{a} + \frac{E_{\lambda\lambda'}}{T}.$$

Note that the preexponential factor $P^{(0)}_{\lambda'\lambda}$ depends weakly on the energy and coordinates of impurity sites. Thus, the electron hopping to the states with the larger energy $E_{\lambda,\lambda'}$ is strongly suppressed.

The conductivity is resulted from a set of hopping over impurity sites with the probability determined by the random quantity defined as *range* $\mathcal{R}$. Due to exponential dependence on $\mathcal{R}$ the hopping with the minimum magnitude of range $\mathcal{R}$ is most favorable.

The total number of electron states per 1 cm^3 in the energy band $\mu \pm E/2$ near chemical potential μ will be equal to $N(E) = g(\mu)E$. The mean distance $\bar{r}$ between impurity sites can be estimated according to formula[3]

$$\frac{4\pi}{3}\bar{r}^3 = \frac{1}{N(E)} = \frac{1}{g(\mu)E} \quad \text{and} \quad \bar{r} = \left(\frac{3}{4\pi g(\mu)E}\right)^{1/3}.$$

Then we have for the range $\mathcal{R}$

$$\mathcal{R} = \frac{2}{a}\left(\frac{3}{4\pi g(\mu)E}\right)^{1/3} + \frac{E}{T}.$$

Let us determine the energy bandwidth which provides the maximal hopping rate, using the condition of the minimum range $\mathcal{R} = \mathcal{R}(E)$

$$\frac{d\mathcal{R}}{dE} = -\frac{2}{3a}\left(\frac{3}{4\pi g(\mu)}\right)^{1/3}\frac{1}{E^{4/3}} + \frac{1}{T} = 0.$$

This equation gives the optimum value $E = \bar{E}$, i.e. saddle point, and characteristic temperature T_0

$$\bar{E} = \frac{1}{4}T_0^{1/4}T^{3/4} \quad \text{where} \quad T_0 = \frac{512}{9\pi}\frac{1}{a^3 g(\mu)}.$$

Then we readily find $\mathcal{R}_{\text{opt}} = (T_0/T)^{1/4}$ and the corresponding characteristic length of electron hopping

$$\bar{r} = \frac{3}{8}a\left(\frac{T_0}{T}\right)^{1/4}.$$

---

[3] Let impurities be chaotically distributed in the space of dimensionality d with concentration n. For the strict estimate of the mean distance $\bar{r}$ between impurities, we consider the density function $P(r)$ of the probability that the nearest impurity is at the distance r from the selected impurity. From the equation

$$P(r) = \left(1 - \int_0^r P(r')\,dr'\right)s_d\, r^{d-1}n \quad \text{we have} \quad P(r) = ns_d r^{d-1}e^{-nv_d r^d},$$

where s_d and $v_d = s_d/d$ is the area and volume of unit sphere in the space of dimensionality d, i.e. $s_3 = 4\pi$ and $s_2 = 2\pi$. The mean distance

$$\bar{r} = \int_0^\infty rP(r)\,dr = \left(\frac{1}{nv_d}\right)^{1/d}\Gamma\left(\frac{d+1}{d}\right).$$

contains the Gamma function as an additional multiplier which here and below is inessential for us.

For validity of exponential approximation, it is implied that the temperature T is small compared with the characteristic temperature T_0.

The probability of electron hopping to small distance decreases due to the Gibbs factor $\exp(E_{\lambda,\lambda'}/T)$ and to large distance decreases due to drastic reduction of the wave function overlapping. This entails that the effective hopping length does not remain constant but depends significantly on the temperature. Since $N(\bar{E})4\pi\bar{r}^3/3=1$ and temperature independent, there is always at least one suitable site for electron hopping at any temperature. The resistance $\rho(T)$ of such impurity-doped semiconductor is mainly governed by the exponential dependence

$$\rho(T) \sim \exp\left(\frac{T_0}{T}\right)^{1/4}.$$

Such temperature behavior of resistance is referred to as the *Mott law* of *variable-range conductivity*. As a rule, such behavior of resistance in impurity-doped semiconductor is typical in the temperature range from tens of millikelvin to several kelvin.

25. The generalization of hopping conductivity to the two-dimensional case reduces to the following. The total number of electron states per 1 cm^2 of a film within the energy bandwidth $\mu \pm E/2$ near chemical potential μ will be equal to $N(E) = g(\mu)E$ as well. The mean distance $\bar{r}$ between impurity sites is estimated as

$$\pi\bar{r}^2 = \frac{1}{N(E)} = \frac{1}{g(\mu)E} \quad \text{or} \quad \bar{r} = \left(\frac{1}{\pi g(\mu)E}\right)^{1/2}.$$

We have for the range $\mathcal{R}(E)$

$$\mathcal{R}(E) = \frac{2}{a}\left(\frac{1}{\pi g(\mu)E}\right)^{1/2} + \frac{E}{T}.$$

The energy bandwidth E yielding the maximum hopping rate is found from the condition

$$\frac{d\mathcal{R}}{dE} = -\frac{1}{a}\left(\frac{1}{\pi g(\mu)}\right)^{1/2}\frac{1}{E^{3/2}} + \frac{1}{T} = 0.$$

This equation gives the optimum value $E = \bar{E}$ and characteristic temperature T_0

$$\bar{E} = \frac{1}{3}T_0^{1/3}T^{2/3} \quad \text{where} \quad T_0 = \frac{27}{\pi a^2 g(\mu)}.$$

Next, we readily find $\mathcal{R}_{\text{opt}} = (T_0/T)^{1/3}$ and corresponding characteristic length of electron hopping

$$\bar{r} = \frac{a}{3}\left(\frac{T_0}{T}\right)^{1/3}.$$

The exponential behavior of resistance $\rho(T)$ in the two-dimensional film of impurity-doped semiconductor is determined by the minimum value $\mathcal{R}(E)$, i.e.

$$\rho(T) \sim \exp\left(\frac{T_0}{T}\right)^{1/3}.$$

Here the exponent in the Mott law equals $p = 1/3$.

26. Let us explain how the density of states $g(\varepsilon)$ near the Fermi level at $T = 0$ can change under interaction between electrons of various impurity sites. Consider two-electron states with energies ε_λ and $\varepsilon_{\lambda'}$ within an arbitrary energy interval Δ near the Fermi level, i.e. $|\varepsilon_\lambda - \mu| \leqslant \Delta$ and $|\varepsilon_{\lambda'} - \mu| \leqslant \Delta$. Let one of the states be occupied with an electron and, therefore, $\varepsilon_\lambda < \mu$ and the second be free $\varepsilon_{\lambda'} > \mu$, the distance between impurity sites being equal to $r_{\lambda\lambda'} = |\boldsymbol{r}_\lambda - \boldsymbol{r}_{\lambda'}|$. If both sites would be occupied, their total energy should increase by the Coulomb interaction energy $U_{\lambda\lambda'} = e^2/\varkappa r_{\lambda\lambda'}$, $\varkappa$ being the permittivity[4] of semiconductor.

Let us in thought transfer an electron from site λ to the infinity. Then the level energy of site λ' will decrease by magnitude $U_{\lambda\lambda'}$ due to lack of the Coulomb repulsion and becomes equal to

$$\widetilde{\varepsilon}_{\lambda'} = \varepsilon_{\lambda'} - U_{\lambda\lambda'} = \varepsilon_{\lambda'} - e^2/\varkappa r_{\lambda\lambda'}\,.$$

This energy $\widetilde{\varepsilon}_{\lambda'}$ should be larger than energy ε_λ. Otherwise, the assumption that impurity site λ is occupied and site λ' is free will be incorrect. Therefore, the following inequality should be satisfied:

$$\left(\varepsilon_{\lambda'} - \frac{e^2}{\varkappa r_{\lambda\lambda'}}\right) - \varepsilon_\lambda > 0 \quad \text{for} \quad \varepsilon_{\lambda'} - \varepsilon_\lambda \leqslant 2\Delta\,.$$

This leads to the conclusion that the inequalities below are valid for an arbitrary value Δ

$$2\Delta \geqslant \varepsilon_{\lambda'} - \varepsilon_\lambda > \frac{e^2}{\varkappa r_{\lambda\lambda'}} \quad \text{or} \quad r_{\lambda\lambda'} \geqslant \frac{e^2}{2\varkappa\Delta}\,.$$

Hence, the spatial separation $r_{\lambda\lambda'}$ between two states lying at the different sides from the Fermi level cannot be less than $e^2/(2\varkappa\Delta)$.

On the other hand, the number $N(\Delta = |\varepsilon - \mu|)$ of electron states which energies differ by not larger than 2Δ is approximately equal to the concentration of impurity sites $N(\Delta) \sim (v_d r_{\lambda\lambda'}^d)^{-1}$ and will be limited by the upper bound

[4] The use of permittivity can be justified since the spatial spacing between impurities is large as compared with the interatomic distance and the impurity charge screening is absent as a result of small electron concentration.

$$N(|\varepsilon - \mu|) \leqslant \frac{1}{v_d}\left(\frac{2\varkappa}{e^2}\right)^d \Delta^d = \frac{1}{v_d}\left(\frac{2\varkappa}{e^2}\right)^d |\varepsilon - \mu|^d.$$

Here $d = 2$ or $d = 3$, depending on the spatial dimensionality and $v_2 = \pi$ or $v_3 = 4\pi/3$. Hence, it also follows the similar limitation for the upper bound of the density of states $g(\varepsilon) = dN(\varepsilon)/d\varepsilon$

$$g(\varepsilon) \leqslant \frac{d}{v_d}\left(\frac{2\varkappa}{e^2}\right)^d |\varepsilon - \mu|^{(d-1)}.$$

The inequality means that the density of states $g(\varepsilon)$ vanishes at $\varepsilon = \mu$ not slower than the law $|\varepsilon - \mu|^{d-1}$.

If $g(\varepsilon)$ decays according to $\sim |\varepsilon - \mu|^\beta$ and goes to zero faster than $|\varepsilon - \mu|^{d-1}$, i.e. $\beta > (d - 1)$, then $N(\Delta = |\varepsilon - \mu|) \sim \Delta^{\beta+1}$. The mean distance $\bar{r}$ between the impurity sites having the level energy difference within Δ can be estimated according to

$$1/\bar{r}^d \sim N(\Delta) \sim \Delta^{\beta+1} \quad \text{or} \quad \bar{r} \sim \Delta^{-\frac{\beta+1}{d}}.$$

The interaction energy U between the electrons at these distances will be about

$$U(\Delta) \sim \frac{e^2}{\varkappa\bar{r}} \sim \Delta^{\frac{\beta+1}{d}}.$$

For $\beta > (d - 1)$, in the limit of energies sufficiently close to the Fermi level when $\Delta \to 0$, we have obligatory $U(\Delta) \ll \Delta$. Since the interaction energy $U(\Delta)$ in this case proves to be much smaller than the typical difference Δ in the energies of states near the Fermi level, we may expect that such weak interaction is inessential and, therefore, should not lead to stronger decrease of density of states $g(\varepsilon)$ near the Fermi level. As a result, we should have $\beta = (d - 1)$ in accordance with the upper bound as an estimate. The behavior $g(\varepsilon) \sim |\varepsilon - \mu|^{d-1}$ holds for since the distance $e^2/(\varkappa|\varepsilon - \mu|)$ runs significantly over the mean distance between impurity sites.

So, if the *Coulomb gap* exists and

$$g(\varepsilon) = g_0(|\varepsilon - \mu|/\varepsilon_0)^\beta = \gamma|\varepsilon - \mu|^\beta, \quad \gamma = g_0\varepsilon_0^{-\beta},$$

the number of states in the vicinity $E = |\varepsilon - \mu|$ around the Fermi level will be approximately equal to

$$N(E) \sim \frac{\gamma}{\beta + 1} E^{\beta+1} = \frac{\gamma}{\beta + 1}|\varepsilon - \mu|^{\beta+1}.$$

The next treatment is completely analogous to the speculation in the two previous problems. The mean distance $\bar{r}$ between impurity sites is estimated according to

$$v_d \bar{r}^d = \frac{1}{N(E)} = \frac{\beta+1}{\gamma}\left(\frac{1}{E}\right)^{\beta+1} \quad \text{or} \quad \bar{r} = \left(\frac{\beta+1}{\gamma v_d}\right)^{\frac{1}{d}}\left(\frac{1}{E}\right)^{\frac{\beta+1}{d}},$$

where v_d is the volume of unit sphere in the d-dimensional space and $v_d = 4\pi/3$ for $d = 3$, $v_d = \pi$ for $d = 2$. Correspondingly, we have for the range $\mathcal{R}$

$$\mathcal{R}(E) = \frac{2\bar{r}(E)}{a} + \frac{E}{T} = \frac{2}{a}\left(\frac{\beta+1}{\gamma v_d}\right)^{\frac{1}{d}}\left(\frac{1}{E}\right)^{\frac{\beta+1}{d}} + \frac{E}{T},$$

where a is the localization radius of an electron at the impurity site and T is the temperature.

The maximum probability of electron hopping from one site to another is determined by the minimum value $\mathcal{R}(E)$. The condition on the minimum reads

$$\frac{d\mathcal{R}}{dE} = -\frac{2}{a}\frac{\beta+1}{d}\left(\frac{\beta+1}{\gamma v_d}\right)^{\frac{1}{d}}\left(\frac{1}{E}\right)^{\frac{\beta+1+d}{d}} + \frac{1}{T} = 0.$$

The equation gives the optimum saddle point $\bar{E}$ and characteristic temperature T_0

$$\bar{E} = \frac{\beta+1}{\beta+d+1} T_0^{\frac{\beta+1}{\beta+1+d}} T^{\frac{d}{\beta+1+d}},$$

$$\text{where} \quad T_0 = \varepsilon_0\left(\frac{2^d(\beta+d+1)^{\beta+d+1}}{(\beta+1)d^d}\right)^{\frac{1}{\beta+1}} \left(g_0 v_d a^d \varepsilon_0\right)^{-1/(\beta+1)}.$$

The dimensional estimate of characteristic temperature equal approximately to

$$T_0 \sim \frac{e^2}{\varkappa a}$$

relates to the energy scale for the Coulomb interaction between the electron and the impurity site at the localization radius a.

The optimum range magnitude equals

$$\mathcal{R}_{\text{opt}} = \left(\frac{T}{T_0}\right)^p \quad \text{where} \quad p = \frac{\beta+1}{\beta+1+d}.$$

Involving that $\beta = d-1$, we get $p = 1/2$ independent of spatial dimensionality d.

And the last point concerns the electrical resistance $\rho(T)$ of weakly doped semiconductor with the presence of the Coulomb gap in the region of sufficiently low $T \lesssim T_0$ temperatures. The temperature behavior of resistance is mainly governed by the exponential dependence

$$\rho(T) \sim \exp\left(\frac{T_0}{T}\right)^{1/2},$$

which is referred to as the *Efros-Shklovskii law*. The law is the same for the bulk $d = 3$ semiconductor as well as for the thin $d = 2$ film. The mean length of electron hopping varies with the temperature according to $\bar{r} \sim T^{-1/2}$.

3.11 Kinetics of Phonons and Thermal Conductivity in Dielectrics

27. Let us write down the transport equation for the phonon distribution function $N_{\boldsymbol{k}}(t, \boldsymbol{r})$

$$\frac{\partial N_{\boldsymbol{k}}}{\partial t} + \frac{\partial \omega_{\boldsymbol{k}}}{\partial \boldsymbol{k}} \frac{\partial N_{\boldsymbol{k}}}{\partial \boldsymbol{r}} = \mathrm{St}\,[N_{\boldsymbol{k}}],$$

where $\partial\omega_{\boldsymbol{k}}/\partial\boldsymbol{k} = \boldsymbol{u}_{\boldsymbol{k}}$ is the phonon velocity. The equilibrium phonon distribution or Planck one

$$N_{0\boldsymbol{k}} = (e^{\hbar\omega_{\boldsymbol{k}}/T} - 1)^{-1}$$

should identically nullify the collision integral $\mathrm{St}\,[N_{0\boldsymbol{k}}] = 0$.

Unlike ordinary particles, in general, neither the number of phonons nor their total quasimomentum conserves in the collisions of excitations in a phonon gas. The energy conservation law alone holds for, meaning that

$$\int \omega_{\boldsymbol{k}} \mathrm{St}\,[N_{\boldsymbol{k}}] \frac{d^3k}{(2\pi)^3} = 0.$$

In accordance, we find for the transport equation

$$\int \frac{d^3k}{(2\pi)^3} \hbar\omega_{\boldsymbol{k}} \frac{\partial N_{\boldsymbol{k}}}{\partial t} + \int \frac{d^3k}{(2\pi)^3} \hbar\omega_{\boldsymbol{k}} \boldsymbol{u}_{\boldsymbol{k}} \frac{\partial N_{\boldsymbol{k}}}{\partial \boldsymbol{r}} = \frac{\partial E}{\partial t} + \mathrm{div}\,\boldsymbol{Q} = 0,$$

where E is the heat energy density of a dielectric and $\boldsymbol{Q}$ is the energy flux density.

We seek for the phonon distribution function as $N_{\boldsymbol{k}} = N_{0\boldsymbol{k}} + \delta N_{0\boldsymbol{k}}$, where $\delta N_{0\boldsymbol{k}}$ is a small correction. As usual, we assume that the temperature varies slowly along the dielectric. In the linear approximation the transport equation takes the form

$$\boldsymbol{u}_{\boldsymbol{k}} \frac{\partial N_{0\boldsymbol{k}}}{\partial \boldsymbol{r}} = (\boldsymbol{u}_{\boldsymbol{k}} \nabla T) \frac{\partial N_{0\boldsymbol{k}}}{\partial T} = \mathrm{St}\,[N_{\boldsymbol{k}}] \approx -\frac{\delta N_{\boldsymbol{k}}}{\tau_{\boldsymbol{k}}}.$$

Hence we find

$$\delta N_{\boldsymbol{k}} = -\tau_{\boldsymbol{k}} (\boldsymbol{u}_{\boldsymbol{k}} \nabla T) \frac{\partial N_{0\boldsymbol{k}}}{\partial T} = -\tau_{\boldsymbol{k}} (\boldsymbol{u}_{\boldsymbol{k}} \nabla T) N_{0\boldsymbol{k}} (1 + N_{0\boldsymbol{k}}) \frac{\hbar\omega_{\boldsymbol{k}}}{T^2}.$$

The energy flux, coincident here with the dissipative heat flow, equals

$$\boldsymbol{Q} = \int \hbar\omega_k \boldsymbol{u}_k N_k \frac{d^3k}{(2\pi)^3} = \int \hbar\omega_k \boldsymbol{u}_k \delta N_k \frac{d^3k}{(2\pi)^3}$$
$$= -\int \boldsymbol{u}_k(\boldsymbol{u}_k\nabla T)\frac{(\hbar\omega_k)^2}{T^2} N_{0k}(1+N_{0k})\tau_k \frac{d^3k}{(2\pi)^3}\,.$$

Vanishing the energy flux at $N_k = N_{0k}$ results formally from the property of phonon spectrum $\omega_k = \omega_{-k}$. Averaging the isotropic phonon spectrum over the solid angle in the space of vector $\boldsymbol{k}$ and using the definition of thermal conductivity from $\boldsymbol{Q} = -\varkappa\nabla T$, we arrive finally at

$$\varkappa = \frac{1}{3}\int u_k^2 \frac{(\hbar\omega_k)^2}{T^2} N_{0k}(1+N_{0k})\tau_k \frac{d^3k}{(2\pi)^3}$$
$$= \frac{1}{3}\int \frac{d^3k}{(2\pi)^3} u_k^2 \tau_k \frac{(\hbar\omega_k/2T)^2}{\sinh^2(\hbar\omega_k/2T)}\,.$$

For a simple sound dispersion $\omega_k = uk$, where u is the sound velocity, the latter relation can readily be rewritten in the form of elementary formula in the kinetic theory of gases

$$\varkappa = \frac{1}{3}C_{ph}(T)u^2\tau_{ph}(T).$$

Here $C_{ph}(T)$ is the phonon specific heat and $\tau_{ph}(T)$ is the effective mean free time of phonons. These quantities are determined by the relations

$$C_{ph}(T) = \int \frac{d^3k}{(2\pi)^3}\frac{(\hbar\omega_k/2T)^2}{\sinh^2(\hbar\omega_k/2T)} = \int d\omega\, g(\omega)\frac{(\hbar\omega/2T)^2}{\sinh^2(\hbar\omega/2T)}\,,$$

$$\tau_{ph}(T) = \frac{\int \frac{d^3k}{(2\pi)^3}\tau(\omega_k)\frac{(\hbar\omega_k/2T)^2}{\sinh^2(\hbar\omega_k/2T)}}{\int \frac{d^3k}{(2\pi)^3}\frac{(\hbar\omega_k/2T)^2}{\sinh^2(\hbar\omega_k/2T)}} = \frac{\int d\omega\, g(\omega)\tau(\omega)\frac{(\hbar\omega/2T)^2}{\sinh^2(\hbar\omega/2T)}}{\int d\omega\, g(\omega)\frac{(\hbar\omega/2T)^2}{\sinh^2(\hbar\omega/2T)}}\,.$$

Here $g(\omega)$ is the density of phonon states per unit volume.

The thermal resistance of a dielectric arises from the processes in which the quasimomentum of phonons does not conserve due to Umklapp processes, defects of spatial structure or phonon scattering at the dielectric boundary. The quantitative determination of relaxation time τ is a complicated problem. The point is that the coupling of phonons results from the anharmonic terms of third and higher orders in displacing the crystal lattice ions with respect to their equilibrium sites. The collision integral should include both the *normal processes* conserving the total quasimomentum of phonons and the *Umklapp processes* in which the total quasimomentum of phonons at their merging or decay equals the reciprocal lattice vector.

Below we only present a description for the temperature behavior of relaxation time τ and thermal conductivity $\varkappa$. For the high temperatures compared with the

Debye temperature Θ_D, the phonon frequencies $\omega \lesssim \Theta_D/\hbar \ll T/\hbar$ and the phonon mean free paths for normal processes l_n and Umklapp processes l_u are of the same order of magnitude. The mean free time of phonons or relaxation time is estimated as

$$\tau \sim \frac{Ma^2}{\hbar}\frac{\theta_D}{T},$$

where M is the atom mass and a is the interatomic distance. Accordingly, involving that the specific heat $C(T)$ at $T \gg \Theta_D$ is constant, the thermal conductivity obeys the behavior $\varkappa \sim 1/T$.

At low $T \ll \Theta_D$ temperatures the thermal resistance in an ideal single crystal of unlimited sizes occurs only due to phonon collisions accompanied with the Umklapp processes. The effective mean free path l_u grows exponentially with lowering the temperature as $l_u \sim \exp(\gamma\Theta_D/T)$, where γ is the number about unity. In essence, the magnitude $\gamma\Theta_D/\hbar$ is the minimum value of phonon frequency for which the Umklapp process is still possible. This gives the dominant dependence for the behavior of thermal conductivity $\varkappa(T) \sim \exp(\gamma\Theta_D/T)$.

In a dielectric of finite sizes, e.g. in the shape of infinite cylinder of radius R, the collisions of phonons with the boundary of a cylinder start to play a key role as the temperature lowers. This takes place when the mean free paths for normal processes $l_n(T)$ and for Umklapp processes $l_u(T)$ prove to be compared with the cylinder radius R. At low $T \ll \Theta_D$ temperatures, as a result of drastically different temperature behavior of mean free paths, being power-like for $l_n(T)$

$$l_n(T) \sim l_n(\Theta_D)\big(\theta_D/T\big)^5,$$

and being exponential for $l_u(T) \sim \exp(\gamma\Theta_D/T)$, there arise two temperatures T_n and T_u when $l_n(T_n) \sim R$ and $l_u(T_u) \sim R$. In addition, inequality $l_n \ll R \ll l_u$ takes place within the temperature range $T_n < T < T_u$. While $l_n \ll R$, a phonon experiences a lot of normal collisions before it reaches the boundary of a dielectric. The phonon motion between two successive collisions with the boundary has a diffusive character with the effective mean free path $l_{\text{eff}} \sim R^2/l_n$. On account of dependence $C(T) \sim T^3$ for the specific heat, we arrive at the estimate

$$\varkappa \sim C(T)ul_{\text{eff}} \sim \frac{C(T)uR^2}{l_n(T)} \sim T^8R^2, \quad (T_n < T < T_u).$$

And, eventually, in the region of ultralow $T < T_n$ temperatures when all the possible mean free paths run over the sizes of a dielectric, the size R of a dielectric plays a role of effective mean free path. Then the estimate of thermal conductivity reduces to

$$\varkappa(T) \sim C(T)uR \sim T^3R \quad (T < T_n).$$

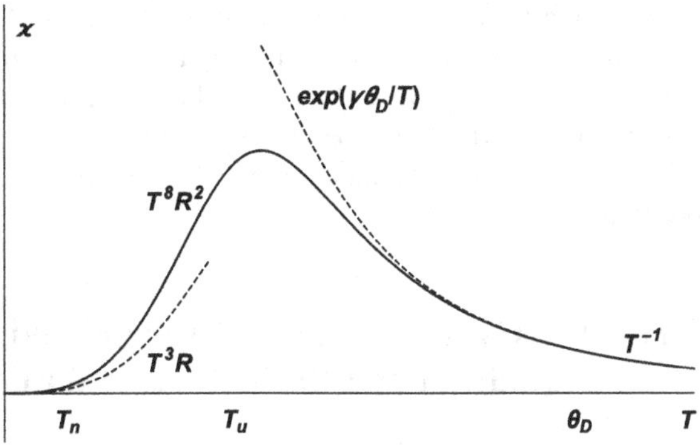

Fig. 3.11 A schematic behavior of the thermal conductivity $\varkappa$ in a dielectric of radius R. In the ultralow $T < T_n$ temperature region one has $\varkappa \sim T^3 R$, then $\varkappa \sim T^8 R^2$ between temperatures T_n and T_u. Next, there is a reduction $\varkappa(T)$ either exponentially or power-like

On the whole, the temperature dependence for the thermal conductivity demonstrates a bell-shaped behavior (Fig. 3.11) with the maximum which position depends on the size of a dielectric.

In conclusion, we discuss the effect of point defects in the crystal lattice on the thermal conductivity of a dielectric. The presence of defects, e.g. impurity atoms, vacancies or interstitial atoms, results in the phonon scattering at defects and hinders the heat flow. Simplest point defect can be imagined as a small region of radius r about interatomic distance a, which properties differ from those of surrounding medium, e.g. by density or atom mass. The scattering of phonons or sound waves with such defect corresponds to the *Rayleigh scattering* when the phonon wavelength λ is significantly larger than the defect size r. If there is a sphere of radius r which density $\rho + \delta\rho$ differs by $\delta\rho$ from density ρ of its environment, the cross-section of sound wave scattering can be estimated as

$$\sigma(k) \sim 4\pi r^2 (kr)^4 (\delta\rho/\rho)^2.$$

Such type of scattering has its inherent elasticity, i.e. phonon frequency remains unchanged and scattering cross-section $\sigma(k)$ is proportional to fourth power of phonon wave vector k. The scattering of a phonon with the impurity changes its quasimomentum. The relaxation time or mean free time τ_k can be estimated, using the relation between mean free path l_k and scattering cross-section $\sigma(k)$

$$l_k = 1/\big(n_i \sigma(k)\big).$$

Here n_i is the number of scattering centers per unit volume, i.e. impurity concentration. The mean free time τ_k can be found from relation $l_k = u\tau_k$, where u is the phonon or sound velocity. Since the typical energy $\hbar\omega$ of thermal phonons is about temperature T and $\omega = uk$, it is readily to see that

$$1/\tau_k \propto n_i \omega^4 \propto n_i T^4.$$

Since the heat capacity $C(T)$ is proportional to T^3 below the Debye temperature, we get to the following temperature behavior of impurity contribution to the thermal conductivity: $\varkappa_i \sim (n_i T)^{-1}$. Thus, the presence of impurities suppresses an exponential growth of thermal conductivity. The latter will behave, obeying the law $\varkappa \sim 1/T$ down to the temperatures when the phonon scattering at the boundaries becomes predominant.

3.12 The Hydrodynamics of Normal and Superfluid Liquids: Sound Oscillations and Energy Dissipation

28. The energy of unit liquid volume $E = E(S, \rho, \boldsymbol{j})$ is a function of the following thermodynamical variables: entropy S, liquid density ρ and mass density flux (momentum per unit volume) $\boldsymbol{j}$. The differential of energy

$$dE(S, \rho, \boldsymbol{j}) = T\,dS + \mu\,d\rho + \boldsymbol{v}\cdot d\boldsymbol{j}$$

determines the conjugated thermodynamical variables: temperature T, chemical potential per unit mass μ and fluid velocity $\boldsymbol{v}$. The conjugated thermodynamic potential $A(T, \mu, \boldsymbol{v})$ and its differential are equal to

$$A(T, \mu, \boldsymbol{v}) = E - TS - \mu\rho - \boldsymbol{v}\cdot\boldsymbol{j},$$
$$dA(T, \mu, \boldsymbol{v}) = -S\,dT - \rho\,d\mu - \boldsymbol{j}\cdot d\boldsymbol{v}.$$

The quantity $P = -A = -E + TS + \mu\rho + \boldsymbol{v}\cdot\boldsymbol{j}$ should be called the *pressure* if we follow the definition of pressure $P = -\partial(EV)/\partial V$ as a minus-sign derivative of total energy EV of a liquid with respect to its volume V under constant total mass ρV, total entropy SV, and total momentum $\boldsymbol{j}V$.

Let us write down the conservation laws for the mass, momentum, energy and entropy growth in the differential form

$$\frac{\partial\rho}{\partial t} + \operatorname{div}\boldsymbol{j} = 0, \quad \frac{\partial j_i}{\partial t} + \frac{\partial\Pi_{ik}}{\partial x_k} = 0,$$
$$\frac{\partial E}{\partial t} + \operatorname{div}\boldsymbol{Q} = 0, \quad \frac{\partial S}{\partial t} + \operatorname{div}\boldsymbol{F} = \frac{R}{T}.$$

In what follows, we should specify a set of the following quantities: mass density flux (momentum density) $\boldsymbol{j}$, energy flux density $\boldsymbol{Q}$, entropy flux density $\boldsymbol{F}$, momentum flux density tensor Π_{ik}, and dissipative Rayleigh function R.

Then we sum all four equations, in advance multiplying each of them by the corresponding factor T, ρ or $\boldsymbol{v}$ and taking into account that

$$\frac{\partial E}{\partial t} = T\frac{\partial S}{\partial t} + \mu\frac{\partial\rho}{\partial t} + \boldsymbol{v}\cdot\frac{\partial\boldsymbol{j}}{\partial t}$$

We find finally after identical transformations

$$R + \operatorname{div} \boldsymbol{Q} = \frac{\partial}{\partial x_i}\big[T F_i + \mu j_i + v_k(\Pi_{ik} + A\delta_{ik})\big]$$
$$- (F_i - S v_i)\frac{\partial T}{\partial x_i} - (j_i - \rho v_i)\frac{\partial \mu}{\partial x_i} - (\Pi_{ik} + A\delta_{ik} - j_k v_i)\frac{\partial v_k}{\partial x_i}\,.$$

If one assumes that the energy dissipation is absent, i.e. $R \equiv 0$, the magnitudes of fluxes $\boldsymbol{Q}$, $\boldsymbol{F}$, $\boldsymbol{j}$ and Π_{ik} must be the functions of thermodynamic variables alone, e.g. T, ρ, and $\boldsymbol{j}$, and be independent of their derivatives with respect to coordinates x_i and time t. In addition, the right-hand side of the above equation must represent the expression in the form of total derivative (divergence). Therefore, it is necessary to put the following for the dissipationless fluxes considered:

$$j_{0i} = \rho v_i\,, \qquad F_{0i} = S v_i\,,$$
$$\Pi_{0ik} = -A\delta_{ik} + j_{0k} v_i = P\delta_{ik} + \rho v_i v_k\,,$$
$$Q_{0i} = T F_{0i} + \mu j_{0i} + v_k(\Pi_{0ik} + A\delta_{ik}) = T F_{0i} + \mu j_{0i} + (\boldsymbol{v}\cdot \boldsymbol{j}_0) v_i\,.$$

For the further applications, we rewrite the last equality in the vector representation

$$\boldsymbol{Q}_0 = T\boldsymbol{F}_0 + \mu \boldsymbol{j}_0 + (\boldsymbol{v}\cdot \boldsymbol{j}_0)\boldsymbol{v}.$$

The term $P\delta_{ik}$ in the momentum flux density tensor means that the pressure in a fluid at rest is isotropic, i.e. pressure is transmitted equally in all directions (Pascal's law).

The non-equilibrium state of a fluid and energy dissipation lead to appearance of additional terms in the fluxes of energy, entropy, and momentum density tensor which depend explicitly that time on the derivatives with respect to the coordinates. If we continue to treat ρ as a fluid density and $\boldsymbol{j}$ as a momentum per unit fluid volume, we can keep the conventional formula for the mass density flow

$$\boldsymbol{j} = \boldsymbol{j}_0 = \rho \boldsymbol{v}.$$

For the other fluxes, the dissipative terms are necessary to be augmented

$$\Pi_{ik} = \Pi_{0ik} + \tau_{ik}\,,$$
$$F_i = F_{0i} + f_i = F_{0i} + q_i/T.$$

Accordingly, we must put

$$Q_i = Q_{0i} + T f_i + v_k \tau_{ki} \equiv Q_{0i} + q_i + v_k \tau_{ki}\,,$$

where vector $\boldsymbol{q}$ is the *dissipative heat flow*, and tensor τ_{ik} is a dissipative part of momentum flux density tensor. The tensor with the reverse sign $\sigma'_{ik} = -\tau_{ik}$ is called the *viscous stress tensor*.

The dissipative function R is given by

$$R = -f_i \frac{\partial T}{\partial x_i} - \tau_{ik} \frac{\partial v_k}{\partial x_i} = -\frac{\boldsymbol{q} \cdot \nabla T}{T} - \tau_{ik} \frac{\partial v_k}{\partial x_i} \geqslant 0$$

and should be positive definite. For the weak non-equilibrium states, the spatial derivatives are small. In first approximation we can restrict ourselves with first-order derivatives. This entails for the dissipative heat

$$\boldsymbol{q} = -\varkappa \nabla T,$$

where $\varkappa$ is the *thermal conductivity*.

The dissipative flux tensor τ_{ik} as a second-order tensor may depend on two independent tensors $\partial v_i / \partial x_k$ and δ_{ik}. The scalar factor in front of δ_{ik}, composed of velocity derivatives, is proportional to div $\boldsymbol{v}$. If the particles of a fluid has no rotational degrees of freedom and has no internal angular momentum, the momentum flux density tensor should be symmetrical $\Pi_{ik} = \Pi_{ki}$. In fact, in such fluid at its uniform rotation as a whole with angular velocity $\boldsymbol{\Omega}$ the fluid remains in the thermodynamic equilibrium state and the dissipative processes should disappear. Since the fluid velocity equals $\boldsymbol{v} = [\boldsymbol{\Omega} \times \boldsymbol{r}]$ under such rotation, the dissipative part of tensor τ_{ik} should not have nonzero antisymmetric combination $(\partial v_i/\partial x_k - \partial v_k/\partial x_i)$ which would be proportional to Ω in this case.

The general form for the symmetrical second-order rank tensor is given by the expression

$$\tau_{ik} = -\eta \left(\frac{\partial v_i}{\partial x_k} + \frac{\partial v_k}{\partial x_i} \right) - \zeta' \delta_{ik} \frac{\partial v_l}{\partial x_l} .$$

As a rule, one subdivides the tensor τ_{ik} into two parts as follows: traceless part representing a purely shear component of viscous stress tensor and the part connected with changing the fluid volume:

$$\tau_{ik} = -\eta \left(\frac{\partial v_i}{\partial x_k} + \frac{\partial v_k}{\partial x_i} - \frac{2}{3} \delta_{ik} \frac{\partial v_l}{\partial x_l} \right) - \zeta \delta_{ik} \frac{\partial v_l}{\partial x_l} .$$

The coefficient η is called the *dynamical* or *shear viscosity* and the second coefficient ζ is referred to as *bulk* or *second viscosity*. The dissipative function will equal

$$R = \frac{\eta}{2} \left(\frac{\partial v_i}{\partial x_k} + \frac{\partial v_k}{\partial x_i} - \frac{2}{3} \delta_{ik} \frac{\partial v_l}{\partial x_l} \right)^2 + \zeta (\operatorname{div} \boldsymbol{v})^2 + \varkappa \frac{(\nabla T)^2}{T} .$$

The requirement of positive definiteness for the dissipative function means straightforwardly that the coefficients of viscosity and thermal conductivity must be strictly positive $\eta > 0$, $\zeta > 0$, and $\varkappa > 0$.

The substitution of tensor Π_{ik} into the momentum conservation equation and the subsequent application of mass conservation law result in the *Navier-Stokes equation*

$$\rho\left(\frac{\partial \boldsymbol{v}}{\partial t} + (\boldsymbol{v}\cdot\nabla)\boldsymbol{v}\right) = -\nabla P + \eta\nabla^2\boldsymbol{v} + \left(\zeta + \frac{\eta}{3}\right)\nabla(\operatorname{div}\boldsymbol{v}).$$

Provided that $\eta = \zeta = 0$, the Navier-Stokes equation reduces to the *Euler equation* for an ideal fluid

$$\rho\left(\frac{\partial \boldsymbol{v}}{\partial t} + (\boldsymbol{v}\cdot\nabla)\boldsymbol{v}\right) = -\nabla P.$$

29. In the sound wave the fluid velocity $\boldsymbol{v}$ is assumed to be small as compared with the sound velocity u and the thermodynamic quantities should be close to the equilibrium magnitudes. So, in first approximation we can treat the linearized hydrodynamical equations, neglecting the energy dissipation processes

$$\frac{\partial \rho}{\partial t} + \rho \operatorname{div}\boldsymbol{v} = 0 \quad \text{and} \quad \rho\frac{\partial \boldsymbol{v}}{\partial t} = -\nabla P.$$

Eliminating the fluid velocity $\boldsymbol{v}$ from these two equation, we obtain the equation relating the density ρ and pressure P

$$\frac{\partial^2 \rho}{\partial t^2} = \nabla^2 P.$$

Since we neglect completely the energy dissipation effects and dissipative function R vanishes, the fluid motion becomes adiabatic and occurs under conservation of total entropy. Adiabaticity of sound oscillations supposes the smallness of oscillation frequency ω in the parameter $\omega\tau \ll 1$, where τ is the typical relaxation time. Relating the pressure oscillations δP with density oscillations $\delta\rho$ under constancy of entropy

$$\delta P = \left(\partial P/\partial\rho\right)_s \delta\rho,$$

we obtain the wave equation

$$\nabla^2 P = \frac{1}{u^2}\frac{\partial^2 P}{\partial t^2}, \qquad u = \left(\frac{\partial P}{\partial \rho}\right)_s^{1/2}$$

which determines the adiabatic sound velocity u and dispersion spectrum $\omega = uk$.

The energy dissipation processes of viscosity and heat conduction in a fluid lead to absorption of sound waves. The loss of sound wave energy E_s results eventually in heating the fluid and increasing its internal energy $E = E(S)$. The latter we consider as a function of entropy S. Differentiating relation $\delta E(S) = -\delta E_s$ with respect to time yields

$$\dot{E}_s = -\dot{E}(S) = -\frac{\partial E}{\partial S}\dot{S} = -T\dot{S} = -\int R\,dV.$$

The energy dissipation power per unit volume R will equal

$$R = \frac{\eta}{2}\left(\frac{\partial v_i}{\partial x_k} + \frac{\partial v_k}{\partial x_i} - \frac{2}{3}\delta_{ik}\frac{\partial v_l}{\partial x_l}\right)^2 + \zeta(\operatorname{div}\boldsymbol{v})^2 + \varkappa\frac{(\nabla T)^2}{T},$$

integration being performed over the whole fluid volume.

Let sound plane wave propagate in the x-axis direction and fluid velocity be given as

$$v_x = v_0\cos(kx - \omega t), \quad v_y = v_z = 0.$$

The two viscosity terms and identity $\frac{\partial v_i}{\partial x_k}\frac{\partial v_k}{\partial x_i} = (\partial v_x/\partial x)^2$ lead to

$$\int\left(\frac{4\eta}{3} + \zeta\right)\left(\frac{\partial v_x}{\partial x}\right)^2 dV = \left(\frac{4\eta}{3} + \zeta\right)k^2 v_0^2 \int \sin^2(kx - \omega t)\, dV.$$

After averaging over time, we obtain per the unit volume

$$\left(\frac{4\eta}{3} + \zeta\right)\frac{k^2 v_0^2}{2}.$$

In order to calculate the contribution from heat conduction, we relate the temperature oscillations δT with the pressure oscillations or sound pressure δP

$$\delta T = \left(\frac{\partial T}{\partial P}\right)_s \delta P = \frac{T}{\rho c_P V}\left(\frac{\partial V}{\partial T}\right)_P \delta P.$$

Here c_P is the *isobaric specific heat* or heat capacity per unit fluid mass under constant pressure. A ratio of sound pressure δP to the fluid velocity v, i.e. *acoustic impedance* equals $\delta P/v = \rho\omega/k = \rho u$. As a result, we find

$$\delta T = u\frac{\alpha_V T}{c_P}v,$$

where α_V is the coefficient of thermal expansion

$$\alpha_V = \frac{1}{V}\left(\frac{\partial V}{\partial T}\right)_P = -\frac{1}{\rho}\left(\frac{\partial \rho}{\partial T}\right)_P.$$

So, the contribution of heat conduction to the dissipative function will be equal to

$$\int \varkappa\frac{(\nabla T)^2}{T}dV = \frac{\varkappa}{T}u^2\frac{\alpha_V^2 T^2}{c_P^2}k^2 v_0^2 \int \sin^2(kx - \omega t)\, dV.$$

After the analogous time average we find

$$\varkappa u^2 \frac{\alpha_V^2 T}{c_P^2}\frac{k^2 v_0^2}{2} = \varkappa\left(\frac{1}{c_V} - \frac{1}{c_P}\right)\frac{k^2 v_0^2}{2}$$

for the dissipative contribution per unit fluid volume. To derive the latter equality, we have used the familiar thermodynamic equality for the difference between the isobaric and the isochoric specific heats

$$c_P - c_V = T\alpha_V^2\left(\frac{\partial P}{\partial \rho}\right)_T = T\alpha_V^2\frac{c_V}{c_P}\left(\frac{\partial P}{\partial \rho}\right)_s = T\alpha_V^2\frac{c_V}{c_P}u^2.$$

Thus the total dissipative function equals

$$R = \left[\frac{4}{3}\eta + \zeta + \varkappa\left(\frac{1}{c_V} - \frac{1}{c_P}\right)\right]\frac{k^2 v_0^2}{2}, \quad k = \frac{\omega}{u}.$$

To find the sound attenuation coefficient $\alpha(\omega)$, we should express $k^2 v_0^2$ in terms of the energy of sound wave. The density of total energy in a fluid is a sum of its kinetic and internal energies and equal to

$$E = \frac{\rho v^2}{2} + \rho\varepsilon(s, \rho),$$

ε being the energy per unit mass. Below we should keep in mind that the sound propagation is an adiabatic process $s = \text{const}$ and the fluid mass remains also unchanged, i.e. $\int \rho\, dV = \text{const}$. For our purposes, it is sufficient to calculate the variation of total energy $E = E + \delta E$ within accuracy up to second-order terms in velocity v and in density perturbation $\delta\rho$. Then we have

$$\delta E \approx \frac{\rho v^2}{2} + \frac{\partial(\rho\varepsilon)_s}{\partial\rho}\delta\rho + \frac{1}{2}\frac{\partial^2(\rho\varepsilon)_s}{\partial\rho^2}(\delta\rho)^2 + \ldots$$

The second derivative is transformed with the aid of definition for the specific enthalpy $w = \varepsilon + P/\rho$

$$\frac{\partial^2(\rho\varepsilon)_s}{\partial\rho^2} = \frac{\partial}{\partial\rho}\left(\varepsilon + \frac{P}{\rho}\right) = \left(\frac{\partial w}{\partial\rho}\right)_s = \frac{\partial w}{\partial P}\left(\frac{\partial P}{\partial\rho}\right)_s = \frac{u^2}{\rho}.$$

Using that $\int \delta\rho\, dV = 0$ and relation $\delta\rho = \delta P/u^2 = \rho v/u$, we find the total energy of sound wave

$$\int \delta E\, dV = \int\left[\frac{\rho v^2}{2} + \frac{u^2}{2\rho}(\delta\rho)^2\right]dV = \int \rho v^2\, dV = \rho v_0^2 \int \cos^2(kx - \omega t)\, dV.$$

Then the mean energy of sound wave per unit volume reads

$$E_s = \rho v_0^2/2.$$

Let us write the equation for the rate of dissipating the energy of sound wave

$$\dot{E}_s = -R = -\left[\frac{4}{3}\eta + \zeta + \varkappa\left(\frac{1}{c_V} - \frac{1}{c_P}\right)\right]\frac{k^2 v_0^2}{2}$$
$$= -\left[\frac{4}{3}\eta + \zeta + \varkappa\left(\frac{1}{c_V} - \frac{1}{c_P}\right)\right]\frac{k^2}{\rho}E_s = -2\alpha u E_s\,.$$

It is readily seen that the intensity of sound wave will decay as $e^{-2\alpha ut}$ and its amplitude as $e^{-\alpha ut}$. Since the distance traveled by the wave is connected with time according to simple relation $x = ut$, the wave amplitude will decay as $e^{-\alpha x}$ with the distance x traversed. Thus, expressing the absorption coefficient α as a function of frequency and substituting $k = \omega/u$, we have the following answer:

$$\alpha(\omega) = \left[\frac{4}{3}\eta + \zeta + \varkappa\left(\frac{1}{c_V} - \frac{1}{c_P}\right)\right]\frac{\omega^2}{2\rho u^3}\,.$$

The sound attenuation is small while $\omega/u \gg \alpha(\omega)$. This is always true in the limit $\omega \to 0$. If all kinetic coefficients are considered to be the same order of magnitude $\eta \sim \zeta \sim \rho u^2\tau$ and $\varkappa \sim \rho c_P u^2\tau$, where τ is some typical mean free time of particles, we obtain the condition of small sound attenuation in the form $\omega\tau \ll 1$.

30. From the phenomenological point of view an existence of two independent types of fluid flows unlike single flow type requires to increase the number of variables describing the state of a fluid. The thermodynamic variables inherent in usual normal fluid should be augmented with both the superfluid velocity $\boldsymbol{v}_s$ and the mass flow of superfluid component $\boldsymbol{j}_s$. The latter variable is conjugated to superfluid velocity. Therefore, the energy density of superfluid liquid $E = E(S, \rho, \boldsymbol{j}, \boldsymbol{v}_s)$ will depend on four variables as follows: entropy S, density ρ, mass flow density or momentum per unit volume $\boldsymbol{j}$, and superfluid velocity $\boldsymbol{v}_s$.

In the first turn let us write the conservation laws of mass, momentum, energy, and entropy in the differential form

$$\frac{\partial\rho}{\partial t} + \operatorname{div}\boldsymbol{j} = 0, \quad \frac{\partial j_i}{\partial t} + \frac{\partial\Pi_{ik}}{\partial x_k} = 0,$$
$$\frac{\partial E}{\partial t} + \operatorname{div}\boldsymbol{Q} = 0, \quad \frac{\partial S}{\partial t} + \operatorname{div}\boldsymbol{F} = 0.$$

The removal of energy dissipation processes from our consideration implies the entropy conservation.

In two-fluid hydrodynamics these four equations must be augmented with one more equation describing the dynamics of superfluid velocity $\boldsymbol{v}_s$. On the analogy with usual fluid this should be an equation on the time derivative of superfluid velocity $\boldsymbol{v}_s$. In order to satisfy the condition of potential or irrotational flow $\operatorname{curl}\boldsymbol{v}_s = 0$, we will write this equation as

$$\frac{\partial\boldsymbol{v}_s}{\partial t} + \nabla\psi = 0.$$

Here $\psi = \psi(S, \rho, \boldsymbol{j}, \boldsymbol{v}_s)$ is a scalar function to be determined along with the fluxes of energy $\boldsymbol{Q}$, entropy $\boldsymbol{F}$, mass $\boldsymbol{j}$, and momentum flux density tensor Π_{ik}.

In order to find the expressions for the fluxes, as a first step, it is practical to choose the reference frame K_0 in which the superfluid component is at rest $\boldsymbol{v}_s = 0$ but the normal component alone flows at relative velocity[5] $\boldsymbol{w} = \boldsymbol{v}_n - \boldsymbol{v}_s$. Since the superfluid component is at rest in this reference frame and does not participate in the fluid motion, we may expect that the fluid properties in the frame K_0 will be the same ones inherent in usual normal fluid. This allows us to determine the corresponding fluxes $\boldsymbol{j}_0$, $\boldsymbol{Q}_0$, $\boldsymbol{F}_0$, and Π_{0ik}.

In what follows, we are going to employ the *Galilean transformation* to determine the fluxes in the laboratory reference frame K in which the superfluid component flows at velocity $\boldsymbol{v}_s$. We relate the corresponding physical quantities in two various reference frames with the aid of the Galilean transformation.

For the energy density $E_0 = E_0(S, \rho, \boldsymbol{j}_0) = E(S, \rho, \boldsymbol{j}, \boldsymbol{v}_s = 0)$ in the reference frame K_0, we write the usual differential

$$dE_0 = T\,dS + \mu\,d\rho + \boldsymbol{w}\cdot d\boldsymbol{j}_0.$$

Here temperature T, chemical potential μ, and relative velocity $\boldsymbol{w}$ are introduced as the thermodynamic quantities conjugated to variables S, ρ and $\boldsymbol{j}_0$. The mass flow vector $\boldsymbol{j}_0$, as a vector in the isotropic fluid, can only be oriented in the direction of relative velocity vector $\boldsymbol{w}$:

$$\boldsymbol{j}_0 = \rho_n \boldsymbol{w}.$$

Here scalar $\rho_n = \rho_n(S, \rho, \boldsymbol{w}^2)$ determines the magnitude of the normal component density and, in general, is a function of entropy, total fluid density, and square of relative velocity $\boldsymbol{w}$.

Accordingly, the expressions for the other fluxes in reference frame K_0 are presented with the following formulas:

$$\Pi_{0ik} = P\delta_{ik} + \rho_n w_i w_k,$$
$$\boldsymbol{Q}_0 = T\boldsymbol{F}_0 + \mu\boldsymbol{j}_0 + (\boldsymbol{w}\cdot\boldsymbol{j}_0)\boldsymbol{w} \quad \text{and} \quad \boldsymbol{F}_0 = S\boldsymbol{w}.$$

The pressure P and its differential dP are specified in terms of the conjugated potential $A = E_0 - TS - \mu\rho - \boldsymbol{w}\cdot\boldsymbol{j}$ according to

$$P = -A = -E_0 + TS + \mu\rho + \boldsymbol{w}\cdot\boldsymbol{j}_0,$$
$$dP = -dA = S\,dT + \rho\,d\mu + \boldsymbol{j}_0\cdot d\boldsymbol{w}.$$

The quantities of our interest in the reference frames K and K_0 are related with the Galilean transformation

[5] The velocity difference $\boldsymbol{w} = \boldsymbol{v}_n - \boldsymbol{v}_s$ is often referred to as counterflow.

$$E = E_0 + \boldsymbol{j}_0 \cdot \boldsymbol{v}_s + \frac{\rho \boldsymbol{v}_s^2}{2},$$
$$\boldsymbol{j} = \boldsymbol{j}_0 + \rho \boldsymbol{v}_s, \qquad \boldsymbol{F} = \boldsymbol{F}_0 + S\boldsymbol{v}_s,$$
$$\Pi_{ik} = \Pi_{0ik} + \rho v_{si} v_{sk} + v_{si} j_{0k} + j_{0i} v_{sk},$$
$$Q_i = Q_{0i} + \big(E_0 + \boldsymbol{j}_0 \cdot \boldsymbol{v}_s + \frac{\rho \boldsymbol{v}_s^2}{2}\big) v_{si} + \frac{\boldsymbol{v}_s^2}{2} j_{0i} + \Pi_{0ik} v_{sk}.$$

Substituting the values of fluxes in the reference frame K_0 and performing the algebraic calculations, we find the value of fluxes in the laboratory frame K expressed via velocities of the normal and superfluid components

$$\boldsymbol{j} = \rho_n \boldsymbol{v}_n + (\rho - \rho_n)\boldsymbol{v}_s = \rho_n \boldsymbol{v}_n + \rho_s \boldsymbol{v}_s,$$
$$\boldsymbol{F} = S\boldsymbol{w} + S\boldsymbol{v}_s = S\boldsymbol{v}_n,$$
$$\Pi_{ik} = P\delta_{ik} + \rho_n v_{ni} v_{nk} + (\rho - \rho_n) v_{si} v_{sk} = P\delta_{ik} + \rho_n v_{ni} v_{nk} + \rho_s v_{si} v_{sk},$$
$$\boldsymbol{Q} = \big(\mu + \frac{\boldsymbol{v}_s^2}{2}\big)\boldsymbol{j} + TS\boldsymbol{v}_n + \rho_n \boldsymbol{v}_n (\boldsymbol{v}_n - \boldsymbol{v}_s) \cdot \boldsymbol{v}_n.$$

As we see, the mass flow density $\boldsymbol{j}$ and the momentum flux density tensor Π_{ik} include the terms associated with the normal component flow as well as with the superfluid component flow. Emphasize that this property does not apply to the entropy transfer. In fact, one can realize from the entropy transfer formula $\boldsymbol{F} = S\boldsymbol{v}_n$ that the entropy flux is connected with the normal component flow alone and the entropy is transferred in the *convective* manner along with the normal component. The superfluid component does not participate in the entropy transfer. Note also that for the equal normal and superfluid velocities $\boldsymbol{v}_n = \boldsymbol{v}_s = \boldsymbol{v}$, the expressions of fluxes resemble those for the corresponding quantities in non-superfluid liquid.

We are now in position to determine the scalar function ψ which gradient determines the time derivative of superfluid velocity. This can be done, for example, using the energy conservation law and the expressions given above for the density of energy and fluxes in the laboratory frame K. Calculating after numerous cancellations results in a simple answer

$$\frac{\partial E}{\partial t} + \operatorname{div} \boldsymbol{Q} = -\rho_s (\boldsymbol{v}_n - \boldsymbol{v}_s) \cdot \left(\frac{\partial \boldsymbol{v}_s}{\partial t} + \nabla \big(\mu + \frac{\boldsymbol{v}_s^2}{2}\big) \right) + \rho \boldsymbol{v}_n \cdot (\boldsymbol{v}_s \times \operatorname{rot} \boldsymbol{v}_s) = 0.$$

The requirement of zero equality for the right-hand side of equation under irrotational superfluid component flow curl $\boldsymbol{v}_s = 0$ leads us to the determination of scalar function $\psi = \mu + \boldsymbol{v}_s^2/2$ and to the equation of motion for the superfluid component velocity

$$\frac{\partial \boldsymbol{v}_s}{\partial t} + \nabla \big(\mu + \frac{\boldsymbol{v}_s^2}{2}\big) = 0.$$

Let us write the other hydrodynamical equations of superfluid liquid

$$\frac{\partial \rho}{\partial t} + \operatorname{div} \boldsymbol{j} = 0, \qquad \boldsymbol{j} = \rho_n \boldsymbol{v}_n + \rho_s \boldsymbol{v}_s,$$

$$\frac{\partial \boldsymbol{j}}{\partial t} + \boldsymbol{v}_s \operatorname{div} \boldsymbol{j} + (\boldsymbol{j} \cdot \nabla)\boldsymbol{v}_s + \boldsymbol{j}_0 \operatorname{div} \boldsymbol{v}_n + (\boldsymbol{v}_n \cdot \nabla)\boldsymbol{j}_0 = -\nabla P,$$

$$\frac{\partial S}{\partial t} + \operatorname{div}(S\boldsymbol{v}_n) = 0, \qquad \boldsymbol{j}_0 = \rho_n(\boldsymbol{v}_n - \boldsymbol{v}_s).$$

The equations are very complicated since the scalar quantities μ, ρ, S, and P entering them are the functions of relative velocity $\boldsymbol{v}_n - \boldsymbol{v}_s$. These quantities cannot be found within the framework of phenomenological approach and, thus, the microscopic theory should be attracted.

31. The sound wave represents a small perturbation of equilibrium state. Thus, the normal and superfluid component velocities $\boldsymbol{v}_n$ and $\boldsymbol{v}_s$ are assumed to be small as compared with the sound velocity and the thermodynamic quantities to be close to the equilibrium ones. Let us write the linearized hydrodynamic equations of superfluid liquid

$$\frac{\partial \rho}{\partial t} + \operatorname{div} \boldsymbol{j} = 0, \quad \frac{\partial \boldsymbol{j}}{\partial t} = -\nabla P; \quad \boldsymbol{j} = \rho_n \boldsymbol{v}_n + \rho_s \boldsymbol{v}_s \text{ and } \rho = \rho_n + \rho_s,$$

$$\frac{\partial(\sigma\rho)}{\partial t} + \sigma\rho \operatorname{div} \boldsymbol{v}_n = 0, \quad \frac{\partial \boldsymbol{v}_s}{\partial t} + \nabla\mu = 0; \quad \sigma = \frac{S}{\rho} \text{ and } \frac{\nabla P}{\rho} = \sigma\nabla T + \nabla\mu.$$

Here, for the further convenience, we have introduced the specific entropy σ instead of entropy density S according to $\sigma = S/\rho$.

Eliminating the momentum density $\boldsymbol{j}$ from the first two equations, we obtain the same equation relating the oscillations of density ρ and pressure P as for a non-superfluid liquid

$$\frac{\partial^2 \rho}{\partial t^2} = \nabla^2 P.$$

Using the thermodynamic relation $d\mu = -\sigma\, dT + dP/\rho$, we have from the second and fourth equations

$$\rho_n \frac{\partial}{\partial t}(\boldsymbol{v}_n - \boldsymbol{v}_s) = -\sigma\rho\nabla T.$$

Applying the first and third equation gives the following equation:

$$\rho_s \operatorname{div}(\boldsymbol{v}_n - \boldsymbol{v}_s) = -\frac{\rho}{\sigma}\frac{\partial \sigma}{\partial t}.$$

Eliminating the difference $(\boldsymbol{v}_n - \boldsymbol{v}_s)$ from the last two equations, we obtain the equation relating the oscillations of entropy σ and temperature T

$$\frac{\partial^2 \sigma}{\partial t^2} = \frac{\sigma^2 \rho_s}{\rho_n} \nabla^2 T.$$

Let us introduce the deviations of pressure δP and temperature δT from the equilibrium magnitudes. The deviations of density $\delta\rho$ and entropy $\delta\sigma$ from their equilibrium magnitudes can be written as

$$\delta\rho = \frac{\partial \rho}{\partial P}\delta P + \frac{\partial \rho}{\partial T}\delta T \quad \text{and} \quad \delta\sigma = \frac{\partial \sigma}{\partial P}\delta P + \frac{\partial \sigma}{\partial T}\delta T.$$

Substituting it into the equation for oscillations yields a set of two equations

$$\frac{\partial \rho}{\partial P}\frac{\partial^2 \delta P}{\partial t^2} - \nabla^2 \delta P = -\frac{\partial \rho}{\partial T}\frac{\partial^2 \delta T}{\partial t^2},$$
$$\frac{\partial \sigma}{\partial T}\frac{\partial^2 \delta T}{\partial t^2} - \frac{\sigma^2 \rho_s}{\rho_n}\nabla^2 \delta T = \frac{\partial \sigma}{\partial P}\frac{\partial^2 \delta P}{\partial t^2}.$$

As usual, we seek for the solution of equations as a plane wave propagating in the x-axis direction at frequency ω and velocity u. In other words, the oscillations of pressure $\delta P(x, t)$ and temperature $\delta T(x, t)$ are proportional to the general factor $\exp[-i\omega(t - x/u)]$. Then,

$$\left(u^2\frac{\partial \rho}{\partial P} - 1\right)\delta P + u^2\frac{\partial \rho}{\partial T}\delta T = 0,$$
$$u^2\frac{\partial \sigma}{\partial P}\delta P + \left(u^2\frac{\partial \sigma}{\partial T} - \frac{\sigma^2 \rho_s}{\rho_n}\right)\delta T = 0.$$

Equating the determinant of the system to zero results in the equation which determines the sound velocity u

$$u^4\left(\frac{\partial \rho}{\partial P}\frac{\partial \sigma}{\partial T} - \frac{\partial \rho}{\partial T}\frac{\partial \sigma}{\partial P}\right) - u^2\left(\frac{\partial \sigma}{\partial T} + \frac{\sigma^2 \rho_s}{\rho_n}\frac{\partial \rho}{\partial P}\right) + \frac{\sigma^2 \rho_s}{\rho_n} = 0.$$

A more obvious form of this equation can be given with using the thermodynamic relations between the specific heat at constant volume (constant density) $c_V = T(\partial\sigma/\partial T)_V$ and at constant pressure $c_P = T(\partial\sigma/\partial T)_P$

$$c_V = T\left(\frac{\partial \sigma}{\partial T}\right)_P - T\frac{\left(\frac{\partial \sigma}{\partial P}\right)_T\left(\frac{\partial \rho}{\partial T}\right)_P}{\left(\frac{\partial \rho}{\partial P}\right)_T}.$$

As a result, we get

$$\left(\frac{u^2}{u_1^2} - 1\right)\left(\frac{u^2}{u_2^2} - 1\right) = 1 - \frac{c_V}{c_P},$$

where we denote

$$\frac{1}{u_1^2} = \frac{c_V}{c_P}\left(\frac{\partial\rho}{\partial P}\right)_T = \left(\frac{\partial\rho}{\partial P}\right)_S,$$
$$\frac{1}{u_2^2} = \frac{c_V}{c_P}\frac{\rho_n(\partial\sigma/\partial T)_P}{\sigma^2\rho_s} = \frac{\rho_n(\partial\sigma/\partial T)_V}{\sigma^2\rho_s}.$$

The biquadratic equation obtained determines two possible sound velocities. The difference $c_P - c_V$ in the specific heats is proportional to the square of thermal expansion coefficient $\alpha_V = -\rho^{-1}(\partial\rho/\partial T)_P$. At low temperatures in superfluid ^{4}He the magnitude α_V is very small and we can make no distinction between the specific heats c_V and c_P, putting them practically equal with the exception of the close vicinity of λ-point of superfluid transition. This circumstance simplifies the solution of equation and its roots are equal to

$$u_1 = \sqrt{\left(\frac{\partial P}{\partial\rho}\right)_\sigma} \quad \text{and} \quad u_2 = \sqrt{\frac{\sigma^2\rho_s}{\rho_n\,(\partial\sigma/\partial T)_\rho}}.$$

The first root determines the velocity of *first sound* u_1 or usual sound which exists above the superfluid transition temperature as well. The pressure or density oscillations propagate at such velocity. On neglecting the thermal expansion coefficient $\alpha_V = 0$ the normal and superfluid component velocities coincide $\boldsymbol{v}_n = \boldsymbol{v}_s$ in the first sound wave, i.e. normal and superfluid components oscillate in-phase. In this sense the first sound is similar to the usual sound in normal fluid.

The second root u_2 determines the *second sound* velocity. The undamped temperature oscillations propagate at this velocity. The existence of such sound waves is a specific attribute of superfluids. At the superfluid transition point the second sound velocity vanishes as well as the superfluid density. For the low temperature $T \to 0$ limit when elementary excitations are phonons, the second sound velocity is $u_2 = u_1/\sqrt{3}$. In approximation $\alpha_V = 0$ the pressure and density in the second sound wave do not oscillate and the total mass flow is absent $\boldsymbol{j} = \rho_n\boldsymbol{v}_n + \rho_s\boldsymbol{v}_s = 0$. Therefore, the superfluid and normal components oscillate in antiphase and compensate mutually the mass flows of the both components.

Nonzero thermal expansion leads to the effect of coupling between first and second sounds. To the extent $\alpha_V \neq 0$ in first sound there also arise temperature oscillations along with the pressure and density oscillations and, strictly speaking, $\boldsymbol{v}_n \neq \boldsymbol{v}_s$. In its turn, in the second sound the pressure and density oscillations appear in addition to the temperature oscillations. The total mass flow in second sound does not vanish strictly, i.e. $\boldsymbol{j} \neq 0$.

32. The velocity of fourth sound will be determined from the linearized system of hydrodynamical equations for the superfluid liquid in which we put $\boldsymbol{v}_n = 0$ and, correspondingly, mass flow $\boldsymbol{j} = \rho_s\boldsymbol{v}_s$

$$\frac{\partial \rho}{\partial t} + \rho_s \mathrm{div}\, \boldsymbol{v}_s = 0, \quad \rho_s \frac{\partial \boldsymbol{v}_s}{\partial t} = -\nabla P \quad \text{and} \quad \rho = \rho_n + \rho_s,$$

$$\frac{\partial S}{\partial t} = 0, \quad \frac{\partial \boldsymbol{v}_s}{\partial t} + \nabla \mu = 0 \quad \text{and} \quad \frac{\nabla P}{\rho} = \sigma \nabla T + \nabla \mu.$$

Here $S = \sigma \rho$ is the entropy density, σ is the specific entropy, P is the pressure, and μ is the chemical potential. Eliminating the velocity $\boldsymbol{v}_s$ from the two first equations and comparing the second and fourth equations, we find

$$\frac{\partial^2 \rho}{\partial t^2} = \nabla^2 P \quad \text{and} \quad \nabla P = \rho_s \nabla \mu.$$

As a result, we arrive at the following relations:

$$\frac{\partial^2 \rho}{\partial t^2} = \rho_s \nabla^2 \mu \quad \text{or} \quad \frac{\partial^2 \rho}{\partial t^2} = \rho_s \left(\frac{\partial \mu}{\partial \rho}\right)_S \nabla^2 \rho.$$

Deriving the last relation, we have used the condition of constancy for the entropy density $S = \text{const}$, resulting from $\boldsymbol{v}_n = 0$, and $\delta\mu = (\partial\mu/\partial\rho)_S\, \delta\rho$ at the small density perturbations. The latter equation, representing the wave equation, determines the square of fourth sound velocity

$$u_4^2 = \rho_s \left(\frac{\partial \mu}{\partial \rho}\right)_S .$$

Below we transform this derivative to more obvious form. Using the identity $d\mu = -\sigma\, dT + dP/\rho$, we have

$$\left(\frac{\partial \mu}{\partial \rho}\right)_S = -\sigma \left(\frac{\partial T}{\partial \rho}\right)_S + \frac{1}{\rho}\left(\frac{\partial P}{\partial \rho}\right)_S .$$

The transformation of the first integral is performed with the aid of Jacobian

$$\left(\frac{\partial T}{\partial \rho}\right)_S = \frac{\partial(T, S)}{\partial(\rho, S)} = \frac{\partial(T, S)}{\partial(\rho, \sigma)} \frac{\partial(\rho, \sigma)}{\partial(\rho, S)} = \left[\rho \left(\frac{\partial T}{\partial \rho}\right)_\sigma - \sigma \left(\frac{\partial T}{\partial \sigma}\right)_\rho\right] \frac{1}{\rho}.$$

Next, we need to transform the first term in the square brackets

$$\left(\frac{\partial T}{\partial \rho}\right)_\sigma = \frac{\partial(T, \sigma)}{\partial(\rho, \sigma)} = \frac{\partial(T, \sigma)}{\partial(T, P)} \frac{\partial(T, P)}{\partial(\sigma, P)} \frac{\partial(\sigma, P)}{\partial(\rho, \sigma)}$$

$$= -\left(\frac{\partial \sigma}{\partial P}\right)_T \left(\frac{\partial T}{\partial \sigma}\right)_P \left(\frac{\partial P}{\partial \rho}\right)_\sigma = -\frac{1}{\rho^2} \left(\frac{\partial \rho}{\partial T}\right)_P \left(\frac{\partial T}{\partial \sigma}\right)_P \left(\frac{\partial P}{\partial \rho}\right)_\sigma .$$

Here, when going over to the last relation, we use a succession of the following equalities:

$$\left(\frac{\partial\sigma}{\partial P}\right)_T = -\frac{\partial^2\mu}{\partial P\partial T} = -\frac{\partial^2\mu}{\partial T\partial P} = -\frac{\partial}{\partial T}\left(\frac{1}{\rho}\right)_P = \frac{1}{\rho^2}\left(\frac{\partial\rho}{\partial T}\right)_P .$$

Collecting all the relations together, we get a simple answer

$$\left(\frac{\partial\mu}{\partial\rho}\right)_S = \frac{1}{\rho}\left(\frac{\partial P}{\partial\rho}\right)_S + \frac{\sigma^2}{\rho}\left(\frac{\partial T}{\partial\sigma}\right)_\rho + \frac{\sigma}{\rho^2}\left(\frac{\partial\rho}{\partial T}\right)_P\left(\frac{\partial T}{\partial\sigma}\right)_P\left(\frac{\partial P}{\partial\rho}\right)_\sigma .$$

The derivative $\partial P/\partial\rho$ under constant entropy density $S = \rho\sigma$ is connected with the derivative $\partial P/\partial\rho$ at constant specific entropy σ as follows:

$$\frac{(\partial P/\partial\rho)_S}{(\partial P/\partial\rho)_\sigma} = \frac{\partial(P,S)}{\partial(\rho,S)}\frac{\partial(\rho,\sigma)}{\partial(P,\sigma)} = \frac{\partial(P,S)}{\partial(P,\sigma)}\frac{\partial(\rho,\sigma)}{\partial(\rho,S)} = 1 + \frac{\sigma}{\rho}\left(\frac{\partial\rho}{\partial\sigma}\right)_P$$

$$= 1 + \frac{\sigma}{\rho}\left(\frac{\partial\rho}{\partial T}\right)_P\left(\frac{\partial T}{\partial\sigma}\right)_P = 1 + \frac{\sigma T}{\rho C_P}\left(\frac{\partial\rho}{\partial T}\right)_P = 1 - \frac{\sigma T}{C_P}\alpha_V .$$

Here C_P is the specific heat at constant pressure and α_V is the thermal expansion coefficient.

As a result, we find that

$$\rho_s\left(\frac{\partial\mu}{\partial\rho}\right)_S = \frac{\rho_s}{\rho}\left(\frac{\partial P}{\partial\rho}\right)_\sigma\left(1 - \frac{2\sigma T}{C_P}\alpha_V\right) + \frac{\rho_s}{\rho}\frac{\sigma^2}{(\partial\sigma/\partial T)_\sigma} .$$

Putting zero magnitude of thermal expansion coefficient α_V as a good approximation at low temperatures and introducing the velocities of first and second sounds

$$u_1 = \sqrt{\left(\frac{\partial P}{\partial\rho}\right)_\sigma} \quad \text{and} \quad u_2 = \sqrt{\frac{\sigma^2\rho_s}{\rho_n\,(\partial\sigma/\partial T)_\rho}} ,$$

we arrive at the final answer for the velocity of fourth sound

$$u_4^2 = \frac{\rho_s}{\rho}u_1^2 + \frac{\rho_n}{\rho}u_2^2 .$$

In superfluid ^{4}He the second term is small compared with the first one practically for all temperatures.

3.13 Kapitza Resistance and Kinetic Phenomena at the Superfluid-Solid Helium Interface

33. A phonon incident onto the interface may either transmit across it to the other medium or not. The phonon transmission coefficient $t(\boldsymbol{k}) = t(k, \vartheta)$ in our case may depend on both wave vector k and angle ϑ between the phonon propagation direction and the normal $\boldsymbol{n}$ to the interface (Fig. 3.12). Let the coordinate z-axis run in the direction of the normal from medium 1 to medium 2. Then we write down the energy flux $Q = Q(T)$ transferred with the phonons from medium 1 to medium 2 across the unit area of plane interface at $z = 0$

$$Q(T) = \int\limits_{(\boldsymbol{u}\cdot\boldsymbol{n})>0} t_{1\to 2}(\boldsymbol{k})(\boldsymbol{u}\cdot\boldsymbol{n})\,\varepsilon_{\boldsymbol{k}} N_{\omega_{\boldsymbol{k}}}(T)\frac{d^3k}{(2\pi)^3}\,, \quad \boldsymbol{u} = \frac{\partial\omega(\boldsymbol{k})}{\partial\boldsymbol{k}}\,.$$

Here $(\boldsymbol{u}\cdot\boldsymbol{n}) = u_z$ is the phonon velocity component normal to the interface, $\varepsilon_{\boldsymbol{k}} = \hbar\omega_{\boldsymbol{k}}$ is the phonon energy, and $N_\omega(T) = [\exp(\hbar\omega/T) - 1]^{-1}$ is the equilibrium phonon distribution function or Planck function at temperature T.

The similar expression can also be written for the opposite energy flux $Q' = Q'(T')$ transferred with the phonons from medium 2 of temperature T' to medium 1. This process will be characterized by the corresponding transmission coefficient $t_{2\to 1}(\boldsymbol{k})$. Under assumptions of the problem, in general, there is no necessity to calculate the flux $Q'(T')$. The point is that for the temperature T of medium 1, there is a thermal equilibrium due to identical temperature of both media. This means that the energy flux $Q'(T')$ is completely compensated with the counterflow $Q(T')$ so that $Q'(T') + Q(T') = 0$. Therefore, it is sufficient to restrict our treatment with the single medium, for example, occupying the left-hand half-space $z < 0$. The result is the following expression for the total resultant energy flux $\Delta Q(T, T')$:

$$\begin{aligned}\Delta Q(T, T') &= Q(T) - Q(T') \\ &= \int\limits_{(\boldsymbol{u}\cdot\boldsymbol{n})>0} t_{1\to 2}(\boldsymbol{k})(\boldsymbol{u}\cdot\boldsymbol{n})\,\varepsilon_{\boldsymbol{k}}\big[N_{\omega_{\boldsymbol{k}}}(T) - N_{\omega_{\boldsymbol{k}}}(T')\big]\frac{d^3k}{(2\pi)^3}\,.\end{aligned}$$

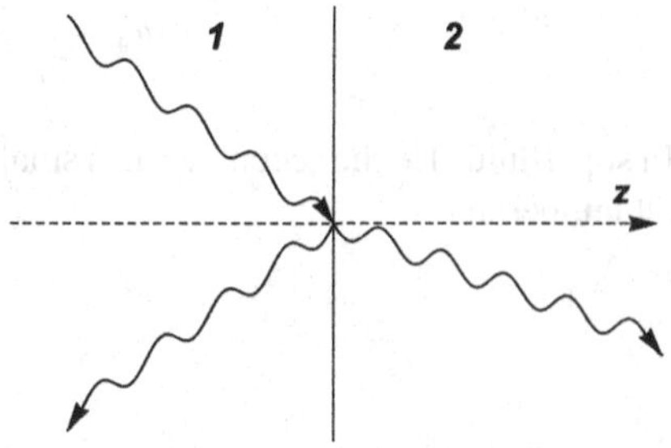

Fig. 3.12 The reflection and refraction of longitudinal sound wave incident onto the interface from medium 1 to medium 2. The densities of media 1 and 2 are ρ and ρ', respectively

For the same sound velocities of the adjacent media, the formula for the transmission coefficient $t_{1\to 2}$ simplifies significantly. In this case, if $u = u'$, the refraction angle equals the incident angle. The transmission coefficient of sound wave is independent of incidence angle and is given by the simple formula

$$t_{1\to 2} = \frac{4\rho\rho'}{(\rho+\rho')^2}\,.$$

The further calculation yields the following integral for the acoustic phonon dispersion $\omega(k) = uk$:

$$\Delta Q(T, T')$$
$$= \frac{4\rho\rho'}{(\rho+\rho')^2}\int_0^{k_D}\frac{k^2\,dk}{8\pi^3}\int_0^{\pi/2}\sin\vartheta\,d\vartheta\int_0^{2\pi}d\varphi\,u\cos\vartheta\left[\frac{\hbar uk}{e^{\hbar uk/T}-1}-\frac{\hbar uk}{e^{\hbar uk/T'}-1}\right]$$
$$= \frac{4\rho\rho'}{(\rho+\rho')^2}\frac{1}{8\pi^2\hbar^3u^2}\left(T^4\int_0^{\hbar\omega_D/T}\frac{x^3\,dx}{e^x-1}-T'^4\int_0^{\hbar\omega_D/T'}\frac{x^3\,dx}{e^x-1}\right),$$

where $\omega_D = uk_D$ is the Debye frequency corresponding to the maximum possible phonon frequency. At low temperature the routine approximation is to put the upper integration limit equal to the infinity. Then we obtain

$$\Delta Q(T, T') = \frac{\pi^2}{30}\frac{\rho\rho'}{(\rho+\rho')^2}\left(\frac{T^4}{\hbar^3u^2}-\frac{T'^4}{\hbar^3u^2}\right).$$

For the small temperature difference $\Delta T = T - T'$, we arrive at the formula required $\Delta Q = \Delta T/R_K$ and thermal Kapitza resistance

$$R_K = \frac{15}{2\pi^2}\frac{(\rho+\rho')^2}{\rho\rho'}\frac{\hbar^3u^2}{T^3}\sim\frac{1}{T^3}\,.$$

The dimensional estimate of the interface thermal conductance R_K^{-1} can be represented as follows:

$$R_K^{-1}\sim\frac{t}{4}C(T)u,\quad C(T) = \frac{2\pi^2}{15}\frac{T^3}{\hbar^3u^3}\,.$$

Here $C(T)$ is the phonon specific heat, u is the sound velocity, and t is a numerical coefficient depending on the conditions of transmitting and reflecting the sound modes of the adjacent media.

34. Let choose the z-axis normal to the plane interface separating the solid and liquid phases. The solid phase occupies the region $z < 0$ and the liquid one does

$z > 0$. Under phase equilibrium at the interface $z = 0$ we have the equalities of pressures $P' = P$ and chemical potentials $\mu'(P' = P_c) = \mu(P = P_c)$, where P_c is the pressure at which the solid and liquid phase coexist.

Let $z = \zeta(x, t) = \zeta e^{ikx - i\omega t}$ represent small displacement of the interface from its plane position $z = 0$ in the x-axis direction with the given wave vector k and frequency ω. Due to various densities of liquid and solid phases the process of melting or crystallization will be accompanied with the liquid flow in the direction to the interface or away from it. The motion of incompressible liquid at velocity $\boldsymbol{v} = \boldsymbol{v}(z)\exp(ikx - i\omega t)$ can readily be described with the aid of velocity potential ϕ according to $\boldsymbol{v} = \nabla\phi$ and satisfying the equation

$$\nabla^2\phi = \frac{\partial^2\phi}{\partial x^2} + \frac{\partial^2\phi}{\partial z^2} = 0$$

due to continuity equation div $\boldsymbol{v} = 0$ for incompressible liquid. Selecting the solution for the liquid phase velocity decaying at $z \to \infty$, we obtain

$$\phi(x, z, t) = -\frac{V}{k}e^{-kz}e^{ikx - i\omega t},$$

where V is the magnitude of the z-component of velocity at $z = 0$.

Let us write the continuity equation of mass flow across the interface $z = \zeta(x, t)$ which moves at rate $\dot{\zeta}$

$$\rho'(v'_\nu - \dot{\zeta})_{z=\zeta} = \rho(v_\nu - \dot{\zeta})_{z=\zeta}\,.$$

Here v'_ν and v_ν are the velocity components normal to the interface. Remaining within the linear approximation in ζ, there is no necessity in distinction between the z and normal components of velocities, i.e. $v'_\nu \approx v'_z$, $v_\nu \approx v_z$. In the same linear approximation of small amplitude ζ we can put $z = 0$ in place of $z = \zeta$ in the condition above. Recalling that velocity is $\boldsymbol{v}' = 0$ in the solid phase, we have

$$-\rho'\dot{\zeta} = \rho(V - \dot{\zeta}) \quad \text{or} \quad V = -\frac{\rho' - \rho}{\rho}\dot{\zeta} = i\omega\frac{\rho' - \rho}{\rho}\zeta(x, t).$$

Next, it is necessary to know the pressure magnitudes in the both phases at the interface. The pressure P in the liquid will be found with the aid of the linearized Euler equation for an ideal fluid

$$\rho\frac{\partial \boldsymbol{v}}{\partial t} = -\nabla P \quad \text{where} \quad \boldsymbol{v} = \nabla\phi.$$

Hence we have

$$P(x, z, t) = P_c - \rho\dot{\phi} = P_c + i\omega\phi = P_c + \frac{\rho' - \rho}{\rho}\frac{\omega^2}{k}e^{-kz}\zeta(x, t),$$

where P_c is the phase equilibrium pressure corresponding to the plane interface at $\zeta(x, t) = 0$. The pressure P' in the immobile solid phase, in general, differs from the phase transition pressure P_c.

The continuity condition for the momentum flux density across the interface on neglecting the squares of interface rate and fluid velocity results in the Young-Laplace formula for the mechanical interface equilibrium. Taking the curvature of interface $z = \zeta(x, t)$ into account, we have

$$\left(P' - P\right)_{z=\zeta} = -\alpha\frac{\partial^2\zeta}{\partial x^2} = \alpha k^2\zeta(x, t).$$

For the linear approximation in oscillation amplitude ζ, it is sufficient to calculate the pressure difference at point $z = 0$. Then we get the following relation:

$$P' = P_c + \left(\frac{\rho' - \rho}{\rho}\frac{\omega^2}{k} + \alpha k^2\right)\zeta(x, t).$$

Finally, we apply an approximation for the infinitely high rate of phase transition when the phase equilibrium, corresponding to the equal chemical potentials of the phases, holds for all time. The small variation of chemical potential $\mu(P)$ per unit mass equals $\delta\mu = \delta P/\rho$ as the pressure varies from P to $P + \delta P$. As above, the pressure magnitudes can be taken at $z = 0$ instead of $z = \zeta$. So, involving equality $\mu(P_c) = \mu'(P_c)$ for the phase transition pressure P_c, we obtain

$$\mu(P) - \mu(P') = \frac{P - P_c}{\rho} - \frac{P' - P_c}{\rho'}$$

$$= \frac{1}{\rho}\frac{\rho' - \rho}{\rho}\frac{\omega^2}{k}\zeta - \frac{1}{\rho'}\left(\frac{\rho' - \rho}{\rho}\frac{\omega^2}{k} + \alpha k^2\right)\zeta = \left(\frac{(\rho' - \rho)^2}{\rho\rho'}\frac{\omega^2}{k} - \frac{\alpha k^2}{\rho'}\right)\zeta = 0.$$

Hence, the relation between the frequency ω and wave vector k or *crystallization wave spectrum* is given by

$$\omega^2 = \frac{\alpha}{\rho_{\text{eff}}}k^3 \quad \text{where} \quad \rho_{\text{eff}} = \frac{(\rho' - \rho)^2}{\rho}.$$

Thus, at sufficiently low temperature the weakly damping waves of melting and crystallization can propagate along the superfluid-crystal ^{4}He interface with the dispersion similar to that of capillary waves propagating along the boundary between two immiscible liquids. The distinction is only in the effective interface density ρ_{eff} which would be equal to $\rho_{\text{eff}} = \rho + \rho'$ in the case of immiscible liquids.

If the effect of gravity force, characterized by the free fall acceleration g, is taken into attention, the crystallization wave spectrum becomes similar to that of *capillary-gravitational waves*

$$\rho_{\text{eff}}\frac{\omega^2}{k} = \alpha k^2 + (\rho' - \rho)g.$$

In the region of sufficiently small wavelengths $\lambda < \lambda_0$ the influence of gravity force can be neglected. The typical length λ_0 is determined by the *capillary length* $\lambda_0 = \sqrt{\alpha/g(\rho' - \rho)}$. In ^{4}He this capillary length equals approximately 1 mm.

35. For solving the problem, we use the results of the previous problem. Let put also that $z = \zeta(x, t) = \zeta e^{ikx - i\omega t}$ is a small displacement of the plane interface from its position at $z = 0$. Next, we use the previous result for the chemical potential difference at the liquid-solid interface

$$\Delta\mu = \mu(P) - \mu(P') = \left(\frac{(\rho' - \rho)^2}{\rho\rho'}\frac{\omega^2}{k} - \frac{\alpha k^2}{\rho'}\right)\zeta(x, t).$$

Equating the difference $\Delta\mu$ to $\dot{\zeta}/K = -i\omega\zeta/K$ yields the following dispersion equation:

$$\rho_{\text{eff}}\frac{\omega^2}{k} = \alpha k^2 - i\omega\frac{\rho'}{K} \quad \text{where} \quad \rho_{\text{eff}} = \frac{(\rho' - \rho)^2}{\rho}.$$

Thus, the finiteness of kinetic growth coefficient K entails the damping of crystallization waves. Assuming the damping to be small, we find

$$\omega(k) \approx \left(\frac{\alpha k^3}{\rho_{\text{eff}}}\right)^{1/2} - \frac{i}{2}\frac{k\rho'}{\rho_{\text{eff}} K} = \omega_0(k) - i\gamma(k).$$

Accordingly, the wave amplitude will decay in time as $e^{-\gamma t}$ and with distance as $e^{-\varkappa x}$ where

$$\varkappa(\omega) = \frac{1}{3}\left(\frac{\omega}{\rho_{\text{eff}}\alpha^2}\right)^{1/3}\frac{\rho'}{K}$$

and $\varkappa(\omega)$ is given according to $\varkappa = \gamma/u$, $u = \partial\omega/\partial k$ being the phase velocity. The measurement of damping allows one to determine the growth coefficient. For $K < K_0 = \rho'/(4\alpha k\rho_{\text{eff}})^{1/2}$, the damping becomes so strong that the interface perturbation with wave vector k has a diffusive aperiodic character.

The dispersion equation of crystallization waves can be treated from the mechanical point of view if it is rewritten with $\zeta(t) \sim \zeta_k e^{-i\omega t}$ as

$$\rho_{\text{eff}}\frac{\ddot{\zeta}_k(t)}{k} + \alpha k^2\zeta_k(t) = -\frac{\rho'}{K}\dot{\zeta}_k(t).$$

Multiplying the both side of equation by $\dot{\zeta}_K(t)$, we have

$$\frac{dH}{dt} = \frac{d}{dt}\left(\frac{1}{2}\rho_{\text{eff}}\frac{\dot{\zeta}_k^2(t)}{k} + \frac{1}{2}\alpha k^2\zeta_k^2(t)\right) = -\frac{\rho'}{K}\dot{\zeta}_k^2(t) = -2R.$$

Here H is the total interface energy per unit area and is a sum of kinetic and potential energies. (In the field of gravity under acceleration g the gravitational energy $(\rho' - \rho)g\zeta_k^2(t)/2$ should be added to the surface tension term.) The energy dissipation rate of total interface energy equals twice the dissipative Rayleigh function R. Since the energy dissipation processes lead to decreasing the energy, the kinetic growth coefficient should be positive, i.e. $K > 0$.

36. Let us consider the liquid-solid phase conversion and let the superfluid-crystal ^{4}He interface grow a constant rate $\dot{\boldsymbol{\zeta}} = \dot{\zeta}\boldsymbol{e}_z$ in the z-axis direction normal to the interface. The z-axis is directed into the liquid phase.

Below we seek for the force exerted to the interface and resulted from the phonon excitations. For this purpose, we should find the momentum transmitted with the phonons of superfluid ^{4}He to the interface per the unit time. The growth of a crystal or its melting will induce the force $\boldsymbol{f}$ additional to the pressure P and proportional to the interface growth rate $\dot{\zeta}$

$$f = -\frac{\rho'}{K}\dot{\zeta}\,.$$

According to the previous problem, the work of this force per unit time

$$-2R = \boldsymbol{f}\dot{\boldsymbol{\zeta}} = -\rho'\dot{\zeta}^2/K$$

determines the energy dissipation power at interface $dH/dt = -2R$ and quantity K has a meaning of interface growth coefficient.

The calculation of force $\boldsymbol{f}$ is better to perform in the reference frame in which the interface is at rest. Then the superfluid phase has a relative velocity $\boldsymbol{v}$ which can be found from the continuity condition for the mass flow across the interface

$$\rho\boldsymbol{v} = -\rho'\dot{\boldsymbol{\zeta}} \quad \text{or} \quad \boldsymbol{v} = -\dot{\boldsymbol{\zeta}}\rho'/\rho.$$

Since the superfluid phase moves as a whole at velocity $\boldsymbol{v}$, it is necessary to recall the energy of elementary excitations (phonons) experiencing the Doppler shift in accordance with the Galilean principle

$$\varepsilon(\boldsymbol{p}) \to \varepsilon(\boldsymbol{p}) + \boldsymbol{p}\boldsymbol{v} \quad \text{or} \quad \frac{\partial\varepsilon(\boldsymbol{p})}{\partial\boldsymbol{p}} \to \frac{\partial\varepsilon(\boldsymbol{p})}{\partial\boldsymbol{p}} + \boldsymbol{v}.$$

The condition for the mirror reflection of phonons from the interface means the following. For such phonon coupling with the interface, the phonon momentum components parallel to the interface plane remain unchanged but the momentum component normal to the interface changes its direction to the opposite one. Thus, after reflection of a phonon from the interface we have for the phonon energy

$$\varepsilon(p) + p_z v \to \varepsilon(p) - p_z v \quad \text{where} \quad v = -\dot{\zeta}\rho'/\rho.$$

Let us turn to calculating the force exerted to the interface. It follows from the symmetry of the problem that the resultant force must be directed along the interface normal or the z-axis. Let us write the momentum flux carried with the phonons per the unit time and unit interface area

$$\Pi_{\text{in}} = \int\limits_{u_z<0} u_z p_z n(\varepsilon_p + p_z v)\frac{d^3p}{(2\pi\hbar)^3}$$

$$\approx \int\limits_{u_z<0} u_z p_z n(\varepsilon_p)\frac{d^3p}{(2\pi\hbar)^3} + v \int\limits_{u_z<0} u_z p_z^2 \frac{\partial n}{\partial \varepsilon_p}\frac{d^3p}{(2\pi\hbar)^3}\,.$$

Here u_z is the z-component of phonon velocity $\boldsymbol{u}$ and $n(\varepsilon) = 1/(e^{\varepsilon/T} - 1)$ is the equilibrium Planck distribution function of phonons at temperature T. The analogous momentum flux, carried away with the phonons after reflection, will be equal to

$$\Pi_{\text{out}} = \int\limits_{u_z>0} u_z p_z n(\varepsilon_p - p_z v)\frac{d^3p}{(2\pi\hbar)^3}$$

$$\approx \int\limits_{u_z>0} u_z p_z n(\varepsilon_p)\frac{d^3p}{(2\pi\hbar)^3} - v \int\limits_{u_z>0} u_z p_z^2 \frac{\partial n}{\partial \varepsilon_p}\frac{d^3p}{(2\pi\hbar)^3}\,.$$

The total force F, exerted with phonons to the interface, is determined by a sum of fluxes Π_{in} and Π_{out} and consists of two terms. The term, independent of the interface growth rate, represents an additional pressure induced with the phonon gas

$$P_{ph}(T) = \int\limits_{u_z<0} u_z p_z n(\varepsilon_p)\frac{d^3p}{(2\pi\hbar)^3} + \int\limits_{u_z>0} u_z p_z n(\varepsilon_p)\frac{d^3p}{(2\pi\hbar)^3} = \frac{\pi^2}{90}\frac{T^4}{u^3\hbar^3}$$

and has no interest here since it is a correction to the chemical potential difference between the solid and liquid phases.

The second term proportional to the growth rate of the solid phase $\dot{\zeta}$

$$f = v \int\limits_{u_z<0} u_z p_z^2 \frac{\partial n}{\partial \varepsilon_p}\frac{d^3p}{(2\pi\hbar)^3} - v \int\limits_{u_z>0} u_z p_z^2 \frac{\partial n}{\partial \varepsilon_p}\frac{d^3p}{(2\pi\hbar)^3}$$

represents the drag force desired. The force prevents from the solid phase growth and has a dissipative character. Note that the first integral equals the second one with the opposite sign. Taking $\varepsilon = up$ into account gives

$$f = 2v \int_0^\infty \frac{2\pi p^2\, dp}{(2\pi\hbar)^3} \int_0^{\pi/2} \sin\theta\, d\theta\, \cos^3\theta\, \frac{up^2}{T} \frac{e^{up/T}}{(e^{up/T} - 1)^2}$$

$$= \left(-\frac{\dot{\zeta}\rho'}{\rho}\right) \frac{T^4}{8\pi^2 u^4 \hbar^3} \int_0^\infty \frac{x^4 e^x}{(e^x - 1)^2}\, dx = -\rho'\dot{\zeta}\, \frac{\pi^2}{30} \frac{T^4}{\rho u^4 \hbar^3}\,.$$

The work $f\dot{\zeta}$, performed with drag force f per unit time, determines the power of interface energy dissipation and relates with the kinetic growth coefficient K as

$$f\dot{\zeta} = -\rho'\dot{\zeta}^2/K.$$

Hence we find the kinetic growth coefficient for the superfluid-crystal ^{4}He interface

$$K(T) = \frac{30}{\pi^2} \frac{\rho u^4 \hbar^3}{T^4}$$

which enhances drastically as the temperature lowers. The kinetic growth coefficient obtained can be expressed in terms of the phonon fraction of the normal component in superfluid ^{4}He

$$K^{-1}(T) = \frac{3}{4} \frac{\rho_{n,ph}(T)}{\rho}\, u \quad \text{where} \quad \rho_{n,ph}(T) = \frac{2\pi^2}{45} \frac{T^4}{u^5 \hbar^3}\,.$$

Thereby, the finiteness of kinetic growth coefficient is due to the presence of normal excitations in the medium and dissipative effects associated with excitations. The dimensionality of inverse growth coefficient or *growth resistance* coincides with the dimensionality of velocity. The magnitude of growth resistance is about typical velocity of elementary excitations multiplied by a ratio of elementary excitation density to the total density of condensed medium.

37. Let a plane sound wave propagate from liquid phase 1 in the half-space $z < 0$ across the interface to the liquid phase 2 in the half-space $z > 0$. In phase 1 we have both the incident and reflected waves. In phase 2 there is the transmitted wave alone. Accordingly, let us write the pressure modulations δP_1 and δP_2 with respect to equilibrium pressure P_0 in phase 1 ($z < 0$) and phase 2 ($z > 0$)

$$\delta P_1(z,t) = \delta P_0 e^{ik_1 z - i\omega t} + \delta P_r e^{-ik_1 z - i\omega t}, \quad k_1 = \omega/u_1\,,$$
$$\delta P_2(z,t) = \delta P_\tau e^{ik_2 z - i\omega t}, \qquad k_2 = \omega/u_2,$$

where δP_0, δP_r, and δP_τ are the amplitudes of incident, reflected, and transmitted waves.

Then we express the chemical potential difference at the interface in terms of pressure modulations

$$\mu_1 - \mu_2 = \mu_1(P_0 + \delta P_1) - \mu_2(P_0 + \delta P_2) \approx \frac{\delta P_1}{\rho_1} - \frac{\delta P_2}{\rho_2}.$$

Next, we involve the continuity condition for the momentum flux across the interface. For the linear approximation in the pressure modulations and interface rate $\dot{\zeta}$, this requirement reduces to a simple condition of mechanical equilibrium, i.e. to the equality of pressures at the interface[6]

$$P_0 + \delta P_1 = P_0 + \delta P_2 \quad \text{or} \quad \delta P_0 + \delta P_r = \delta P_\tau .$$

The last boundary condition we use is a continuity of mass flow J across the interface. In addition, we will need the relation of mass flow J with the chemical potential difference

$$J = \rho_1(v_1 - \dot{\zeta}) = \rho_2(v_2 - \dot{\zeta}),$$

$$J = \xi\left(\frac{\rho_1\rho_2}{\rho_1 - \rho_2}\right)^2\left(\frac{\delta P_1}{\rho_1} - \frac{\delta P_2}{\rho_2}\right).$$

Here $\dot{\zeta}$ is the interface growth rate and v_1, v_2 are the normal velocity components in the both liquids. In the same lowest linear approximation these components coincide with the values of the z-components of the velocities at point $z = 0$.

In order to connect the pressure modulations with the fluid velocity, we will use the linearized Euler equation

$$\rho\partial \boldsymbol{v}/\partial t = -\nabla P \quad \text{or} \quad -i\omega\rho\boldsymbol{v} = -\nabla P.$$

Applying this relation for liquids 1 and 2, we obtain the following relations at $z = 0$:

$$\rho_1 u_1 v_1 = Y_1 v_1 = \delta P_0 - \delta P_r ,$$

$$\rho_2 u_2 v_2 = Y_2 v_2 = \delta P_\tau .$$

The quantity $Y = \rho u$, being a product of density ρ by sound velocity u, represents the *acoustic impedance*.

Now we are in position to solve the problem required. Using the relations obtained, we have

$$\xi\frac{\rho_2}{\rho_2 - \rho_1}\delta P_\tau = v_1 - \dot{\zeta} = \frac{\delta P_0 - \delta P_r}{Y_1} - \dot{\zeta},$$

$$\xi\frac{\rho_1}{\rho_2 - \rho_1}\delta P_\tau = v_2 - \dot{\zeta} = \frac{\delta P_\tau}{Y_2} - \dot{\zeta}.$$

[6] For the linear approximation in small interface distortion, there is no necessity to discern the perturbed interface position $\zeta(t) \sim \zeta e^{-i\omega t}$ from unperturbed one $z = 0$.

Hence, from $\delta P_0 + \delta P_r = \delta P_\tau$ we find the acoustic coefficients of reflection r and transmission τ for the sound wave

$$\tau = \frac{\delta P_\tau}{\delta P_0} = \frac{2Y_2}{Y_1 + Y_2 + Y_1 Y_2 \xi},$$
$$r = \frac{\delta P_r}{\delta P_0} = \frac{Y_2 - Y_1 - Y_1 Y_2 \xi}{Y_1 + Y_2 + Y_1 Y_2 \xi}.$$

Let us discuss two limiting cases $\xi = 0$ and $\xi = \infty$ for the kinetic growth coefficient. The first case $\xi = 0$ implies no phase conversion at the interface and, in essence, corresponds to the boundary between two immiscible liquids. This case is traditionally discussed in the acoustic problems.

In the second limiting case $\xi = \infty$ (i.e. $Y_{1,2}\xi \gg 1$) we see that the sound incident onto the interface cannot transmit to the second phase $\tau = 0$ and reflects completely from the interface $r = -1$ with an additional phase equal to π. As a result of the fast phase conversion at the interface, the phase equilibrium $\mu_1 = \mu_2$ is kept all the time. Accordingly, the modulations of sound wave pressure cannot be transmitted to the second phase in any way and the pressure in the second phase remains unchanged and equal to the phase equilibrium pressure P_0. Thus in the second phase there is no any excitation of pressure oscillations, entailing no transmitted sound wave, i.e. $\tau = 0$.

The fraction

$$\epsilon = 1 - |r|^2 - \frac{Y_1}{Y_2}|\tau|^2 = \frac{2Y_1 Y_2^2(\xi + \xi^*)}{|Y_1 + Y_2 + Y_1 Y_2 \xi|^2}.$$

determines a loss of sound energy flux at the interface in the process of sound reflection and transmission at the interface. This fraction of sound wave energy dissipates due to irreversible processes associated with the conversion of the substance from one phase to another. The energy dissipation is related to the real part of kinetic growth coefficient $\xi = \xi_\omega$ and is absent only in two cases. The first is the case of two immiscible liquids when there is no phase conversion $\xi = 0$. The second is the case of the infinite growth coefficient $\xi = \infty$ when the phase conversion rate is so high that the phase equilibrium has no time to break down.

3.14 Collisionless Plasma: Permittivity and the Longitudinal and Transverse Oscillations

38. Let $n(\boldsymbol{r}, \boldsymbol{p}, t)$ be the distribution function of particles, i.e. $dN = n(\boldsymbol{r}, \boldsymbol{p}, t)d^3r d\Gamma_p$ is the number of particles which the coordinates and momenta lie within the intervals $\boldsymbol{r} \div \boldsymbol{r} + d\boldsymbol{r}$ and $\boldsymbol{p} \div \boldsymbol{p} + d\boldsymbol{p}$ at the time moment t. Two spin projections lead to $d\Gamma_p = 2d^3p/(2\pi\hbar)^3$. Let us write the transport equation

$$\frac{\partial n}{\partial t}+\boldsymbol{v}\frac{\partial n}{\partial \boldsymbol{r}}+\boldsymbol{F}\frac{\partial n}{\partial \boldsymbol{p}}=\mathrm{St}\,[n].$$

Here $\boldsymbol{v}$ is the particle velocity and $\boldsymbol{F}=q(\boldsymbol{E}+[\boldsymbol{v}\times\boldsymbol{B}]/c)$ is the Lorentz force exerted to the particle of charge q. Treating the plasma as *collisionless*, we should neglect the collision term $\mathrm{St}\,[n]$ which equals $\delta n/\tau$ for the small deviations $\delta n = n - n_0$ from the equilibrium distribution function $n_0(\boldsymbol{p})$, τ being the typical collision time. A possibility of such approximation arises provided that

$$\frac{\partial\,\delta n}{\partial t}\sim\omega\delta n\gg\frac{\delta n}{\tau}\quad\text{or}\quad \boldsymbol{v}\frac{\partial\,\delta n}{\partial \boldsymbol{r}}\sim v\frac{\delta n}{L}\gg\frac{\delta n}{\tau}.$$

Here ω is a typical frequency and L is the typical length at which the electric $\boldsymbol{E}$ and magnetic $\boldsymbol{B}$ fields vary in time and space. Thus, it is necessary to realize the high-frequency situation $\omega\tau\gg 1$ or large mean free path of plasma particles $l=v\tau\gg L$.

So, in the collisionless plasma we have the following transport equations for the distribution functions of electrons with charge $(-e)$ and ions with charge Ze:

$$\frac{\partial n}{\partial t}+\boldsymbol{v}\frac{\partial n}{\partial \boldsymbol{r}}-e\big(\boldsymbol{E}+\frac{1}{c}[\boldsymbol{v}\times\boldsymbol{B}]\big)\frac{\partial n}{\partial \boldsymbol{p}}=0,$$
$$\frac{\partial n_i}{\partial t}+\boldsymbol{v}\frac{\partial n_i}{\partial \boldsymbol{r}}+Ze\big(\boldsymbol{E}+\frac{1}{c}[\boldsymbol{v}\times\boldsymbol{B}]\big)\frac{\partial n_i}{\partial \boldsymbol{p}}=0.$$

The charge density $\rho(\boldsymbol{r},t)$ and current density $\boldsymbol{j}(\boldsymbol{r},t)$ are expressed with the aid of distribution functions as

$$\rho=e\int(Zn_i-n)\,d\Gamma_{\boldsymbol{p}}\quad\text{and}\quad \boldsymbol{j}=e\int\boldsymbol{v}(Zn_i-n)\,d\Gamma_{\boldsymbol{p}}.$$

In order to get a closed set of equations, we should include the Maxwell equations into consideration. The Maxwell equations determine the evolution of electric and magnetic fields under electric charges and currents

$$\operatorname{div}\boldsymbol{E}=4\pi\rho,\qquad \operatorname{div}\boldsymbol{B}=0,$$
$$\operatorname{rot}\boldsymbol{E}=-\frac{1}{c}\frac{\partial\boldsymbol{B}}{\partial t},\qquad \operatorname{rot}\boldsymbol{B}=\frac{4\pi}{c}\boldsymbol{j}+\frac{1}{c}\frac{\partial\boldsymbol{E}}{\partial t}.$$

All together, the equations above constitute the closed *self-consistent system of Vlasov equations* for determining the distribution functions and the fields $\boldsymbol{E}$ and $\boldsymbol{B}$.

Since in our approximation we have eliminated the motion of ions, the ion distribution function remains equilibrium $n_i = n_{i0}$. The electron distribution function is taken as $n = n_0 + \delta n$. Due to electroneutrality and the lack of electrical currents in equilibrium we have $e\int(Zn_{i0}-n_0)d^3p=0$ and $e\int\boldsymbol{v}(Zn_{i0}-n_0)d^3p=0$. Then

$$\rho = -e\int \delta n\, d\Gamma_{\boldsymbol{p}} \quad \text{and} \quad \boldsymbol{j} = -e\int \boldsymbol{v}\delta n\, d\Gamma_{\boldsymbol{p}}.$$

Let us analyze small deviations from equilibrium $\boldsymbol{E}_0 = \boldsymbol{B}_0 = 0$. We find in the linear approximation in all deviations from the equilibrium values

$$\frac{\partial\, \delta n}{\partial t} + \boldsymbol{v}\frac{\partial\, \delta n}{\partial \boldsymbol{r}} - e\big(\boldsymbol{E} + \frac{1}{c}[\boldsymbol{v}\times \boldsymbol{B}]\big)\frac{\partial n_0}{\partial \boldsymbol{p}} = 0.$$

Since $\partial n_0(\varepsilon_{\boldsymbol{p}})/\partial \boldsymbol{p} = \boldsymbol{v}\partial n_0/\partial\varepsilon$, the term with magnetic field vanishes completely. Then,

$$\frac{\partial\, \delta n}{\partial t} + \boldsymbol{v}\frac{\partial\, \delta n}{\partial \boldsymbol{r}} = e(\boldsymbol{v}\boldsymbol{E})\frac{\partial n_0}{\partial \varepsilon}.$$

If the electric field has a single harmonic

$$\boldsymbol{E}(\boldsymbol{r}, t) = \boldsymbol{E}_{\boldsymbol{k},\omega}\exp(i\boldsymbol{k}\boldsymbol{r} - i\omega t),$$

then $\delta n(\boldsymbol{r}, \boldsymbol{p}, t) = \delta n_{\boldsymbol{k},\omega}(\boldsymbol{p})\exp(i\boldsymbol{k}\boldsymbol{r} - i\omega t)$ due to linearity of equations. Accordingly, we obtain

$$(\omega - \boldsymbol{k}\boldsymbol{v})\delta n_{\boldsymbol{k},\omega}(\boldsymbol{p}) = ie(\boldsymbol{v}\boldsymbol{E}_{\boldsymbol{k},\omega})\partial n_0/\partial\varepsilon.$$

When solving this equation, division by zero occurs and a singular point should appear as a simple pole. In this regard we make the following comment. In general, equation $xy(x) = 1$ has a number of solutions resulting directly from identity $x\delta(x) = 0$

$$y(x) = \mathcal{P}\frac{1}{x} + b\,\delta(x).$$

Here $\mathcal{P}\frac{1}{x}$ is the *Caushy principal value* for $1/x$ and b is an arbitrary number. Correspondingly, equation for $\delta n_{\boldsymbol{k},\omega}$ has no unique solution. For selection of the solution having the *physical meaning*, we draw our attention that the perturbation of distribution function $\delta n_{\boldsymbol{k},\omega}$ by electric field $\boldsymbol{E}_{\boldsymbol{k},\omega}$ represents the *retarded response* to external influence. Then, according to the physical principles, the response $\delta n_{\boldsymbol{k},\omega}$ must be analytical function of frequency in the upper half-plane of complex variable ω. This means straightforwardly that the pole singularity $\delta n_{\boldsymbol{k},\omega}$ can only be in the lower half-plane of complex variable ω. The following solution,

$$\delta n_{\boldsymbol{k},\omega}(\boldsymbol{p}) = \frac{ie(\boldsymbol{v}\boldsymbol{E}_{\boldsymbol{k},\omega})}{\omega - \boldsymbol{k}\boldsymbol{v} + i\delta}\frac{\partial n_0}{\partial\varepsilon},$$
$$= ie(\boldsymbol{v}\boldsymbol{E}_{\boldsymbol{k},\omega})\frac{\partial n_0}{\partial\varepsilon}\Big(\mathcal{P}\frac{1}{\omega - \boldsymbol{k}\boldsymbol{v}} - i\pi\delta(\omega - \boldsymbol{k}\boldsymbol{v})\Big)$$

satisfies this physical condition since it provides us a pole at point $\omega = \boldsymbol{k}\boldsymbol{v} - i\delta$ in the lower half-plane of complex variable ω. The latter equality corresponds to *Sokhotsky's formula*

$$\frac{1}{x+i\delta}=\mathcal{P}\frac{1}{x}-i\pi\,\delta(x).$$

It is readily to find the electron component polarization $\boldsymbol{P}_{\boldsymbol{k},\omega}$ with the aid of relation $\boldsymbol{j}(\boldsymbol{r},t)=\partial\boldsymbol{P}(\boldsymbol{r},t)/\partial t$ between the current and the polarization or for the Fourier transform $\boldsymbol{j}_{\boldsymbol{k},\omega}=-i\omega\boldsymbol{P}_{\boldsymbol{k},\omega}$. Thus,

$$\omega\boldsymbol{P}_{\boldsymbol{k},\omega}=e^2\int\boldsymbol{v}(\boldsymbol{v}\boldsymbol{E}_{\boldsymbol{k},\omega})\frac{\partial n_0/\partial\varepsilon}{\omega-\boldsymbol{k}\boldsymbol{v}+i\delta}\,d\Gamma_{\boldsymbol{p}}.$$

Rewriting in terms of the Fourier transforms, we obtain the susceptibility $\chi_{\alpha\beta}(\omega,\boldsymbol{k})$ using the relation

$$P_\alpha=\frac{e^2}{\omega}E_\beta\int\frac{v_\alpha v_\beta}{\omega-\boldsymbol{k}\boldsymbol{v}+i\delta}\frac{\partial n_0}{\partial\varepsilon}\,d\Gamma_{\boldsymbol{p}}\equiv\chi_{\alpha\beta}E_\beta$$

and next we find the permittivity $\varepsilon_{\alpha\beta}$ according to

$$\varepsilon_{\alpha\beta}(\omega,\boldsymbol{k})=\delta_{\alpha\beta}+4\pi\chi_{\alpha\beta}(\omega,\boldsymbol{k})=\delta_{\alpha\beta}+\frac{4\pi e^2}{\omega}\int\frac{v_\alpha v_\beta}{\omega-\boldsymbol{k}\boldsymbol{v}+i\delta}\frac{\partial n_0}{\partial\varepsilon}\,d\Gamma_{\boldsymbol{p}}.$$

It is simpler to calculate the integral using the following trick. Let us introduce the unit vectors $\boldsymbol{l}$ and $\boldsymbol{e}$ in the direction of velocity vector $\boldsymbol{v}$ and wave vector $\boldsymbol{k}$, i.e. $\boldsymbol{v}=v\boldsymbol{l}$ and $\boldsymbol{k}=k\boldsymbol{e}$. Then one has

$$\varepsilon_{\alpha\beta}(\omega,\boldsymbol{k})=\delta_{\alpha\beta}+\frac{4\pi e^2}{\omega^2}\int d\Gamma_{\boldsymbol{p}}\,sv^2\frac{\partial n_0}{\partial\varepsilon}\int\frac{d\Omega_{\boldsymbol{p}}}{4\pi}\frac{l_\alpha l_\beta}{s-\boldsymbol{l}\boldsymbol{e}+i\delta},\quad s=\frac{\omega}{vk}.$$

The integration over the solid angle of vector $\boldsymbol{p}$ should result in some second-rank tensor depending on two indexes α and β. In our disposal there are two independent second-rank tensor $\delta_{\alpha\beta}$ and $e_\alpha e_\beta$. Therefore, the integral over solid angle $\Omega_{\boldsymbol{p}}$ should have the following structure:

$$\int\frac{d\Omega_{\boldsymbol{p}}}{4\pi}\frac{l_\alpha l_\beta}{s-\boldsymbol{l}\boldsymbol{e}+i\delta}=A(s)\delta_{\alpha\beta}+B(s)e_\alpha e_\beta.$$

The unknown scalar functions $A(s)$ and $B(s)$ are to be determined.

To find $A(s)$ and $B(s)$, we calculate the integral twice. Once for the components $\alpha=\beta=z$ and the second for the convolution. Let us choose the z-axis in the direction of vector $\boldsymbol{e}$. Then we find

$$A\cdot 1+B=\int\frac{d\Omega}{4\pi}\frac{l_z^2}{s-\boldsymbol{l}\boldsymbol{e}+i\delta}=\frac{1}{2}\int_0^\pi\sin\vartheta\,d\vartheta\frac{\cos^2\vartheta}{s-\cos\vartheta+i\delta}=-sW(s),$$

$$3A+B=\int\frac{d\Omega}{4\pi}\frac{l_\alpha^2}{s-\boldsymbol{le}+i\delta}=\frac{1}{2}\int\limits_0^\pi \sin\vartheta\,d\vartheta\frac{1}{s-\cos\vartheta+i\delta}=\frac{1-W(s)}{s},$$

where function $W(s)$ equals

$$W(s)=1-\frac{s}{2}\ln\frac{s+1}{s-1+i\delta}.$$

For $s>1$ or $v<\omega/k$, function $W(s)$ is real. For $s<1$ or $v>\omega/k$, function $W(s)$ has an imaginary part $\mathrm{Im}\,W=\pi s/2$. This singularity appears at the particle velocity v exceeding the phase velocity ω/k and leads to the imaginary component in the permittivity and to the specific damping in plasma, *Landau damping*.

A set of two equations for $A(s)$ and $B(s)$ is readily solved

$$A(s)=\frac{1-W(s)+s^2W(s)}{2s}=\frac{s}{2}\left(1+\frac{1-s^2}{2s}\ln\frac{s+1}{s-1+i\delta}\right),$$

$$B(s)=\frac{-1+W(s)-3s^2W(s)}{2s}=-\frac{s}{2}\left(3+\frac{1-3s^2}{2s}\ln\frac{s+1}{s-1+i\delta}\right).$$

The result for permittivity $\varepsilon_{\alpha\beta}$ reads

$$\varepsilon_{\alpha\beta}(\omega,\boldsymbol{k})=\left(1+\frac{4\pi e^2}{\omega^2}\int d\Gamma_p\, sv^2A(s)\frac{\partial n_0}{\partial\varepsilon}\right)\delta_{\alpha\beta}+\left(\frac{4\pi e^2}{\omega^2}\int d\Gamma_p\, sv^2B(s)\frac{\partial n_0}{\partial\varepsilon}\right)\frac{k_\alpha k_\beta}{k^2}.$$

Let us separate permittivity $\varepsilon_{\alpha\beta}$ into the longitudinal and transverse components, i.e. represent it in the form:

$$\varepsilon_{\alpha\beta}=\varepsilon_l\frac{k_\alpha k_\beta}{k^2}+\varepsilon_t\left(\delta_{\alpha\beta}-\frac{k_\alpha k_\beta}{k^2}\right).$$

The meaning of such representation is the following. If the electric field $\boldsymbol{E}_l=E\boldsymbol{k}/k$ is *longitudinal*, i.e. being directed along vector $\boldsymbol{k}$, it is easy to see that the vector of electric induction $\boldsymbol{D}$ is also oriented in the direction $\boldsymbol{E}_l$ and fully determined by the longitudinal component of permittivity

$$\boldsymbol{D}=\varepsilon_l\boldsymbol{E}_l.$$

For the transverse electric field, one has $\boldsymbol{kE}_t=0$. It is readily to check that the vector of electric induction $\boldsymbol{D}$ is completely determined by the transverse permittivity component alone

$$\boldsymbol{D}=\varepsilon_t\boldsymbol{E}_t.$$

Finally, we have for the longitudinal and transverse components of permittivity

$$\varepsilon_l = 1 + \frac{4\pi e^2}{k^2}\int d\Gamma_p \frac{A(s)+B(s)}{s}\frac{\partial n_0}{\partial \varepsilon}$$
$$= 1 - \frac{4\pi e^2}{k^2}\int\left(1 - \frac{s}{2}\ln\frac{s+1}{s-1+i\delta}\right)\frac{\partial n_0}{\partial \varepsilon}\nu(\varepsilon)d\varepsilon$$

and

$$\varepsilon_t = 1 + \frac{4\pi e^2}{k^2}\int d\Gamma_p \frac{A(s)}{s}\frac{\partial n_0}{\partial \varepsilon}$$
$$= 1 + \frac{2\pi e^2}{k^2}\int\left(1 + \frac{1-s^2}{2s}\ln\frac{s+1}{s-1+i\delta}\right)\frac{\partial n_0}{\partial \varepsilon}\nu(\varepsilon)d\varepsilon.$$

Here $\nu(\varepsilon)$ is the density of states, i.e. $d\Gamma_p = \nu(\varepsilon)d\varepsilon$, $v = v(\varepsilon)$ is the electron velocity, and parameter s equals $s = \omega/vk$.

39. Let us start from the analyzing the longitudinal component of permittivity $\varepsilon_l(\omega, \boldsymbol{k})$. In the region of high $\omega \gg vk$ frequencies we can decompose the integrand in $1/s \ll 1$ and find approximately

$$\varepsilon_l(\omega \gg vk) = 1 + \frac{4\pi e^2}{k^2}\int_0^\infty\left(\frac{1}{3s^2} + \frac{1}{5s^4} + \cdots\right)\frac{\partial n_0}{\partial \varepsilon}\nu(\varepsilon)d\varepsilon$$
$$= 1 + \frac{4\pi e^2}{3\omega^2}\int_0^\infty v^2\left(1 + \frac{3}{5}\frac{k^2v^2}{\omega^2} + \cdots\right)\frac{\partial n_0}{\partial \varepsilon}\nu(\varepsilon)d\varepsilon.$$

For the further transformations, we employ the following relation:

$$\int \varepsilon^r\frac{\partial n_0}{\partial \varepsilon}d\Gamma_p = \int_0^\infty \varepsilon^r\frac{\partial n_0}{\partial \varepsilon}\nu(\varepsilon)d\varepsilon = -\left(r + \frac{1}{2}\right)\int_0^\infty \varepsilon^{r-1}n_0(\varepsilon)\nu(\varepsilon)d\varepsilon$$
$$= -\left(r + \frac{1}{2}\right)\int \varepsilon^{r-1}n_0(\varepsilon)\,d\Gamma_p \quad (r > -1/2).$$

To derive this formula, we integrate by parts and take into account that for spectrum $\varepsilon = mv^2/2$, the density of states is $\nu(\varepsilon) \sim \varepsilon^{1/2}$. So, we obtain

$$\varepsilon_l(\omega \gg vk) \approx 1 - \frac{\Omega^2}{\omega^2}\left(1 + \frac{k^2\langle v^2\rangle}{\omega^2}\right), \quad \Omega^2 = \frac{4\pi n e^2}{m}.$$

The frequency Ω is called the *plasma frequency* and referred to as an important characteristic of plasma. The electron density is defined as usual $n = \int d\Gamma_p\, n_0(\varepsilon)$

and the angular brackets hereinafter denote the following average with the distribution function n_0:

$$\langle(\ldots)\rangle = \frac{\int d\Gamma_p\,(\ldots)n_0(\varepsilon)}{\int d\Gamma_p\, n_0(\varepsilon)}\,.$$

The calculation for the imaginary part of longitudinal permittivity component can be performed in the analytical form

$$\operatorname{Im}\varepsilon_l = -\pi\frac{4\pi e^2}{k^2}\int\limits_{v>\omega/k}\frac{\omega}{2vk}\frac{\partial n_0}{\partial\varepsilon}d\Gamma_p = \omega\frac{\Omega^2}{k^2}\frac{m^3}{nk}\frac{n_0(m\omega^2/2k^2)}{2\pi\hbar^3}\,.$$

For the low-frequency region $\omega \ll vk$, we get using the integrand decomposition in $s \ll 1$

$$\varepsilon_l(\omega \ll vk) \approx 1 - \frac{4\pi e^2}{k^2}\int d\Gamma_p\left(1 + i\frac{\pi s}{2}\right)\frac{\partial n_0}{\partial\varepsilon}$$
$$= 1 + \frac{\Omega^2}{k^2}\left\langle\frac{1}{v^2}\right\rangle + i\omega\frac{\Omega^2}{k^2}\frac{m^3}{nk}\frac{n_0(\varepsilon=0)}{2\pi\hbar^3}\,.$$

The static permittivity, equal to

$$\varepsilon_l(0,k) = 1 + \frac{1}{k^2 r_D^2} \quad\text{and}\quad r_D^{-2} = \Omega^2\left\langle\frac{1}{v^2}\right\rangle,$$

determines the screening of electric charge in plasma. In fact, for a point-like charge of magnitude q, the Fourier transform of potential φ_k equals

$$\varphi_k = \frac{4\pi q}{k^2\varepsilon(0,\,k)} = \frac{4\pi q}{k^2 + r_D^{-2}}\,.$$

The corresponding spatial behavior of potential corresponds to the *Debye screening*

$$\varphi(r) = \frac{q}{r}e^{-r/r_D}$$

with the *Debye radius* r_D. The simple calculation yields

$$r_D^2 = \begin{cases} \frac{T}{4\pi ne^2}\,, & T \gg \varepsilon_F \quad \text{(non-degenerated plasma)} \\ \frac{\varepsilon_F}{6\pi ne^2} = \frac{1}{4\pi e^2\nu(\varepsilon_F)}\,, & T \ll \varepsilon_F \quad \text{(degenerate plasma),} \end{cases}$$

where ε_F is the Fermi energy and T is the temperature.

Consider a possibility for existing the longitudinal plasma oscillations. Let electric field $\boldsymbol{E}_l(\boldsymbol{r},t) \sim \boldsymbol{E}_l\exp(i\boldsymbol{kr} - i\omega t)$ be longitudinal, i.e. $\boldsymbol{E}_l = E\boldsymbol{k}/k$. Then one has curl $\boldsymbol{E}_l = i[\boldsymbol{k} \times \boldsymbol{E}_l] = 0$ and, therefore, $\partial \boldsymbol{B}/\partial t = -i\omega \boldsymbol{B} = 0$. The magnetic field is absent $\boldsymbol{B} = 0$ and zero value curl $\boldsymbol{B} = 0$ entails $\partial \boldsymbol{D}/\partial t = -i\omega \boldsymbol{D} = 0$, i.e. $\boldsymbol{D} = 0$. On the other hand, the vector of electric induction equals $\boldsymbol{D} = \varepsilon_l(\omega, \boldsymbol{k})\boldsymbol{E}_l = 0$ for the longitudinal field. Nontrivial solutions are possible only provided that

$$\varepsilon_l(\omega, \boldsymbol{k}) = 0.$$

The latter determines the dispersion of longitudinal plasma oscillations. Substituting the longitudinal permittivity into this equation yields

$$1 - \frac{\Omega^2}{\omega^2}\left(1 + \frac{k^2\langle v^2\rangle}{\omega^2}\right) = 0 \quad \text{or} \quad \omega^2 \approx \Omega^2 + k^2\langle v^2\rangle \ \text{ and } \ \omega \approx \Omega + \frac{k^2\langle v^2\rangle}{2\Omega}.$$

The second term, depending on wave vector k, is small as compared with the plasma frequency Ω. Hence, there exist longitudinal and weakly dispersive *plasma oscillations*[7] or *plasmons* as quanta of plasma oscillations.

The transverse oscillations of electric field $\boldsymbol{E}_t(\boldsymbol{r}, t) \sim \boldsymbol{E}_t \exp(i\boldsymbol{kr} - i\omega t)$ imply the equality $\boldsymbol{kE}_t = 0$. From $\boldsymbol{D}_t = \varepsilon_t \boldsymbol{E}_t$ it also follows $\boldsymbol{kD}_t = 0$ for the vector of induction. According to the Maxwell equations, we find for the Fourier transforms of electric and magnetic fields

$$\begin{cases} \text{curl}\, \boldsymbol{E}_t = -\frac{1}{c}\frac{\partial \boldsymbol{B}}{\partial t}, & \text{div}\, \boldsymbol{B} = 0, \\ \text{curl}\, \boldsymbol{B} = \frac{1}{c}\frac{\partial \boldsymbol{D}_t}{\partial t}, & \text{div}\, \boldsymbol{D}_t = 0, \end{cases} \Rightarrow \begin{cases} \boldsymbol{k} \times \boldsymbol{E}_t = \frac{\omega}{c}\boldsymbol{B}, & \boldsymbol{kB} = 0, \\ \boldsymbol{k} \times \boldsymbol{B} = -\frac{\omega}{c}\boldsymbol{D}_t, & \boldsymbol{kD}_t = 0. \end{cases}$$

Hence, eliminating the magnetic field $\boldsymbol{B}$ leads to

$$\boldsymbol{k} \times [\boldsymbol{k} \times \boldsymbol{E}_t] = \boldsymbol{k}(\boldsymbol{kE}_t) - k^2\boldsymbol{E}_t = \frac{\omega}{c}[\boldsymbol{k} \times \boldsymbol{B}] = -\frac{\omega^2}{c^2}\boldsymbol{D}_t = -\frac{\omega^2}{c^2}\varepsilon_t \boldsymbol{E}_t.$$

For an existence of nontrivial solution $\boldsymbol{E}_t \neq 0$, it is necessary to satisfy the following dispersion equation:

$$k^2 = \frac{\omega^2}{c^2}\varepsilon_t(\omega, k).$$

Treating the high-frequency $\omega \gg vk$ transverse oscillations, we see that the spatial dispersion in permittivity ε_t is insignificant because $v \ll c$ and it is possible to put $\varepsilon_t(\omega, k \approx \varepsilon_t(\omega, 0) = 1 - \Omega^2/\omega^2$. This gives

$$\omega^2 = \Omega^2 + c^2k^2.$$

At $\omega \gg \Omega$ the plasma effect is negligible and $\omega = ck$ as in vacuum. The frequencies $\omega < \Omega$ correspond to the imaginary values of wave vector k. From the physical

[7] Also known as the *Langmuir waves*.

point of view this means that the oscillations at such frequencies decay and cannot propagate deep into plasma.

In conclusion, we write the imaginary part of the transverse permittivity component

$$\operatorname{Im}\varepsilon_t = -\pi\frac{2\pi e^2}{k^2}\int\limits_{v>\omega/k}\frac{v^2k^2-\omega^2}{2\omega vk}\frac{\partial n_0}{\partial\varepsilon}d\Gamma_p = \frac{\pi\Omega^2}{2\omega k}\left\langle\frac{1}{v}\right\rangle\frac{\int\limits_{m\omega^2/2k^2}^{\infty}n_0(\varepsilon)d\varepsilon}{\int\limits_0^{\infty}n_0(\varepsilon)d\varepsilon}.$$

In the lower $\omega \ll vk$ frequency region the real part of the transverse permittivity ε_t equals approximately

$$\operatorname{Re}\varepsilon_t(\omega \ll vk) \approx 1 - \frac{1}{k^2 r_D^2}.$$

The main contribution to the transverse permittivity ε_t is connected with its imaginary part which can be estimated as

$$\operatorname{Im}\varepsilon_t(\omega \ll vk) \approx \frac{\pi}{2}\frac{\Omega^2}{\omega k}\left\langle\frac{1}{v}\right\rangle \sim \frac{\Omega}{\omega k r_D}.$$

Substituting the approximate value ε_t into the dispersion equation entails

$$k^2 = \frac{\omega^2}{c^2}\left(i\frac{\pi\Omega^2}{2\omega k}\left\langle\frac{1}{v}\right\rangle\right).$$

Hence, we find the dispersion law for the long wave $kr_D \ll 1$ transverse oscillations

$$\omega = -i\frac{2}{\pi}\frac{c^2k^3}{\Omega^2\langle v^{-1}\rangle}.$$

The purely imaginary value of frequency means that such plasma oscillations are aperiodic and strongly damping.

40. Let two-dimensional plasma layer occupy the plane $z = 0$. As a first step, we write the collisionless transport equation on the electron distribution function $n(\boldsymbol{r}, \boldsymbol{p}, t)$, where $\boldsymbol{r}$ is the radius-vector lying in the layer plane

$$\frac{\partial n}{\partial t} + \boldsymbol{v}\frac{\partial n}{\partial \boldsymbol{r}} + \boldsymbol{F}\frac{\partial n}{\partial \boldsymbol{p}} = 0.$$

Here $\boldsymbol{v} = \boldsymbol{p}/m$ is the electron velocity and $F = -e(\boldsymbol{E} + [\boldsymbol{v} \times \boldsymbol{B}]/c)$ is the Lorentz force exerted to an electron of charge $(-e)$.

Let us turn to small deviations $\delta n(\boldsymbol{r}, \boldsymbol{p}, t)$ of distribution function from equilibrium value $n_0(\varepsilon_p)$ taking into account that $\partial n_0(\varepsilon_p)/\partial \boldsymbol{p} = \boldsymbol{v}\partial n_0/\partial\varepsilon$. Then we have

$$\frac{\partial\,\delta n}{\partial t} + \boldsymbol{v}\frac{\partial\,\delta n}{\partial \boldsymbol{r}} = e(\boldsymbol{v}\boldsymbol{E})\frac{\partial n_0}{\partial \varepsilon},$$

where $\boldsymbol{E} = \boldsymbol{E}(\boldsymbol{r}, z = 0, t)$ is the electric field in the layer. If the electric field contains a harmonic $\boldsymbol{E}(\boldsymbol{r}, z, t) = \boldsymbol{E}_{\boldsymbol{k},\omega}(z)\exp(i\boldsymbol{k}\boldsymbol{r} - i\omega t)$, then also $\delta n(\boldsymbol{r}, \boldsymbol{p}, t) = \delta n_{\boldsymbol{k},\omega}(\boldsymbol{p})\exp(i\boldsymbol{k}\boldsymbol{r} - i\omega t)$ due to linearity of equation. Since the response to electric field must have the retarded character, we find that

$$\delta n_{\boldsymbol{k},\omega}(\boldsymbol{p}) = \frac{ie\big(\boldsymbol{v}\boldsymbol{E}_{\boldsymbol{k},\omega}(z=0)\big)}{\omega - \boldsymbol{k}\boldsymbol{v} + i\delta}\frac{\partial n_0}{\partial \varepsilon}.$$

Accordingly, the density of polarization charges, induced by the electric field perturbation, will be equal to

$$\rho_{\boldsymbol{k},\omega} = -e\int \delta n_{\boldsymbol{k},\omega}\frac{2d^2p}{(2\pi\hbar)^2} = -ie^2\int \frac{\big(\boldsymbol{v}\boldsymbol{E}_{\boldsymbol{k},\omega}(z=0)\big)}{\omega - \boldsymbol{k}\boldsymbol{v} + i\delta}\frac{\partial n_0}{\partial \varepsilon}\frac{2\,d^2p}{(2\pi\hbar)^2}.$$

Let us analyze the following integral:

$$\boldsymbol{I}(\omega, \boldsymbol{k}) = \int \frac{\boldsymbol{v}}{\omega - \boldsymbol{k}\boldsymbol{v} + i\delta}\frac{\partial n_0}{\partial \varepsilon}\frac{2\,d^2p}{(2\pi\hbar)^2}.$$

The integration over angle φ between vectors $\boldsymbol{k}$ and $\boldsymbol{v}$ will lead to some vector in the direction of wave vector $\boldsymbol{k}$,

$$\boldsymbol{I}(\omega, \boldsymbol{k}) = \frac{\boldsymbol{k}}{k^2}\int \frac{\partial n_0}{\partial \varepsilon}A(s)\frac{2\,d^2p}{(2\pi\hbar)^2}.$$

Here $s = \omega/vk$ and function $A(s)$ is determined by the expression

$$A(s) = \int_0^{2\pi} \frac{d\varphi}{2\pi}\frac{\cos\varphi}{s - \cos\varphi + i\delta} = \begin{cases} -1 + s/\sqrt{s^2-1}\,, & s > 1, \\ -1 - is/\sqrt{1-s^2}\,, & s < 1. \end{cases}$$

The origin of imaginary part in $A(s)$ is directly associated with the *Landau damping* mechanism when the particle velocity v exceeds the wave phase velocity ω/k.

After replacing the integration over momentum with that over energy $\varepsilon = p^2/2m$, the polarization charge density $\rho_{\boldsymbol{k},\omega}$ is given by the formula

$$\rho_{\boldsymbol{k},\omega} = -\frac{i\big(\boldsymbol{k}\boldsymbol{E}_{\boldsymbol{k},\omega}(z=0)\big)}{\pi k^2 a_B}\int_0^\infty \frac{\partial n_0}{\partial \varepsilon}A(s)\,d\varepsilon,$$

$a_B = \hbar^2/me^2$ being the Bohr radius.

The polarization charges in the layer with the bulk density $\rho_V(\boldsymbol{r}, z) = \rho(\boldsymbol{r})\delta(z)$ is a reason of appearing the polarization potential

$$\varphi(\boldsymbol{r}, z) = \int \frac{\rho(\boldsymbol{r}')\delta(z')}{\sqrt{(\boldsymbol{r}-\boldsymbol{r}')^2 + (z-z')^2}}\, d^2r'\, dz' = \int \frac{\rho(\boldsymbol{r}')\, d^2r'}{\sqrt{(\boldsymbol{r}-\boldsymbol{r}')^2 + z^2}}.$$

Using the property of convolution and that

$$\int \frac{e^{i\boldsymbol{k}\boldsymbol{r}}\, d^2r}{\sqrt{r^2+z^2}} = 2\pi \int\limits_0^\infty \frac{r\, J_0(kr)\, dr}{\sqrt{r^2+z^2}} = \frac{2\pi}{k} e^{-k|z|},$$

where J_0 is the Bessel function of zero order, we disclose the following relation between the Fourier transforms of the potential and density of polarization charges appearing in the infinitely thin layer[8]

$$\varphi_{\boldsymbol{k},\omega}(z) = \frac{2\pi}{k} e^{-k|z|} \rho_{\boldsymbol{k},\omega}\,.$$

Since we know the polarization charge potential, we can find the corresponding electric field $\boldsymbol{E}_{\rm p}$ at point $z = 0$ or polarization according to $\boldsymbol{E}_{\rm p} = -4\pi\boldsymbol{P}$ since $\boldsymbol{D} = \boldsymbol{E}_{\rm p} + 4\pi\boldsymbol{P} = 0$. We are interested in the field component parallel to the electron layer

$$\boldsymbol{E}_{\rm p}(z) = -\frac{\partial\varphi_{\boldsymbol{k},\omega}(z)}{\partial\boldsymbol{r}} \quad \text{or} \quad \boldsymbol{E}_{{\rm p}\,\boldsymbol{k},\omega}(z) = -i\boldsymbol{k}\varphi_{\boldsymbol{k},\omega}(z) = -\frac{2\pi i\boldsymbol{k}}{k} e^{-k|z|}\rho_{\boldsymbol{k},\omega}.$$

Hence, putting $z = 0$, we obtain the following expression for the polarization vector $\boldsymbol{P}_{\boldsymbol{k},\omega}$

$$4\pi\boldsymbol{P}_{\boldsymbol{k},\omega} = \frac{2\pi i\boldsymbol{k}}{k}\rho_{\boldsymbol{k},\omega} = \frac{2\boldsymbol{k}(\boldsymbol{k}\boldsymbol{E}_{\boldsymbol{k},\omega})}{k^3 a_B} \int\limits_0^\infty \frac{\partial n_0}{\partial\varepsilon} A(s)\, d\varepsilon$$

and for the susceptibility $\chi_{\alpha\beta}(\omega, \boldsymbol{k})$

$$4\pi\chi_{\alpha\beta}(\omega, \boldsymbol{k}) = 4\pi\chi_l(\omega, k)\frac{k_\alpha k_\beta}{k^2} = \frac{2}{k a_B}\left(\int\limits_0^\infty \frac{\partial n_0}{\partial\varepsilon} A(s)\, d\varepsilon\right)\frac{k_\alpha k_\beta}{k^2}.$$

Thus, we arrive at the following formula for the longitudinal permittivity $\varepsilon_l(\omega, k)$ of thin plasma layer

[8] Note here that at $z = 0$ the factor $2\pi/k$, connecting the Fourier transforms of potential and charge density, differs from the routine factor $4\pi/k^2$ in the three-dimensional bulk case.

$$\varepsilon_l(\omega, k) = 1 + 4\pi\chi_l(\omega, k) = 1 + \frac{2}{ka_B}\int_0^\infty \frac{\partial n_0}{\partial \varepsilon} A(s)\, d\varepsilon.$$

Let us start first from most interesting case of zero temperature when the electron component represents a degenerate Fermi gas. Then we can put $\partial n_0/\partial\varepsilon = -\delta(\varepsilon - \varepsilon_F)$ and obtain immediately the expression for the longitudinal permittivity

$$\varepsilon_l(\omega, k) = \begin{cases} 1 + \frac{2}{ka_B}\left(1 - \frac{\omega}{\sqrt{\omega^2 - v_F^2k^2}}\right), & \omega > v_F k, \\ 1 + \frac{2}{ka_B}\left(1 + \frac{i\omega}{\sqrt{v_F^2k^2 - \omega^2}}\right), & \omega < v_F k. \end{cases}$$

The static value of permittivity $\varepsilon(0, k) = 1 + 2/ka_B$ means that the electric field of point-like charge q in a thin metal film starts to be screened at the distances about the Bohr radius a_B. However, unlike the bulk metal, the screening here is much less effective and becomes non-exponential. The electric field from the charge q inserted into the metal layer decays in the power-like manner as qa_B^2/r^3 at the large distances.

The spectrum of plasma oscillations or plasmons is determined from the dispersion equation $\varepsilon_l(\omega, k) = 0$. Assuming $\omega > v_F k$, we get the equation

$$1 + \frac{ka_B}{2} = \frac{\omega}{\sqrt{\omega^2 - v_F^2k^2}} \quad \text{or} \quad \omega^2 = k\frac{v_F^2}{a_B}\frac{(1 + ka_B/2)^2}{1 + ka_B/4}.$$

Hence, in the natural long-wavelength limit $ka_B \ll 1$ we have the following dispersion relation for plasmons with the square-root behavior as a function of wave vector

$$\omega(k) = v_F\sqrt{k/a_B} = \sqrt{2\pi ne^2k/m}.$$

Deriving the latter equality, we take into account that, for two-dimensional electron gas, the Fermi momentum p_F and Fermi velocity v_F are related with the electron density n as $p_F = mv_F = \hbar(2\pi n)^{1/2}$.

For the finite temperatures, the plasmon spectrum remains still unchanged in the long-wavelength $ka_B \ll 1$ limit. Using the expansion $A(s) \approx 1/2s^2$ at $s \gg 1$ yields

$$\varepsilon_l = 1 + \frac{2e^2k}{\hbar^2\omega^2}\int_0^\infty \varepsilon\frac{\partial n_0}{\partial\varepsilon}\, d\varepsilon = 1 - \frac{2e^2k}{\hbar^2\omega^2}\int_0^\infty n_0(\varepsilon)\, d\varepsilon = 1 - \frac{2\pi ne^2k}{m\omega^2}.$$

The condition $\varepsilon_l = 0$ results in the same square-root spectrum of plasmons. At the finite temperatures the Landau damping mechanism leads to insignificant damping of two-dimensional plasmons.

41. Let us employ results of the previous problem and write the general expression for the longitudinal permittivity of two-dimensional electron layer

$$\varepsilon_l(\omega, k) = 1 + \frac{2\pi e^2}{k}\int \frac{\partial n_0(\varepsilon)}{\partial \varepsilon} A(s) \frac{\nu\, d^2 p}{(2\pi\hbar)^2}\,.$$

Here $n_0(\varepsilon) = 1/\left(1 + e^{(\varepsilon-\mu)/T}\right)$ is the Fermi distribution function, $\nu = 2 \times 2$. The function $A(s)$, equal to

$$A(s) = \begin{cases} -1 + s/\sqrt{s^2 - 1}\,, & s > 1, \\ -1 - is/\sqrt{1 - s^2}\,, & s < 1, \end{cases}$$

can be taken out of integral sign since its argument is independent of momentum p. Replacing the integration over momentum with that over energy and introducing the density of states $g(\varepsilon)$, we obtain for the permittivity

$$\varepsilon_l(\omega, k) = 1 + \frac{2\pi e^2}{k} A\left(\frac{\omega}{vk}\right) \int_{-\infty}^{\infty} \frac{\partial n_0(\varepsilon)}{\partial \varepsilon} g(\varepsilon)\, d\varepsilon = 1 - \frac{2\pi e^2}{k} A\left(\frac{\omega}{vk}\right)$$

$$\times \frac{\nu}{2\pi\hbar^2 v^2} \int_{-\infty}^{\infty} \frac{d\varepsilon}{4T} \frac{|\varepsilon|}{\cosh^2 \frac{\varepsilon-\mu}{2T}} = 1 - \frac{4e^2 T}{\hbar^2 v^2 k} A\left(\frac{\omega}{vk}\right) \ln\left(2\cosh\frac{\mu}{2T}\right).$$

The equation $\varepsilon_l(\omega, k) = 0$ provides us the plasmon dispersion. Solving the equation is simple

$$\omega^2 = \frac{\varkappa k}{1 + v^2 k/4\varkappa} + v^2 k^2, \quad \varkappa(T) = \frac{2e^2 T}{\hbar^2} \ln\left(2\cosh\frac{\mu}{T}\right).$$

In most interesting long-wavelength region $k \ll \varkappa/v^2$ the plasma oscillations (plasmons) have the same square-root behavior as a function of wave vector k

$$\omega(k) = \sqrt{\varkappa k}.$$

This dispersion is the same as for the ordinary two-dimensional systems but having the temperature dependence $\varkappa(T)$ at $\mu \lesssim T$.

3.15 Kinetics of First-Order Phase Transitions: Melting and Solidification

42. Let the x-axis run inside the lake and we measure the distances from the air-ice interface (Fig. 2.4). The temperature distribution $T(x, t)$ deep into the ice phase obeys the heat equation

$$\rho C_p \frac{\partial T}{\partial t} = \varkappa \frac{\partial^2 T}{\partial x^2} \quad \text{for} \quad 0 < x < x_0(t),$$

where $x_0(t)$ is the position of the water-ice interface at the time moment t.

For the unambiguous solution of heat equation, it is necessary to set the boundary conditions which should take place at the interfaces between the phases at $x = 0$ and $x = x_0(t)$. First of all, there should be the equal temperatures of the both phases at each of the interfaces

$$T(0, t) = T_0 \quad \text{and} \quad T(x_0(t), t) = T_m\,.$$

Next, we should take into account that the first-order phase transition occurs at the water-ice interface with releasing the freezing heat equal to the melting heat. If the interface shifts by distance dx_0, the amount of heat equal to $\rho L dx_0$ releases in the ice element, withdrawing into the depth of ice with the aid of heat conduction mechanism. Thus, in the process of phase conversion the amount of heat $Q(t)$, equal to $\rho L dx_0/dt$, releases for unit time per the unit interface area. Since the temperature of water remains constant $T(x > x_0, t) = T_m$ because of initial assumption about its high (infinite) heat conduction, the release of heat $Q(t)$ during the phase conversion at interface $x = x_0(t)$ compensates completely with the heat flow $q = -\varkappa \partial T/\partial x$ in the direction to the ice. As a result, we arrive at the following boundary condition:

$$q\big|_{x=x_0(t)} + Q(t) = 0 \quad \text{or} \quad \varkappa \frac{\partial T}{\partial x}\bigg|_{x=x_0(t)} = \rho L \frac{dx_0}{dt}\,.$$

From the mathematical point of view we have the *Stefan problem*, i.e. solution of partial differential equation with the boundary conditions given at the mobile boundaries.

The solution of heat equation under condition $T(0, t) = T_0$ is expressed in terms of *error function*

$$\operatorname{erf}(\eta) = \frac{2}{\sqrt{\pi}} \int_0^{\eta} e^{-\xi^2} d\xi,$$

according to

$$T(x, t) = T_0 + A \operatorname{erf}\left(\frac{x}{2\sqrt{\chi t}}\right),$$

where $\chi = \varkappa/\rho C_p$ is the thermal diffusivity of ice. Unknown coefficient A should satisfy the following condition at the interface $x = x_0(t)$:

$$T(x_0(t), t) = T_0 + A \operatorname{erf}\left(\frac{x_0(t)}{2\sqrt{\chi t}}\right) = T_m\,.$$

Since the right-hand side of this equation is time independent, the equation can only be satisfied provided that the argument of function *erf* is also time independent. Accordingly, we have the following growth of the ice thickness in time:

$$x_0(t) = 2B\sqrt{\chi t},$$

where unknown constant B to be determined below.

To find the constant B, we use the second boundary which reduces to the following equation:

$$\frac{\varkappa}{\sqrt{\pi}}\frac{A}{\sqrt{\chi t}}e^{-B^2} = \rho L B\sqrt{\frac{\chi}{t}}.$$

So, we arrive at a set of two equations to determine unknown A and B

$$A\,\mathrm{erf}\,(B) = T_m - T_0\,,$$
$$\frac{A}{\sqrt{\pi}}e^{-B^2} = \frac{\rho L\chi}{\varkappa}B = \frac{L}{C_p}B.$$

The latter leads to the equation which solves the problem on determining unknown constant B

$$Be^{B^2}\mathrm{erf}\,(B) = \frac{1}{\sqrt{\pi}}\frac{C_p}{L}(T_m - T_0).$$

Below we examine only the case of sufficiently neighboring temperatures T_m and T_0. Assuming $B \ll 1$, we have approximately

$$\frac{2}{\sqrt{\pi}}B^2 \approx \frac{1}{\sqrt{\pi}}\frac{C_p}{L}(T_m - T_0).$$

So, we find at $T_m - T_0 \ll 2L/C_p$

$$B \approx \sqrt{\frac{C_p}{2L}(T_m - T_0)} \quad \text{and} \quad A \approx \sqrt{\frac{\pi L}{2C_p}(T_m - T_0)}.$$

The temperature of ice decreases approximately in accordance with the linear law as a function of distance x in the direction to the air

$$T(x,t) \approx T_0 + \sqrt{\frac{\pi L}{2C_p}(T_m - T_0)}\frac{x}{\sqrt{\chi t}} \qquad (0 < x < x_0).$$

Regardless of the neighboring temperature approximation chosen, the ice thickness increases in time according to the square-root law $x_0(t) \sim \sqrt{t}$. Note some limitation of the approximations selected. In the first turn this concerns the sufficiently short time when the rate of new phase $dx_0/dt \sim 1/\sqrt{t}$ is unlimitedly large. As a consequence, the equilibrium inside the phases ceases to run and the applicability of heat equations is limited.

In conclusion, emphasize that the neglect of a difference in the densities of water and ice allows us not to consider the inflow or outflow of substance in the region of interface during the process of phase transition.

43. Let us direct the x-axis to the melt and measure the distance from the furnace bottom (Fig. 2.5). If the densities of the metal and the melt are different, under phase transition there appears a flow of liquid melt in the direction to the solid metal phase due to necessity of supplying the substance to the denser phase $\rho_2 > \rho_1$.

The heat equation in the moving liquid, assumed incompressible, takes the form

$$\rho_1 C_1 \left(\frac{\partial T_1}{\partial t} + v_1 \frac{\partial T_1}{\partial x} \right) = \varkappa_1 \frac{\partial^2 T_1}{\partial x^2} \quad \text{at} \quad x > x_0(t).$$

Here v_1 is the flow velocity of liquid melt in the x-axis direction and $x_0(t)$ is the position of the metal-melt interface at the time moment t. To find the velocity of the melt, one can use the continuity condition for the mass flow across the interface $x = x_0(t)$ growing at rate $\dot{x}_0(t)$

$$\rho_1(v_1 - \dot{x}_0) = \rho_2(v_2 - \dot{x}_0),$$

where v_1 and v_2 are the velocities of liquid and solid phases. Keeping in mind that the solid phase is immobile, i.e. $v_2 = 0$, we find the flow velocity for the melt

$$v_1 = -\frac{\rho_2 - \rho_1}{\rho_1} \dot{x}_0 .$$

The temperature distribution $T_2(x, t)$ in the depth of metallic phase obeys the usual heat conduction equation

$$\rho_2 C_2 \frac{\partial T_2}{\partial t} = \varkappa_2 \frac{\partial^2 T_2}{\partial x^2} \quad \text{at} \quad 0 < x < x_0(t).$$

According to the problem statement, the temperatures T_1 and T_2 at the upper side of the melt and at the lower side of metal near the furnace bottom are equal to

$$T_1(\infty, t) = T_l \quad \text{and} \quad T_2(0, t) = T_0 ,$$

Let us turn to analyzing the boundary conditions for the metal-melt interface at $x = x_0(t)$. First of all, the temperature of the both phases must be equal

$$T_1(x_0(t), t) = T_2(x_0(t), t) = T_c .$$

The requirement for the continuity of energy flux across the interface $x = x_0(t)$ results in the following boundary condition:

$$\rho_1(v_1 - \dot{x}_0)\left(\frac{(v_1 - \dot{x}_0)^2}{2} + w_1\right) + q_1 = \rho_2(v_2 - \dot{x}_0)\left(\frac{(v_2 - \dot{x}_0)^2}{2} + w_2\right) + q_2,$$

where $q_{1,2} = -\varkappa_{1,2}\partial T_{1,2}/\partial x$ are the dissipative heat flows and $w_{1,2}$ are the enthalpies (heat functions) of the melt and the metal, respectively. Denoting the mass flow of substance across the interface as $J = \rho_1(v_1 - \dot{x}_0) = -\rho_2\dot{x}_0$, we will have for the difference in the dissipative heat flows at the interface

$$q_1 - q_2 = J\left(w_2 - w_1 + J^2\frac{\rho_1^2 - \rho_2^2}{2\rho_1^2\rho_2^2}\right) = -\rho_2\left(w_2 - w_1 + \frac{\rho_1^2 - \rho_2^2}{2\rho_2^2}\dot{x}_0^2\right)\dot{x}_0.$$

The enthalpy (heat function) equals $w = \mu + Ts$ where μ is the chemical potential and s is the entropy. For the enthalpy difference $w_1 - w_2 = (\mu_1 - \mu_2) + T_1s_1 - T_2s_2$ at the interface $x = x_0$, we pay attention that the temperatures of the melt T_1 and the metal T_2 coincide with the crystallization temperature T_c. In addition, the condition of the melt-metal phase equilibrium, determining the crystallization temperature, implies the equality of chemical potentials for the both phases, i.e. $\mu_1(T_c) = \mu_2(T_c)$. This results in the following enthalpy difference at the interface:

$$w_1(T_c) - w_2(T_c) = T_m(s_1 - s_2) = L,$$

the latter being obviously the latent heat of phase transition (heat of crystallization) L. Thus, we have at the interface of the melt with the metal

$$q_1 - q_2 = \rho_2\left(L - \frac{\rho_1^2 - \rho_2^2}{2\rho_2^2}\dot{x}_0^2\right)\dot{x}_0.$$

Next, we apply the condition of small rate of the melt formation when the rate-squared term $\dot{x}_0^2$ in the brackets can be neglected as compared with the latent heat L. This gives the following approximate boundary condition:

$$\varkappa_2\frac{\partial T_2}{\partial x}\bigg|_{x=x_0(t)} - \varkappa_1\frac{\partial T_1}{\partial x}\bigg|_{x=x_0(t)} = \rho_2 L\frac{dx_0}{dt}$$

which becomes exact under the equal densities of the melt and the metal ($\rho_1 = \rho_2$).

So, after establishing the equations for the temperature distribution in the melt and in the metal and setting the boundary conditions at the interfaces, we can turn to the mathematical treatment of the problem. The solution of the heat equation in the metal is similar to that in the previous problem. The solution can be expressed in terms of error function

$$\mathrm{erf}\,(\eta) = \frac{2}{\sqrt{\pi}} \int_0^{\eta} e^{-\xi^2} d\xi,$$

according to

$$T_2(x,t) = T_0 + A_2\,\mathrm{erf}\left(\frac{x}{2\sqrt{\chi_2 t}}\right), \quad \text{at} \quad 0 < x < x_0(t)$$

and corresponding consequence from the boundary condition at $x = x_0(t)$:

$$T_0 + A_2\,\mathrm{erf}\left(\frac{x_0(t)}{2\sqrt{\chi_2 t}}\right) = T_c\,.$$

Here $\chi_2 = \varkappa_2/\rho_2 C_2$ is the thermal diffusivity of a metal. The last condition can only be satisfied provided that the melt growth rate obeys the following time law $x_0(t) = 2B\sqrt{t}$, where B is unknown constant.

Analyzing the equation which determines the temperature distribution in the melt, we should have

$$\rho_1 C_1 \left(\frac{\partial T_1}{\partial t} - \frac{\rho_2 - \rho_1}{\rho_1}\frac{B}{\sqrt{t}}\frac{\partial T_1}{\partial x}\right) = \varkappa_1 \frac{\partial^2 T_1}{\partial x^2} \quad \text{for} \quad x > x_0(t).$$

The general solution of the equation in the melt region under $T_1(\infty, t) = T_l$ at the external side of the melt is expressed via the complimentary error function

$$\mathrm{erfc}\,(\eta) = \frac{2}{\sqrt{\pi}} \int_{\eta}^{\infty} e^{-\xi^2} d\xi = 1 - \mathrm{erf}\,(\eta),$$

according to

$$T_1(x,t) = T_l + A_1\,\mathrm{erfc}\left(\frac{x}{2\sqrt{\chi_1 t}} + \frac{\rho_2 - \rho_1}{\rho_1}\frac{B}{\sqrt{\chi_1}}\right), \qquad x > x_0(t).$$

Here $\chi_1 = \varkappa_1/\rho_1 C_1$ is the thermal diffusivity of a melt.

Then we must satisfy two boundary conditions at the melt-metal interface $x = x_0(t)$. The first condition yields

$$T_l + A_1\,\mathrm{erfc}\left(\frac{\rho_2}{\rho_1}\frac{B}{\sqrt{\chi_1}}\right) = T_c$$

and the second takes the following form:

$$-\frac{\varkappa_1}{\sqrt{\pi\chi_1}}A_1e^{-\frac{\rho_2^2}{\rho_1^2}\frac{B^2}{\chi_1}}+\frac{\varkappa_2}{\sqrt{\pi\chi_2}}A_2e^{-\frac{B^2}{\chi_2}}=\rho_2LB.$$

Using the values for the constant A_1 and A_2, we arrive at the final equation to determine the unknown quantity B

$$\frac{\varkappa_1}{\sqrt{\pi\chi_1}}\frac{\exp\left(-\frac{\rho_2^2}{\rho_1^2}\frac{B^2}{\chi_1}\right)}{\operatorname{erfc}\left(\frac{\rho_2}{\rho_1}\frac{B}{\sqrt{\chi_1}}\right)}(T_l-T_c)+\frac{\varkappa_2}{\sqrt{\pi\chi_2}}\frac{\exp\left(-\frac{B^2}{\chi_2}\right)}{\operatorname{erf}\left(\frac{B}{\sqrt{\chi_2}}\right)}(T_c-T_0)=\rho_2LB.$$

For $T_l = T_c$, this equation is fully analogous to that analyzed in the previous problem.

Let us turn to the symmetrical case of the liquid and solid phases with the same physical parameters, i.e. $\varkappa_1=\varkappa_2=\varkappa$, $\rho_1=\rho_2=\rho$, and $C_1=C_2=C$. Putting $B=\sqrt{\chi}b$ gives

$$(T_c-T_0)+(T_l-T_c)\frac{\operatorname{erf}(b)}{\operatorname{erfc}(b)}=\sqrt{\pi}\frac{L}{C}be^{b^2}\operatorname{erf}(b).$$

Below, more detailed, we examine the case of approximately same temperatures T_0, T_c, and T_l. For $b \ll 1$, the following expansion takes place:

$$\frac{2L}{C}\left(1+\frac{2C(T_l-T_c)}{\pi L}\right)b^2+\frac{2}{\sqrt{\pi}}(T_l-T_c)b-(T_c-T_0)\approx 0.$$

Supposing $(T_l-T_c)\ll L/C$, we find the constant b as a root of quadratic equation

$$b\approx-\frac{C(T_l-T_c)}{2\sqrt{\pi}L}+\sqrt{\frac{C^2(T_l-T_c)^2}{4\pi L^2}+\frac{C(T_c-T_0)}{2L}}.$$

The plus sign in the front of square root is chosen to have a positive value b. Here we give the limiting values

$$b\approx\begin{cases}\sqrt{\frac{C(T_c-T_0)}{2L}}, & (T_l-T_c)^2\ll\frac{L(T_c-T_0)}{C},\\[2mm] \sqrt{\pi}\,\frac{T_c-T_0}{T_l-T_c}, & (T_l-T_c)^2\gg\frac{L(T_c-T_0)}{C}.\end{cases}$$

On the whole, the dependence of parameter b or B on the physical parameters of the both phases proves to be rather complicated and the numerical analysis is required. However, regardless of physical parameters of the both phases the metal thickness grows in time, obeying the general square-root law $x_0(t)\sim\sqrt{t}$.

44. First, we will direct the x-axis toward to the liquid. Let solid phase at time moment t occupy the region $0<x<x_0(t)$, where $x_0(t)$ is the position of the liquid-solid interface. The temperature distribution in the solid is governed by the heat

equation

$$\frac{\partial T}{\partial t} = \chi \frac{\partial^2 T}{\partial x^2} .$$

In accordance with the problem statement we have at the lower $x = 0$ boundary of solid phase

$$T(0, 0) = T_0 \quad \text{and} \quad \varkappa \frac{\partial T(x, t)}{\partial x}\bigg|_{x=0} = Q = \text{const.}$$

The following boundary conditions are determined for the upper boundary $x = x_0(t)$ between the solid and liquid phases:

$$T(x_0, t) = T_0 \quad \text{and} \quad \varkappa \frac{\partial T(x, t)}{\partial x}\bigg|_{x=x_0(t)} = \rho L \dot{x}_0 .$$

These boundary conditions at the interface specify the equality of solid phase temperature to the solidification temperature as well as the heat flow associated with the heat of crystallization L.

The general solution of heat equation is attempted with expanding into a series

$$T(x, t) = T_0 + a_1 x + a_2(x^2 + 2\chi t) + a_3(x^3 + 6\chi x t) + a_4(x^4 + 12\chi x^2 t + 12\chi^2 t^2) + \ldots$$

which satisfies directly the initial condition $T(0, 0) = T_0$. The expansion coefficients a_1 and a_3 can be found from the boundary condition at $x = 0$. In fact,

$$Q = \varkappa \frac{\partial T(x, t)}{\partial x}\bigg|_{x=0} \approx \varkappa a_1 + 6\varkappa a_3 \chi t .$$

Since the heat flow Q is time independent, this condition, obviously, is satisfied at

$$a_1 = Q/\varkappa \quad \text{and} \quad a_3 = 0.$$

The two remaining coefficients will be determined with the aid of conditions at the liquid-solid interface. From the other two conditions

$$a_1 x_0 + a_2(x_0^2 + 2\chi t) + a_4(x_0^4 + 12\chi x_0^2 t + 12\chi^2 t^2) + \ldots = 0,$$

$$\frac{\rho L}{\varkappa}\dot{x}_0 = a_1 + 2a_2 x_0 + 4a_4(x_0^3 + 6\chi x_0 t) + \ldots,$$

and representing the interface displacement as a series in powers t

$$x_0(t) = b_1 t + b_2 t^2 + b_3 t^3 + \ldots$$

we will compare the terms of the same powers of t and get the following values for the expansion coefficients:

$$b_1 = \frac{Q}{\rho L}, \quad a_2 = -\frac{a_1 b_1}{2\chi} = -\frac{1}{2\chi}\frac{Q^2}{\varkappa \rho L}, \quad b_2 = \frac{\varkappa}{\rho L} a_2 b_1 = -\frac{1}{2\chi}\left(\frac{Q}{\rho L}\right)^3,$$

$$a_4 = -\frac{a_1 b_2 + a_2 b_1^2}{12\chi^2} = \frac{1}{12\chi^3}\frac{Q^4}{\varkappa(\rho L)^3},$$

$$b_3 = \frac{2}{3}\frac{\varkappa}{\rho L}(2a_2 b_2 + 12\chi a_4 b_1) = \frac{5}{6\chi^2}\left(\frac{Q}{\rho L}\right)^5.$$

As a result, we obtain the proper expansions for the solid accumulation law and temperature distribution over the solid phase

$$x_0(t) = \frac{Q}{\rho L} t - \frac{1}{2\chi}\left(\frac{Q}{\rho L}\right)^3 t^2 + \frac{5}{6\chi^2}\left(\frac{Q}{\rho L}\right)^5 t^3 + \dots,$$

$$T(x,t) = T_0 + \frac{Q}{\varkappa} x - \frac{1}{2\chi}\frac{Q^2}{\rho L}(x^2 + 2\chi t)$$

$$+ \frac{1}{12\chi^3}\frac{Q^4}{\varkappa(\rho L)^3}(x^4 + 12\chi x^2 t + 12\chi^2 t^2) + \dots$$

Unlike the two previous problems, the initial rate of solidifying the liquid is finite under constancy of external heat flow. The rate is proportional to the magnitude of external heat flow Q and inversely proportional to the heat of crystallization L. The temperature at the lower boundary of solid phase decreases with the growth of the solid and becomes smaller than the solidification temperature. The expansion obtained is a good approximation while the time of solidification is not too long, i.e.

$$t \ll \chi\left(\frac{\rho L}{Q}\right)^2.$$

In the limit of infinite thermal diffusivity of solid phase $\chi = \infty$ (but the finite magnitude of thermal conductivity $\varkappa$) we have the simple dependences

$$x_0(t) = \frac{Q}{\rho L} t \quad \text{and} \quad T(x,t) = T_0 + \frac{Q}{\varkappa} x - \frac{Q^2}{\varkappa \rho L} t$$

for the growth of solid phase and the temperature distribution in it.

3.16 Macroscopic Quantum Tunneling

45. Let us write action S for the total physical system: particle + medium

$$S[q(t), x_\alpha(t)] = \int \left[\frac{M\dot{q}^2}{2} - U_0(q) + \sum_\alpha \left(\frac{m\dot{x}_\alpha^2}{2} - \frac{m\omega_\alpha^2 x_\alpha^2}{2} \right) - q \sum_\alpha C_\alpha x_\alpha \right] dt.$$

The last term is responsible for the particle-medium coupling. The equations of motion for the particle and phonons read

$$M\ddot{q} + U_0'(q) = -\sum_\alpha C_\alpha x_\alpha,$$
$$m\ddot{x}_\alpha + m\omega_\alpha^2 x_\alpha = -qC_\alpha.$$

The solution of the linear equation for $x_\alpha(t)$ can be represented as a sum of the forced and free oscillations

$$x_\alpha(t) = -C_\alpha \int_{-\infty}^{t} \frac{\sin \omega_\alpha (t-s)}{m\omega_\alpha} q(s)\, ds$$
$$+ x_\alpha(0) \cos \omega_\alpha t + \frac{\dot{x}_\alpha(0)}{\omega_\alpha} \sin \omega_\alpha t = \int_{-\infty}^{\infty} K_\alpha^R(t-s) q(s)\, ds + x_\alpha^{(\mathrm{fr})}(t).$$

Here we have extended integration over all possible time interval and introduced the *retarded response function* K_α^R

$$K_\alpha^R(t) = -\vartheta(t) \frac{C_\alpha \sin \omega_\alpha t}{m\omega_\alpha}.$$

Then we define the corresponding Fourier transform

$$K_\alpha^R(\omega) = \int_{-\infty}^{\infty} K_\alpha^R(t) e^{i\omega t} dt = -\frac{C_\alpha}{m\big[\omega_\alpha^2 - (\omega + i\delta)^2\big]}$$

which has the poles $\omega = \pm\omega_\alpha - i\delta$ only in the lower half-plane of complex variable ω. The quantities $x_\alpha(0)$ and $\dot{x}_\alpha(0)$ can be interpreted as a position and velocity of a phonon which the phonon would have in the lack of its coupling with the particle at the same time, for example, at $t = 0$.

The total force $F(t)$, acting on the particle, is determined by the relation

$$F(t) = \int_{-\infty}^{\infty} K^R(t-s) q(s)\, ds + f(t),$$

$$f(t) = -\sum_\alpha C_\alpha \left(x_\alpha(0) \cos \omega_\alpha t + \frac{\dot{x}_\alpha(0)}{\omega_\alpha} \sin \omega_\alpha t \right).$$

The total response function $K^R(t)$ here equals

$$K^R(t) = -\sum_\alpha C_\alpha K^R_\alpha(t) = \vartheta(t)\sum_\alpha C^2_\alpha \frac{\sin\omega_\alpha t}{m\omega_\alpha}.$$

Its Fourier transform is readily found and can be expressed in terms of spectral density $J(\Omega)$

$$K^R(\omega) = \sum_\alpha \frac{C^2_\alpha}{m\left[\omega^2_\alpha - (\omega + i\delta)^2\right]} \equiv \frac{2}{\pi}\int\limits_0^\infty \frac{\Omega}{\Omega^2 - (\omega + i\delta)^2} J(\Omega)\, d\Omega.$$

Finally, we obtain the following equation of particle motion

$$M\ddot{q} + U'_0(q) = \int\limits_{-\infty}^{\infty} K^R(t-s)q(s)\, ds + f(t).$$

Let us separate response function K^R into two parts. The first one K_0 is frequency-independent

$$K_0 = \frac{2}{\pi}\int\limits_0^\infty \frac{J(\Omega)}{\Omega} d\Omega = \sum_\alpha \frac{C^2_\alpha}{m\omega_\alpha}$$

and, on the contrary, the second one K^R_1 is frequency-dependent

$$K^R_1(\omega) = \frac{2}{\pi}\int\limits_0^\infty \frac{J(\Omega)}{\Omega}\frac{\omega^2}{\Omega^2 - (\omega + i\delta)^2} d\Omega$$

so that $K^R(\omega) = K_0 + K^R_1(\omega)$. If $J(\Omega) = \eta\Omega$, one can readily see that

$$K^R_1(\omega) = i\omega\eta \quad \text{and} \quad K^R(\omega) = K_0 + i\omega\eta.$$

Performing the inverse Fourier transform, we find the time representation for the response function

$$K^R(t) = K_0\delta(t) - \eta\delta'(t).$$

Let us substitute $K^R(t)$ into the equation of particle motion and calculate the integrals

$$\begin{aligned} M\ddot{q} + U'_0(q) &= K_0\int\limits_{-\infty}^{\infty}\delta(t-s)q(s)\, ds - \eta\int\limits_{-\infty}^{\infty}\delta'(t-s)q(s)\, ds + f(t) \\ &= K_0 q(t) - \eta\dot{q}(t) + f(t). \end{aligned}$$

We write this equation in the form of the *Langevin equation* with the damping coefficient η and the friction force proportional to particle velocity $\dot{q}$

$$M\ddot{q} + \eta\dot{q} + U'(q) = f(t) \quad \text{and} \quad U(q) = U_0(q) - K_0 q^2/2.$$

Thus, the particle-medium coupling results, firstly, in renormalizing the particle potential energy $U_0(q) \to U(q)$, secondly, in appearing the friction force, and thirdly, in giving rise of random force $f(t)$ which properties depend on the state of a medium.

Let us now turn to examining the properties of random force $f(t)$. Since for the thermodynamically equilibrium system the thermal averages of oscillator coordinate and velocity are equal to zero, i.e. $\langle x_\alpha(0)\rangle = 0$ and $\langle \dot{x}_\alpha(0)\rangle = 0$, the average value of random force equals zero as well

$$\langle f(t)\rangle = 0.$$

The force correlator of second order is determined with a double sum

$$\langle f(t)f(t')\rangle = \sum_{\alpha\beta} C_\alpha C_\beta \langle \Big(x_\alpha(0)\cos\omega_\alpha t + \frac{\dot{x}_\alpha(0)}{\omega_\alpha}\sin\omega_\alpha t\Big) \times \Big(x_\beta(0)\cos\omega_\beta t' + \frac{\dot{x}_\beta(0)}{\omega_\beta}\sin\omega_\beta t'\Big)\rangle.$$

Since oscillators or phonons are independent of each other, the product of the average of two quantities for various oscillators is equal to the product of the averages of these quantities, e.g. $\langle x_\alpha(0)x_\beta(0)\rangle = \langle x_\alpha(0)\rangle\langle x_\beta(0)\rangle = 0$ at $\alpha \neq \beta$. Therefore, the terms with $\alpha = \beta$ alone remain in a sum

$$\langle f(t)f(t')\rangle = \sum_\alpha C_\alpha^2 \Big[\langle x_\alpha(0)^2\rangle \cos\omega_\alpha t \cos\omega_\alpha t' + \frac{\langle x_\alpha(0)\dot{x}_\alpha(0)\rangle}{\omega_\alpha}\sin\omega_\alpha(t+t') + \frac{\langle \dot{x}_\alpha(0)\dot{x}_\alpha(0)\rangle}{\omega_\alpha^2}\sin\omega_\alpha t \sin\omega_\alpha t'\Big].$$

Next, we remind that the thermodynamical average for the product of oscillator coordinate and velocity taken at the same time equals zero, i.e. $\langle x_\alpha(0)\dot{x}_\alpha(0)\rangle = 0$. The average square of velocity is straightforwardly expressed via the average square of coordinate $\langle \dot{x}_\alpha^2(0)\rangle = \omega_\alpha^2\langle x_\alpha^2(0)\rangle$ due to equality of the averages for the kinetic and potential energies. Involving that

$$\langle x_\alpha^2(0)\rangle = \frac{\hbar}{2m\omega_\alpha}\coth\frac{\hbar\omega_\alpha}{2T}$$

for the thermodynamically equilibrium oscillator at temperature T, we find for the time force correlator

$$\langle f(t)f(t')\rangle = \sum_\alpha C_\alpha^2 \frac{\hbar}{2m\omega_\alpha} \coth\frac{\hbar\omega_\alpha}{2T}\cos\omega_\alpha(t-t')$$

$$= \frac{\hbar}{\pi}\int_0^\infty J(\Omega)\coth\frac{\hbar\Omega}{2T}\cos\Omega(t-t')\,d\Omega = \frac{\eta}{\pi}\int_0^\infty \hbar\Omega\coth\frac{\hbar\Omega}{2T}\cos\Omega(t-t')\,d\Omega.$$

Because the integrand function is even, it is not difficult to find the Fourier transform of time force correlator or *spectral density*. In fact,

$$\langle f(t)f(t')\rangle = \int \langle f(t)f(t')\rangle_\Omega\, e^{-i\Omega(t-t')}\frac{d\Omega}{2\pi},$$

$$\langle f(t)f(t')\rangle_\Omega = \eta\hbar\Omega\coth\frac{\hbar\Omega}{2T} = \begin{cases} 2\eta T, & \text{classical limit } \hbar = 0,\\ \eta\hbar|\Omega|, & \text{quantum limit } T = 0.\end{cases}$$

In the time representation the force correlator $\langle f(t)f(t')\rangle$ equals

$$\langle f(t)f(t')\rangle = \begin{cases} 2\eta T\,\delta(t-t'), & \text{classical limit } \hbar = 0\\ -\frac{\eta\hbar}{\pi(t-t')^2}, & \text{quantum limit } T = 0\end{cases}$$

in the two limiting cases. The random force proves to be uncorrelated in the classical limit and correlated in the quantum limit. Comparing the answer with the results of problem 1, we emphasize that the obtained spectral correlator of random force corresponds completely to the statement of the *fluctuation-dissipation theorem* which relates the fluctuations of particle coordinate with the energy dissipation properties of thermodynamically equilibrium medium.

46. The solution of the problem is similar to that of the previous problem. We write action $S = S[q(t), x_\alpha(t)]$ for the system: particle + medium

$$S = \int\left[\frac{M\dot q^2}{2} - U_0(q) + \sum_\alpha\left(\frac{m\dot x_\alpha^2}{2} - \frac{m\omega_\alpha^2 x_\alpha^2}{2}\right) - \gamma(q)\sum_\alpha C_\alpha x_\alpha\right]dt.$$

The equations of particle and phonon motion read

$$M\ddot q + U_0'(q) = -\gamma'(q)\sum_\alpha C_\alpha x_\alpha,$$

$$m\ddot x_\alpha + m\omega_\alpha^2 x_\alpha = -\gamma(q)C_\alpha.$$

The solution of linear equation with respect of $x_\alpha(t)$ can be represented as a sum of forced and free oscillations

$$x_\alpha(t) = -C_\alpha \int\limits_{-\infty}^{t} \frac{\sin\omega_\alpha(t-s)}{m\omega_\alpha}\gamma(q_s)\,ds + x_\alpha(0)\cos\omega_\alpha t + \frac{\dot{x}_\alpha(0)}{\omega_\alpha}\sin\omega_\alpha t$$

$$= C_\alpha \int\limits_{-\infty}^{\infty} K_\alpha^R(t-s)\gamma(q_s)\,ds + x_\alpha^{(\mathrm{fr})}(t).$$

Here we have enlarged the integration over all possible time interval and introduced the retarded response function $K_\alpha^R(t)$

$$K_\alpha^R(t) = -\vartheta(t)\frac{\sin\omega_\alpha t}{m\omega_\alpha}.$$

Let us determine the Fourier transform for $K_\alpha^R(t)$

$$K_\alpha^R(\omega) = \int\limits_{-\infty}^{\infty} K_\alpha^R(t)e^{i\omega t}dt = -\frac{1}{m\big[\omega_\alpha^2 - (\omega+i\delta)^2\big]}$$

which has the poles $\omega = \pm\omega_\alpha - i\delta$ only in the lower half-plane of complex variable ω. The quantities $x_\alpha(0)$ and $\dot{x}_\alpha(0)$ can be treated as a position and velocity of a phonon which the latter would have at the same time, e.g. $t = 0$, if there is no phonon-particle coupling.

The total force $F(t)$, acting on the particle, is given by the relation

$$F(t) = \gamma'(q)\int\limits_{-\infty}^{\infty} K^R(t-s)\gamma(q_s)\,ds + f(t),$$

$$f(t) = -\gamma(q)\sum_\alpha C_\alpha\left(x_\alpha(0)\cos\omega_\alpha t + \frac{\dot{x}_\alpha(0)}{\omega_\alpha}\sin\omega_\alpha t\right).$$

The total response function $K^R(t)$ equals

$$K^R(t) = -\sum_\alpha C_\alpha^2 K_\alpha^R(t) = \vartheta(t)\sum_\alpha C_\alpha^2\frac{\sin\omega_\alpha t}{m\omega_\alpha}.$$

Its Fourier transform is readily found and can be expressed via the spectral density $J(\Omega)$

$$K^R(\omega) = \sum_\alpha \frac{C_\alpha^2}{m\big[\omega_\alpha^2 - (\omega+i\delta)^2\big]} \equiv \frac{2}{\pi}\int\limits_0^\infty \frac{\Omega}{\Omega^2 - (\omega+i\delta)^2}J(\Omega)\,d\Omega.$$

Finally, we arrive at the following equation of particle motion:

$$M\ddot{q} + U_0'(q) = \gamma'(q) \int_{-\infty}^{\infty} K^R(t-s)\gamma(q_s)\, ds + f(t).$$

The response function K^R separates into two parts. One of which K_0 is frequency-independent

$$K_0 = \frac{2}{\pi}\int_0^{\infty} \frac{J(\Omega)}{\Omega} d\Omega = \sum_{\alpha} \frac{C_\alpha^2}{m\omega_\alpha}$$

and the other K_1^R depends on the frequency

$$K_1^R(\omega) = \frac{2}{\pi}\int_0^{\infty} \frac{J(\Omega)}{\Omega} \frac{\omega^2}{\Omega^2 - (\omega + i\delta)^2} d\Omega,$$

so that $K^R(\omega) = K_0 + K_1^R(\omega)$. For $J(\Omega) = \eta\Omega$, we have

$$K_1^R(\omega) = i\omega\eta \quad \text{and} \quad K^R(\omega) = K_0 + i\omega\eta.$$

Performing the inverse Fourier transform, we obtain the time representation of response function

$$K^R(t) = K_0\delta(t) - \eta\delta'(t).$$

Then we substitute $K^R(t)$ into the equation of particle motion and calculate the integrals

$$M\ddot{q} + U_0'(q) = K_0\gamma'(q_t) \int_{-\infty}^{\infty} \delta(t-s)\gamma(q_s)\, ds$$

$$- \eta\gamma'(q_t) \int_{-\infty}^{\infty} \delta'(t-s)\gamma(q_s)\, ds + f(t) = K_0\gamma'(q)\gamma(q) - \eta\gamma'^2(q)\dot{q}(t) + f(t).$$

Let us write this equation in the form of the *Langevin equation* with the nonlinear drag coefficient $\mu(q)$ and friction force proportional to the particle velocity $\dot{q}$

$$M\ddot{q} + \mu(q)\dot{q} + U'(q) = f(t) \quad \text{and} \quad \begin{cases} U(q) = U_0(q) - \frac{1}{2}K_0\gamma^2(q), \\ \mu(q) = \eta\left(\frac{\partial\gamma(q)}{\partial q}\right)^2. \end{cases}$$

The particle-medium coupling results, firstly, in renormalizing the particle potential energy $U_0(q) \to U(q)$, secondly, in appearing the friction force, and thirdly, in giving rise of random force $f(t)$. The drag coefficient $\mu(q)$ and interaction vertex $\gamma(q)$ are unambiguously connected with each other.

47. Let a particle be at the energy level $E = E_n$. Since the semiclassical condition is assumed, we can use the exponential approximation for the tunneling rate

$$\Gamma(E) = \nu(E)e^{-A(E)/\hbar}.$$

Here $\nu(E)$ is the attempt frequency of tunneling and the exponent is given by the well-known WKB formula

$$A(E) = 2\int_{q_1}^{q_2} \sqrt{2M\big(U(q) - E\big)}\,dq \equiv \oint \sqrt{2M\big(U(q) - E\big)}\,dq.$$

The turning points $q_1 = q_1(E)$ and $q_2 = q_2(E)$ are the points of entry under the potential barrier and the exit from it. The semiclassical condition $A(E) \gg \hbar$ supposes also the large number of energy levels in the potential well. Let us write $A(E)$ as

$$A(E) = \oint \sqrt{2M\big(\tilde{E} - \tilde{U}(q)\big)}\,dq,$$

where $\tilde{E} = -E$ and the *inverted potential* $\tilde{U}(q) = -U(q)$. The last relation for $A(E)$ can be interpreted as a motion of a classical particle in the inverted potential $\tilde{U}(q)$ (Fig. 3.13).

Let a particle, bound by the motion in the potential well, be in the thermal equilibrium at temperature T. Then the probability to find the particle at the energy level $E = E_n$ is determined with the Gibbs distribution

$$w(E_n) = Z^{-1}e^{-\beta E_n}, \quad Z = \sum_n e^{-\beta E_n}, \quad \beta = 1/T.$$

The probability or decay rate $\Gamma(T)$ is given by the following sum over the energy levels:

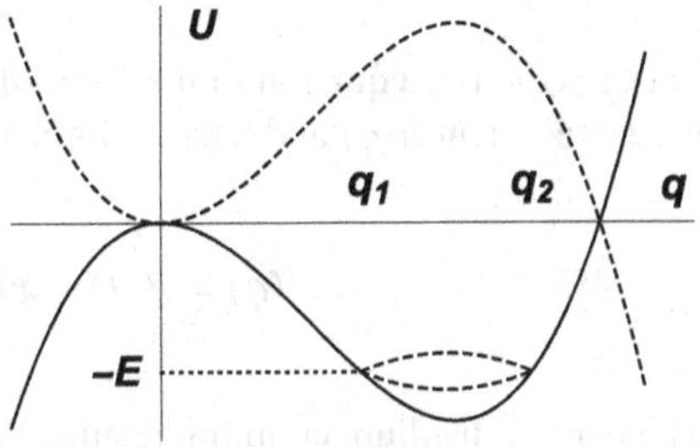

Fig. 3.13 The periodic path in the inverted (solid line) potential $\tilde{U} = -U(q)$. The turning points q_1 and q_2 correspond to the underbarrier motion at energy $E^* = -E$

$$\Gamma(T) = \frac{1}{Z}\sum_n \Gamma(E_n)e^{-\beta E_n} = \frac{1}{Z}\int dE\, g(E)e^{-\frac{A(E)}{\hbar}-\beta E}$$
$$= \frac{1}{Z}\int dE\, g(E)e^{-S(E)/\hbar},$$

where $g(E) = \sum_n \delta(E - E_n)$ is the density of states and $S(E) = A(E) + \hbar\beta E$.

The main contribution to the integral due to its exponential character is gained from the stationary point $E = E^*$, where derivative is $\partial S(E)/\partial E = 0$. This leads to the following equation:

$$\hbar\beta = -\frac{\partial A}{\partial E} = \frac{\partial A}{\partial \tilde{E}} = \oint dq \sqrt{\frac{M}{2(\tilde{E} - \tilde{U}(q))}} = \tau(\tilde{E}).$$

The last integral is exactly the period for the classical particle motion in the inverted potential $\tilde{U}(q)$ if the particle has energy $\tilde{E}$. In other words, the motion period required must be equal to the inverse temperature with accuracy to the Planck constant

$$\tau(E^*) = \hbar\beta = \hbar/T.$$

In essence, this condition means the optimum combination of thermal activation and quantum tunneling. Accordingly, the decay rate of metastable state $\Gamma = \Gamma(T)$ will be determined by magnitude $S(E)$ at $E = E^*$ within the exponential accuracy

$$\Gamma(T) \sim \exp\bigl(-S(E^*)/\hbar\bigr).$$

Let us discuss the answer from another point of view. For this purpose, we represent $S(E)$ as follows:

$$S(E) = A(E) + \hbar\beta E = \oint \sqrt{2M\bigl(\tilde{E} - \tilde{U}(q)\bigr)}\, dq - \int_{-\hbar\beta/2}^{\hbar\beta/2} \tilde{E}\, d\tau.$$

The first term in this expression is commonly called the *abbreviated action*. The integration over parameter τ, having the sense of time, can be chosen within the interval $[0, \hbar\beta]$. For the latter relation, we can assign the meaning of some action $S_{\text{eff}}[q_\tau, \dot{q}_\tau]$ with the Euclidean Lagrangian L_E

$$S_{\text{eff}}[q(\tau), \dot{q}(\tau)] = \int_{-\hbar\beta/2}^{\hbar\beta/2} L_E[q(\tau), \dot{q}(\tau)]\, d\tau$$

$$= \int_{-\hbar\beta/2}^{\hbar\beta/2} [\frac{M}{2}\left(\frac{dq}{d\tau}\right)^2 - \tilde{U}(q)]\, d\tau \equiv \int_{q(-\hbar\beta/2)=q(\hbar\beta/2)} [\frac{M}{2}\left(\frac{dq}{d\tau}\right)^2 + U(q)]\, d\tau.$$

Thus, in order to find the exponent $S[E^*]$ in the expression for the decay rate, one can calculate the magnitude of action $S_{\text{eff}}[q_\tau, \dot{q}_\tau]$ on the closed path which period $\tau(E)$ equals $\hbar/T$. The action $S_{\text{eff}}[q_\tau, \dot{q}_\tau]$ is referred to as the *effective* or, infrequently, *Euclidean action* and the parameter τ is the *imaginary time*. In fact, if we perform the transformation $\tau = it$, called the *Wick rotation*, we disclose the following relations between the effective action and the conventional action

$$S_{\text{eff}} = \int \left[\frac{M}{2}\left(\frac{dq}{d\tau}\right)^2 + U(q) \right] d\tau \overset{\tau=it}{\longrightarrow} -i \int \left[\frac{M}{2}\left(\frac{dq}{dt}\right)^2 - U(q) \right] dt = -iS,$$

$$e^{iS/\hbar} \overset{t=-i\tau}{\longrightarrow} e^{-S_{\text{eff}}/\hbar}.$$

Let us recall that the transition amplitude in quantum mechanics is proportional to $\exp(iS/\hbar)$. From this point of view the underbarrier motion of a particle can be interpreted as a motion of classical particle along the axis of imaginary time τ within interval $\hbar/T$ in the inverted potential $\tilde{U}$.

The temperature dependence for decay rate $\Gamma(T)$ is governed by the behavior of motion period $\tau(E)$ in the inverted potential $\tilde{U}(q)$. For the potential with the smooth behavior at the points of local minimum $q = 0$ and maximum $q = q_0$ corresponding to the barrier height $U_0 = U(q_0)$, the period motion varies from $\tau(E = 0) = \infty$ at $E = 0$ to the minimal value $\tau(E = U_0) = 2\pi/\omega_0$, where $\omega_0 = \left(|U''(q_0)|/M\right)^{1/2}$. Accordingly, there exist no nontrivial periodic paths in the inverted potential at temperatures $T > T_q = \hbar\omega_0/2\pi$. There is a single path $q(\tau) \equiv q_0$, called the *classical*, which satisfies the condition of periodicity $q(-\hbar\beta/2) = q(\hbar\beta/2)$ and extremeness. Since for such path $\dot{q}(\tau) = 0$, the extremum value of effective action on the classical path is obvious and equal to the classical exponent within accuracy to the Planck constant $\hbar$

$$S(E^*) = \hbar U(q_0)/T = \hbar U_0/T.$$

As a result, for temperatures $T > T_q$, the decay of metastable state will occur via the classical thermal activation, obeying the *Arrhenius law*

$$\Gamma(T) \sim e^{-U_0/T}, \qquad T > T_q = \hbar\left(|U''(q_0)|/M\right)^{1/2}/2\pi.$$

At temperature $T = T_q$ determined by the potential barrier curvature, there occurs a crossover to the quantum decay regime. The probability of quantum decay reduces as the temperature lowers, approaching at the minimum value at zero temperature

$$\Gamma(0) \sim \exp\left(-\frac{2}{\hbar} \int_0^{q_c} \sqrt{2MU(q)}\, dq \right).$$

Here q_c is the exit point of a particle from under the barrier at $E = 0$, determined by condition $U(q_c) = 0$.

48. For the system of particle + medium, we write the effective action S_{eff} determined in the imaginary time τ within interval $[-\hbar\beta/2,\ \hbar\beta/2]$, where $\beta = 1/T$ is the inverse temperature

$$S_{\text{eff}} = \int \left[\frac{M\dot{q}_\tau^2}{2} + U_0(q_\tau) + \sum_\alpha \left(\frac{m\dot{x}_\alpha^2}{2} + \frac{m\omega_\alpha^2 x_\alpha^2}{2} \right) + \gamma(q_\tau) \sum_\alpha C_\alpha x_\alpha \right] d\tau.$$

In order to analyze the tunneling probability, it is necessary to find the extremum value of effective action for the particle and phonon paths with the period $\hbar\beta$. The extremum paths of particle and phonons obey the equations

$$\delta S_{\text{eff}}/\delta q(\tau) = -M\ddot{q} + U_0'(q) + \gamma'(q) \sum_\alpha C_\alpha x_\alpha = 0,$$

$$\delta S_{\text{eff}}/\delta x_\alpha(\tau) = -m\ddot{x}_\alpha + m\omega_\alpha^2 x_\alpha + \gamma(q_\tau) C_\alpha = 0.$$

Since we examine the time-periodic paths, it is useful to employ the decomposition of paths into a Fourier series

$$x_\alpha(\tau) = \frac{T}{\hbar} \sum_n x_{\alpha n} e^{-i\omega_n \tau}, \quad \omega_n = 2\pi n T/\hbar, \quad n = 0, \pm 1, \pm 2, \ldots$$

$$x_{\alpha n} = \int_{-\hbar/2T}^{\hbar/2T} x_\alpha(\tau) e^{i\omega_n \tau} d\tau,$$

where the frequencies ω_n are called the *Matsubara frequencies*. Then the solution for the phonon path $x_\alpha(\tau)$ can be expressed as

$$x_\alpha(\tau) = C_\alpha \int_{-\hbar/2T}^{\hbar/2T} \mathcal{K}_\alpha(\tau - s)\gamma(q_s)\, ds.$$

Here the Fourier transform of the *Matsubara response function* $\mathcal{K}_\alpha(\tau)$ is determined by the relation

$$\mathcal{K}_\alpha(\omega_n) = \int_{-\hbar/2T}^{\hbar/2T} \mathcal{K}_\alpha(\tau) e^{i\omega_n \tau} d\tau = -\frac{1}{m(\omega_\alpha^2 + |\omega_n|^2)}.$$

Next, we find the effective equation which the particle motion obeys

$$-M\ddot{q} + U_0'(q) - \gamma'(q_\tau) \int_{-\hbar/2T}^{\hbar/2T} \mathcal{K}(\tau - s)\gamma(q_s)\, ds = 0.$$

Here we have introduced the total response function $\mathcal{K}(\tau)$

$$\mathcal{K}(\tau) = -\sum_\alpha C_\alpha^2 \mathcal{K}_\alpha(\tau)$$

having the following Fourier transform:

$$\mathcal{K}(\omega_n) = \sum_\alpha \frac{C_\alpha^2}{m(\omega_\alpha^2 + |\omega_n|^2)} \equiv \frac{2}{\pi} \int_0^\infty \frac{\Omega}{\Omega^2 + |\omega_n|^2} J(\Omega)\, d\Omega.$$

Then, from the Matsubara response function $\mathcal{K}(\omega_n)$ we select the frequency-independent part $K_0 = \mathcal{K}(\omega_n = 0)$

$$K_0 = \frac{2}{\pi} \int_0^\infty \frac{J(\Omega)}{\Omega} d\Omega = \sum_\alpha \frac{C_\alpha^2}{m\omega_\alpha}$$

and frequency-dependent part $\mathcal{K}_1(\omega_n)$ so that $\mathcal{K}(\omega_n) = K_0 + \mathcal{K}_1(\omega_n)$ and

$$\mathcal{K}_1(\omega_n) = -\frac{2}{\pi} \int_0^\infty \frac{J(\Omega)}{\Omega} \frac{|\omega_n|^2}{\Omega^2 + |\omega_n|^2} d\Omega_{|J=\eta\Omega} = -\eta|\omega_n|.$$

Let us compare the Matsubara response function $\mathcal{K}(\omega_n)$ with the retarded response $K^R(\omega)$. It is seen that the static parts of the responses are completely the same but the dynamical parts are also connected with each other. On the whole, we reveal the following *general* relation between the Matsubara and the retarded responses:

$$\mathcal{K}(\omega_n) = K^R(\omega = i|\omega_n|).$$

Emphasize that the retarded response is also *real* on the positive segment of imaginary axis of complex variable ω.

In order to find the time behavior of Matsubara response $\mathcal{K}(\tau)$, we will perform the inverse transform of function $\mathcal{K}(\omega_n) = K_0 - \eta|\omega_n|$. It is obvious that

$$\mathcal{K}(\tau) = K_0\delta(\tau) + \mathcal{K}_1(\tau).$$

The calculation $\mathcal{K}_1(\tau)$ will be considered separately

$$\mathcal{K}_1(\tau) = -\eta\frac{T}{\hbar}\sum_n |\omega_n| e^{-i\omega_n\tau} = -\eta\frac{iT}{\hbar}\frac{\partial}{\partial\tau}\sum_n \operatorname{sgn}(\omega_n) e^{-i\omega_n\tau}$$

$$= -\eta\frac{iT}{\hbar}\frac{\partial}{\partial\tau}\left(\sum_{n=0}^\infty e^{-i\omega_n\tau} - \sum_{n=0}^\infty e^{i\omega_n\tau}\right).$$

From the mathematical point of view the function $\mathcal{K}_1(\tau)$ is a generalized function or distribution. For the regularization $\mathcal{K}_1(\tau)$, the infinitely small quantity $\pm i\delta$ can be added to the exponent in τ in order to have the convergent series. For brevity, we below omit the infinitely small quantity. Summing two geometric progressions yields

$$\mathcal{K}_1(\tau) = -\eta \frac{iT}{\hbar}\frac{\partial}{\partial\tau}\left(\frac{1}{1-e^{-2\pi iT\tau/\hbar}} - \frac{1}{1-e^{2\pi iT\tau/\hbar}}\right)$$
$$= -\eta\frac{T}{\hbar}\frac{\partial}{\partial\tau}\cot\left(\frac{\pi T\tau}{\hbar}\right) = \eta\frac{1}{\pi}\frac{(\pi T/\hbar)^2}{\sin^2(\pi T\tau/\hbar)}.$$

Let us return to the effective equation of particle motion, originating from the elimination of the medium variables x_α

$$\frac{\delta S_{\text{eff}}}{\delta q_\tau} = -M\ddot{q} + U_0'(q) - \gamma'(q_\tau)\int\limits_{-\hbar/2T}^{\hbar/2T}\left[K_0\delta(\tau-s) + \mathcal{K}_1(\tau-s)\right]\gamma(q_s)\,ds$$
$$= -M\ddot{q} + U'(q) - \gamma'(q_\tau)\int\limits_{-\hbar/2T}^{\hbar/2T}\mathcal{K}_1(\tau-s)\gamma(q_s)\,ds = 0.$$

Here we have denoted the renormalized potential $U(q) = U_0(q) - K_0\gamma^2(q)/2$. The renormalization of the potential is a direct consequence of the interaction between the particle and the medium in which the particle moves.

The equation found represents the variation for the effective action of the particle

$$S_{\text{eff}}[q_\tau] = \int\limits_{-\hbar/2T}^{\hbar/2T}\left(\frac{M\dot{q}_\tau^2}{2} + U(q_\tau)\right)d\tau - \frac{1}{2}\iint\limits_{-\hbar/2T}^{\hbar/2T} d\tau\, ds\, \gamma(q_\tau)\mathcal{K}_1(\tau-s)\gamma(q_s).$$

It is important that the variables of the medium are succeeded to eliminate completely. Since $\mathcal{K}_1(\tau-s)$ has a singular behavior at point $\tau = s$, we transform the second integral to more convenient form in which this singularity does not manifest. Let us write

$$-2\iint d\tau\, ds\gamma_\tau\mathcal{K}_1(\tau-s)\gamma_s = \iint d\tau\, ds\,(\gamma_\tau-\gamma_s)^2\mathcal{K}_1(\tau-s)$$
$$-\iint d\tau\, ds\gamma_\tau^2\mathcal{K}_1(\tau-s) - \iint d\tau\, ds\gamma_s^2\mathcal{K}_1(\tau-s)$$
$$= \iint d\tau\, ds\,(\gamma_\tau-\gamma_s)^2\mathcal{K}_1(\tau-s).$$

Here, when going over to the latter equation, we take into account that the two last integrals vanish. The point is that the integration of periodic function performed over its full period gives the Fourier transform at zero argument. In fact, for example,

$$\int_{-\hbar/2T}^{\hbar/2T} ds\, \mathcal{K}_1(\tau - s) = \int_{-\hbar/2T}^{\hbar/2T} ds\, \mathcal{K}_1(s) = \mathcal{K}_1(\omega_n = 0) \equiv 0.$$

Taking this equality and explicit form $\mathcal{K}_1$, we arrive at the effective action for the particle as

$$S_{\text{eff}}[q_\tau] = \int_{-\hbar/2T}^{\hbar/2T} \left(\frac{M\dot{q}_\tau^2}{2} + U(q_\tau)\right) d\tau$$
$$+ \frac{\eta}{4\pi} \iint_{-\hbar/2T}^{\hbar/2T} \left[\gamma(q_\tau) - \gamma(q_s)\right]^2 \frac{(\pi T/\hbar)^2}{\sin^2[\pi T(\tau - s)/\hbar]} d\tau\, ds.$$

In such equivalent representation the singularity of denominator at point $\tau = s$ is compensated with vanishing the nominator at the same point.

As we have seen from the previous problems, there is a simple relation between the drag coefficient $\mu(q)$ and the interaction vertex $\gamma(q)$

$$\mu(q) = \eta \left(\frac{\partial \gamma(q)}{\partial q}\right)^2 = \left(\sqrt{\eta}\, \gamma'(q)\right)^2.$$

For convenience, we introduce a factor $\eta^{1/2}$ into definition $\gamma(q)$ according to $\eta^{1/2}\gamma \to \gamma$. Then, we arrive at the following form of the effective action which describes the motion of a particle of mass M in the classically accessible region of potential $U(q)$ with the drag coefficient $\mu(q)$

$$S_{\text{eff}}[q_\tau] = \int_{-\hbar/2T}^{\hbar/2T} \left(\frac{M\dot{q}_\tau^2}{2} + U(q_\tau)\right) d\tau$$
$$+ \frac{1}{4\pi} \iint_{-\hbar/2T}^{\hbar/2T} \left[\gamma(q_\tau) - \gamma(q_s)\right]^2 \frac{(\pi T/\hbar)^2}{\sin^2[\pi T(\tau - s)/\hbar]} d\tau\, ds.$$

The vortex $\gamma(q)$ is fixed with the drag coefficient $\mu(q)$ in accordance with

$$\mu(q) = \gamma'^2(q) \quad \text{or} \quad \gamma(q) = \int^q \mu^{1/2} dq.$$

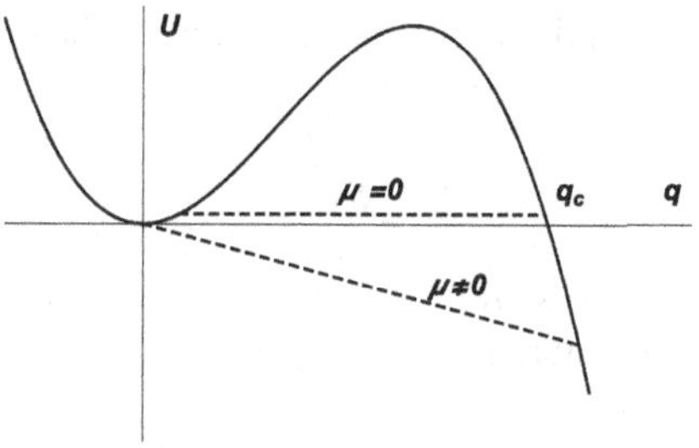

Fig. 3.14 The dashed lines show the extremum tunneling paths of a particle at $T = 0$ in the dissipationless $\mu = 0$ and energy dissipation $\mu \neq 0$ media. In the dissipative case the particle energy does not conserve and the exit point from the barrier lies to the right from point q_c. The particle energy losses in the tunneling process are compensated with the smaller magnitude of the potential energy at which the particle escapes from the barrier

If $\mu(q) \sim q^k$, then $\gamma \sim q^{(k+2)/2}$. In the important case of the Ohmic $k = 0$ friction we have $\gamma = \sqrt{\mu}\, q$.

In order to determine the decay rate of metastable state $\Gamma(T) \sim \exp(-A)$, we should find the minimum value $A = S_{\text{eff}}/\hbar$ calculated for the periodic path $q(-\hbar/2T) = q(\hbar/2T)$. From the positive definiteness of the friction term we conclude immediately that the energy dissipation in the medium results in decreasing the quantum tunneling probability and decay rate of metastable state.

Let us estimate the contribution of friction term into the decay probability at $T = 0$. Let typical barrier size be q_c (Fig. 3.14) and typical time of underbarrier motion be τ_c. Then the approximate order-of-magnitude estimate of the integral with the friction term equals

$$S_\mu \sim \gamma^2(q_c)\frac{1}{\tau_c^2}\tau_c^2 \sim \gamma^2(q_c) \sim \frac{\gamma^2(q_c)}{q_c^2}q_c^2 \sim \gamma'^2(q_c)q_c^2 \sim \mu(q_c)q_c^2\,.$$

If we introduce the typical velocity of underbarrier motion as $\dot{q}_c \sim q_c/\tau_c$, the latter expression can be interpreted as a work of friction force or energy dissipated under barrier $[\mu(q_c)\dot{q}_c]q_c$ multiplied by the tunneling time τ_c.

On the whole, the magnitude of effective action can be estimated as a sum of dissipationless term S_0 at $\mu = 0$ and dissipative term S_μ. Then we have

$$\Gamma(T) \sim \exp(-A) \;\; \text{where} \;\; A \sim \frac{S_0 + S_\mu}{\hbar}\,.$$

For high temperatures, the minimal extremum of action realizes at the classical stationary path $q(\tau) = q_0$ where point q_0 satisfying $U'(q_0) = 0$ corresponds to the maximum of the potential barrier height U_0. The classical path gives the Arrhenius answer

$$S_{\text{eff}} = \frac{\hbar U_0}{T} \quad \text{and} \quad A = \frac{S_{\text{eff}}}{\hbar} = \frac{U_0}{T}\,.$$

Comparing the exponents in these two cases, we obtain the estimate

$$T_q \sim \frac{\hbar U_0}{S_0 + S_\mu}$$

for the thermal-quantum crossover temperature. The energy dissipation in the medium reduces the temperature region of quantum decay.

49. Let us write effective action S_{eff} in imaginary time τ, corresponding to the equation of particle motion $\mu\dot{q}_t + U'(q_t) = 0$ in the real time t

$$S_{\text{eff}}[q_\tau] = \int\limits_{-\hbar/2T}^{\hbar/2T} \frac{M\omega^2 q_\tau^2}{2}\left(1 - \frac{q_\tau}{q_c}\right)d\tau$$

$$- \frac{\mu}{2\pi}\iint\limits_{-\hbar/2T}^{\hbar/2T} \gamma(q_\tau)\frac{(\pi T/\hbar)^2}{\sin^2[\pi T(\tau - s)/\hbar]}\gamma(q_s)\,d\tau\,ds.$$

In order to find the decay probability $\Gamma(T) \sim \exp(-A)$ within the exponential accuracy, it is necessary to calculate the extremum value S_{eff} for the closed symmetrical path $q(-\hbar/2T) = q(\hbar/2T)$. By that, $A = S_{\text{eff}}/\hbar$. We solve the problem with the aid of expanding the path $q(\tau)$ into a Fourier series over Matsubara frequencies $\omega_n = 2\pi nT/\hbar$

$$q(\tau) = \frac{T}{\hbar}\sum_n q_n\, e^{i\omega_n\tau}, \qquad (n = 0,\, \pm 1,\, \pm 2,\, \ldots).$$

Substituting this expansion into the expression for the action gives

$$S_{\text{eff}}[q_n] = \frac{T}{\hbar}\sum_n \frac{M\omega^2}{2} q_n q_{-n}$$

$$- \left(\frac{T}{\hbar}\right)^2 \sum_{n,m} \frac{M\omega^2}{2q_c} q_n q_m q_{-(m+n)} + \frac{\mu}{2}\frac{T}{\hbar}\sum_n |\omega_n| q_n q_{-n}\,.$$

Since $q(\tau)$ is a real function and the extremum path can be taken as an even time function $q(\tau) = q(-\tau)$, the expansion coefficients q_n possess the symmetric property as $q_{-n} = q_n$. Varying S_{eff} with respect to q_n, we get the equation determining the extremum path

$$\frac{\delta S_{\text{eff}}}{\delta q_n} = M\omega^2 q_n - \frac{3}{2}\frac{M\omega^2}{q_c}\frac{T}{\hbar}\sum_m q_m q_{m+n} + \mu|\omega_n| q_n = 0.$$

The solution is attempted as $q_n = B\exp(-b|n|)$. Using that

$$\sum_{m=-\infty}^{\infty} e^{-b|m+n|}e^{-b|m|} = \big(|n| + \coth b\big)e^{-b|n|}, \quad (b > 0),$$

we obtain a simple equation to determine unknown B and b

$$M\omega^2 - \frac{3}{2}\frac{M\omega^2}{q_c}\frac{T}{\hbar}B\big(|n| + \coth b\big) + 2\pi\mu\frac{T}{\hbar}|n| = 0.$$

Hence, we find that

$$B = \frac{4\pi\mu q_c}{3M\omega^2} = \frac{2}{3}q_c\frac{\hbar}{T_q}, \quad \left(T_q = \frac{\hbar M\omega^2}{2\pi\mu}\right),$$
$$\coth b = \frac{\hbar M\omega^2}{2\pi\mu T} = \frac{T_q}{T}.$$

It is seen from the solution that the parameter b tends to the infinity at temperature $T = T_q$. This means that all the harmonics with $n \neq 0$ vanish and $q(\tau) \equiv BT_q/\hbar = 2q_c/3 = q_0$. The point q_0 corresponds to the coordinate at which the potential barrier reaches its maximum value U_0

$$U_0 = U(q_0) = \frac{2M\omega^2 q_c^2}{27}, \quad q_0 = \frac{2}{3}q_c.$$

In other words, at temperatures $T \geqslant T_q$ there exist no nontrivial time-dependent paths but there exists a single classical path $q(\tau) \equiv q_0$ with $S_{\text{eff}} = \hbar U_0/T$.

Let us turn now to calculating the extremum S_{eff}. Here we are in need of attracting two uncomplicated sums

$$\sum_n e^{-2b|n|} = \coth b \quad \text{and} \quad \sum_n |n|e^{-2b|n|} = \frac{1}{2}(\coth^2 b - 1).$$

Substituting the solution $q_n = B\exp(-b|n|)$ into $S_{\text{eff}}[q_n]$ yields

$$S_{\text{eff}} = \frac{T}{\hbar}B^2\left[\frac{M\omega^2}{2}\coth b - B\frac{T}{\hbar}\frac{M\omega^2}{2q_c}\frac{3\coth^2 b - 1}{2} + \frac{\pi\mu T}{\hbar}\frac{\coth^2 b - 1}{2}\right]$$
$$= \frac{\pi\mu T^2B^2}{2\hbar^2}\left(\coth^2 b - \frac{1}{3}\right) = \frac{2\pi\mu q_c^2}{9}\left(1 - \frac{1}{3}\frac{T^2}{T_q^2}\right) = \frac{3}{2}\frac{\hbar U_0}{T_q}\left(1 - \frac{1}{3}\frac{T^2}{T_q^2}\right).$$

Finally, we have for the exponent A in the expression of decay rate

$$A = A(T) = \begin{cases} \frac{3}{2}\frac{U_0}{T_q}\left(1 - \frac{1}{3}\frac{T^2}{T_q^2}\right), & T < T_q, \\ \frac{U_0}{T}, & T > T_q. \end{cases}$$

At temperature T_q there occurs a crossover from the quantum decay regime to the classical thermal activation governed by the Arrhenius formula.

To conclude, let us find the explicit formula for the extremum path

$$q(\tau) = \frac{BT}{\hbar}\sum_n e^{-b|n|}e^{2\pi i n T\tau/\hbar} = \frac{2q_c}{3}\frac{T}{T_q}\frac{1 - e^{-2b}}{1 - 2e^{-2b}\cos(2\pi T\tau/\hbar) + e^{-2b}}$$

$$= \frac{2q_c}{3}\frac{(1 - e^{-2b})\tanh b}{1 - 2e^{-b}\cos(2\pi T\tau/\hbar) + e^{-2b}}.$$

The behavior at zero temperature $T = 0$ can be obtained by expanding the exponentials, tangents, and cosine in the limit of small arguments

$$q(\tau) = q_c\frac{4/3}{1 + (2\pi T_q\tau/\hbar)^2}.$$

Emphasize the following physical aspect. In a medium with energy dissipation the exit point $q = 4q_c/3$ where the particle escapes from the potential barrier lies farther than the usual exit point $q = q_c$ under dissipationless tunneling.

3.17 Macroscopic Quantum Nucleation

50. The tunneling as a dynamic process is associated, first of all, with the kinetic energy of a particle. The growth of a nucleus with density $\rho' = \rho + \Delta\rho$, different from ρ of the surrounding liquid, will be accompanied with the fluid flow toward the nucleus or outwards, depending on the sign $\Delta\rho$ in order to compensate the density difference. The fluid flow will inevitably be accompanied by an appearance of kinetic energy of a liquid, additional to the nucleus energy $U(R) = 4\pi\sigma R^2(1 - R/R_c)$.

So, let the nucleus radius vary in time as $R(t)$. We determine the kinetic energy of a nucleus as a kinetic energy of the whole liquid

$$K = \frac{1}{2}\int \rho(\boldsymbol{r})v^2(\boldsymbol{r})\,d^3r.$$

To find the velocity distribution in the incompressible liquid, we employ the continuity equations inside and outside the nucleus

$$\operatorname{div}\boldsymbol{v}(\boldsymbol{r}) = 0 \quad \text{and} \quad \operatorname{div}\boldsymbol{v}'(\boldsymbol{r}) = 0.$$

Because of spherical symmetry of a nucleus the velocity distribution will also have the spherical symmetry. We find immediately

$$v(r) = v(R)\frac{R^2}{r^2}, \quad r > R(t),$$
$$v'(r) = 0, \quad r < R(t).$$

Due to incompressibility of new phase inside the nucleus the flow velocity is identically zero. Otherwise, we would have the unlimited magnitude of velocity at the nucleus center $r = 0$.

To determine the fluid velocity $v(R)$ at the nucleus boundary $r = R(t)$, we use the continuity condition for the mass flux of substance across the nucleus boundary $r = R$ growing at rate $\dot{R}$

$$\rho'(0 - \dot{R}) = \rho\big(v(R) - \dot{R}\big).$$

Hence, we find the fluid velocity at the nucleus boundary

$$v(R) = -\frac{\rho' - \rho}{\rho}\dot{R} = -\frac{\Delta\rho}{\rho}\dot{R}$$

and the kinetic energy of the nucleus

$$K = \frac{1}{2}\int\limits_{r>R(t)} \rho\left(-\frac{\Delta\rho}{\rho}\dot{R}\right)^2 \frac{R^4}{r^4} 4\pi r^2\, dr = 4\pi\frac{(\Delta\rho)^2}{\rho}R^3\frac{\dot{R}^2}{2} = M(R)\frac{\dot{R}^2}{2}.$$

Here we have defined the nucleus mass $M(R) = 4\pi\rho_{\text{eff}}R^3$ with the effective density $\rho_{\text{eff}} = (\Delta\rho)^2/\rho$. As a result, we arrive at the *Rayleigh-Plesset Lagrangian L* describing the nucleus evolution

$$L(R, \dot{R}) = \frac{M(R)\dot{R}^2}{2} - U(R) = 4\pi\rho_{\text{eff}}R^3\frac{\dot{R}^2}{2} - 4\pi\sigma R^2\left(1 - \frac{R}{R_c}\right).$$

In order to find the nucleation probability of critical nucleus or decay rate of metastable liquid phase, we apply the conception of effective action

$$S_{\text{eff}}[R(\tau)] = \int\limits_{-\hbar/2T}^{\hbar/2T} \left(\frac{1}{2}M(R)\left(\frac{dR}{d\tau}\right)^2 + U(R)\right)d\tau$$

specified in the imaginary time τ and then calculate the extremum of action at periodic path $R(-\hbar/2T) = R(\hbar/2T)$.

Let us start from examining the quantum path. Unlike the tunneling of particles with the constant mass when the motion period can reach the infinite magnitude, in the problem with the coordinate-dependent mass the motion period remains always finite even for the path corresponding to the bottom $E = 0$ of potential well. In fact, from the equation

$$\frac{M(R)\dot{R}^2}{2} - U(R) = E = 0 \quad \text{or} \quad \left(\frac{dR}{d\tau}\right)^2 = \frac{2\sigma}{\rho_{\text{eff}} R}\left(1 - \frac{R}{R_c}\right)$$

we obtain the following law of motion

$$\frac{|\tau(x)|}{\tau_c} = \frac{\pi}{2} - \arcsin\sqrt{x} + \sqrt{x(1-x)}\,.$$

Here we denote $x = R/R_c$ and $\tau_c = \sqrt{M(R_c)/8\pi\sigma} = \sqrt{\rho_{\text{eff}} R_c^3/2\sigma}$. The motion period remains finite and equals $\pi\tau_c$. Thus, while the temperature T of metastable liquid remains sufficiently low, at least, smaller than $\hbar/\pi\tau_c$, the extremum value of effective action is achieved at the boundary $E = 0$ of the possible energy interval. Therefore, the extremum value of effective action proves to be temperature independent. Substituting the dependence found for path $R(\tau)$ gives the simple answer

$$S_0 = 2\int_0^{R_c} \sqrt{2M(R)U(R)}\, dR = \frac{5\sqrt{2}\,\pi^2}{16}\sqrt{\sigma\rho_{\text{eff}}}\, R_c^{7/2}, \qquad T < \frac{\hbar}{\pi\tau_c}\,.$$

On the other hand, there always exists a classical path $R(\tau) \equiv 2R_c/3$ corresponding to extremum $U'(R) = 0$. This entails the following extremum value of action

$$S_{\text{cl}} = \frac{\hbar U_0}{T}, \qquad U_0 = \frac{4}{27}4\pi\sigma R_c^2,$$

where U_0 is the potential barrier height. It is easy to see that for temperature T_q,

$$T_q = \frac{128}{135\pi}\hbar\left(\frac{2\sigma}{\rho_{\text{eff}} R_c^3}\right)^{1/2} = \frac{128}{135}\frac{\hbar}{\pi\tau_c} < \frac{\hbar}{\pi\tau_c}\,,$$

the values of action $S_{\text{eff}}(0)$ for the quantum path and action S_{cl} for the classical path become equated. We should choose the smallest value from two possible ones. Since the motion period for the quantum path $\pi\tau_c$ proves to be smaller than $\hbar/T_q$, the temperature T_q will be the thermal-quantum crossover temperature. As a result, we have for the decay rate of metastable phase within the exponential accuracy

$$\Gamma(T) = B\exp(-A), \qquad A(T) = \begin{cases} S_0/\hbar, & T < T_q, \\ U_0/T, & T > T_q. \end{cases}$$

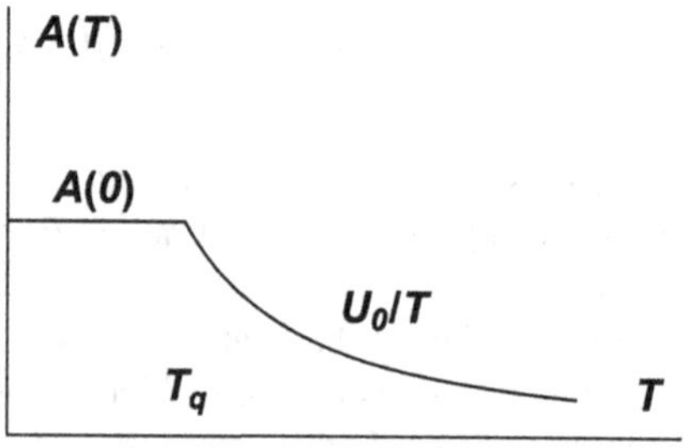

Fig. 3.15 The temperature behavior of exponent $A(T)$

In Fig. 3.15 the behavior of exponent $A(T)$ is shown as a function of temperature.

The preexponential factor B can be estimated as a tunneling attempt frequency ν multiplied by the number N of possible nucleation centers in volume V, i.e. as $B \sim \nu N$. For the order-of-magnitude estimate of the attempt frequency or heterophase fluctuation frequency, we take the frequency of motion in the potential $U(R)$ at $R \ll R_c$ according to the equation

$$4\pi \rho_{\text{eff}} R^3 (\nu R)^2 \sim 4\pi \sigma R^2.$$

We estimate ν taking the typical spatial scale of heterophase fluctuations R as several interatomic distances a

$$\nu \sim \sqrt{\sigma/(\rho_{\text{eff}}\, a^3)}\,.$$

The number of possible nucleation centers is estimated as $N \sim V/(4\pi R_c^3/3)$.

51. As usual, at low temperatures the mean free path of excitations $l(T)$ grows drastically as the temperature lowers. Depending on the temperature, the hydrodynamic and ballistic regimes of nucleus growth are possible. The calculation of energy dissipation rate of nucleus $Q = -dE/dt$ differs in kind for these two limiting cases. Let us start from the hydrodynamical regime $R \gg l(T)$. In this case we can apply the hydrodynamical formula for the density of energy dissipation function integrated over the whole volume of the system

$$Q = Q_\eta + Q_\varkappa = \int \frac{\eta}{2}\left(\frac{\partial v_i}{\partial x_k} + \frac{\partial v_k}{\partial x_i}\right)^2 d^3r + \int \varkappa \frac{(\nabla T)^2}{T}\, d^3r.$$

For calculating the viscous contribution Q_η, we use the velocity distribution $\boldsymbol{v}(\boldsymbol{r})$ found in the previous problem

$$\boldsymbol{v}(\boldsymbol{r}) = \begin{cases} -\frac{\Delta\rho}{\rho}\dot{R}\frac{R^2}{r^2}\frac{\boldsymbol{r}}{r}, & r > R(t), \\ 0, & r < R(t). \end{cases}$$

Integrating leads to

$$Q_\eta = 16\pi\eta\left(\frac{\Delta\rho}{\rho}\right)^2 R\dot{R}^2 = \mu_\eta(R)\dot{R}^2.$$

The corresponding drag force that slows down the nucleus growth is analogous to the Stokes force for a ball of radius R traveling at velocity $v(R) = -(\Delta\rho/\rho)\dot{R}$ in a viscous liquid.

For calculating the heat conduction contribution, it is required to find the spatial temperature distribution which becomes inhomogeneous resulting from nonzero latent heat $L(T)$. For this purpose, we use the *general equation of heat transfer* for the incompressible liquid with specific heat C_p at constant pressure

$$C_p\left(\frac{\partial T}{\partial t} + \boldsymbol{v}\nabla T\right) = \varkappa\nabla^2 T + \frac{\eta}{2}\left(\frac{\partial v_i}{\partial x_k} + \frac{\partial v_k}{\partial x_i}\right)^2.$$

We will solve this equation in the limit of quasi stationary growth $\dot{R} \to 0$. Expecting naturally that the temperature variation δT is proportional to the growth rate $\dot{R}$, all the terms of the equation, with the exception of term with $\nabla^2 T$, will include the terms proportional to $\dot{R}^2$ or $\ddot{R}$. The latter terms are of higher order of smallness in growth rate $\dot{R}$ and thus can be neglected. Therefore, in first approximation the general equation of heat transfer reduces to the *stationary heat equation*

$$\nabla^2 T = \frac{1}{r^2}\frac{\partial}{\partial r}\left(r^2\frac{\partial T}{\partial r}\right) = 0$$

which has the following general solution on taking the spherical symmetry into consideration

$$T(r) = \begin{cases} T + \delta T(t)\frac{R}{r}, & r > R(t), \\ T', & r < R(t). \end{cases}$$

An unknown temperature jump $\delta T(t)$ at the nucleus boundary $r = R$ will be determined from the continuity condition for the energy flux across the boundary

$$(v' - \dot{R})\rho'[\frac{(v' - \dot{R})^2}{2} + w'] + q' = (v - \dot{R})\rho[\frac{(v - \dot{R})^2}{2} + w] + q,$$

where the dashed quantities refer to the nucleus. Accordingly, $v' = 0$, $v = v(R)$, fluxes q and q' are the dissipative heat flows, w and w' are the enthalpies (heat functions) per unit mass. Putting $q' \sim \nabla T' = 0$ and involving the smallness of velocities, we obtain in the linear approximation in $\dot{R}$

$$-\dot{R}\rho' w' = (v - \dot{R})\rho w + q.$$

Then, using the continuity of mass flow at the nucleus boundary as $-\rho'\dot{R} = \rho(v - \dot{R})$, we get the heat flow at the boundary $r = R$

$$q(r = R) = \rho'(w - w')\dot{R}.$$

The enthalpy (heat function) equals $w = \mu + Ts$ where μ is the chemical potential and s is the entropy. For the difference $w - w' = (\mu - \mu') + (Ts - T's')$, in order to stay, on the whole, within the framework of the linear approximation in $\dot{R}$, it is necessary to take the values of chemical potentials and entropy at the phase equilibrium, i.e. at $\dot{R} \equiv 0$. Then, putting $\mu' = \mu$ and $T' = T$, we find

$$w - w' \approx T(s - s') = L,$$

$L = L(T)$ being the latent heat of phase transition. Eventually, the heat flow at the nucleus boundary equals

$$q(r = R) \approx \rho' L\dot{R}.$$

Depending on the sign L, the heat either is carried away from the growing nucleus or delivered to the nucleus. The magnitude of heat flow is proportional to the nucleus growth rate.

We are now in the position to determine the temperature jump δT at the nucleus boundary. From the relation

$$q = -\varkappa\nabla T = -\varkappa\left(-\delta T\frac{R}{r^2}\right)_{r=R} = \frac{\varkappa\,\delta T}{R} = \rho' L\dot{R}$$

we find δT and the temperature distribution

$$T(r) = \begin{cases} T + \frac{\rho' L}{\varkappa}\frac{R^2}{r}\dot{R}, & r > R(t), \\ T' = T + \frac{\rho' LR}{\varkappa}\dot{R}, & r < R(t). \end{cases}$$

To determine the temperature T' inside the nucleus, we have applied the condition of temperature continuity at the boundary between two media. This condition is used for the heat equations under hydrodynamical description of heat flows.

Releasing the heat in the phase transition $L > 0$ results in the local increase of the temperature beside the growing nucleus. The large heat conduction in the medium surrounding the nucleus moderates this effect. On the contrary, the negative latent heat $L < 0$ leads to the local decrease of temperature. The calculation of heat conduction contribution to the energy dissipation rate becomes now very simple

$$Q_\varkappa = \int \varkappa\frac{(\nabla T)^2}{T}\,d^3r = 4\pi\frac{L^2}{\varkappa T}R^3\dot{R}^2 = \mu_\varkappa(R)\dot{R}^2.$$

The viscosity and heat conduction in the metastable medium around the nucleus produce the drag force that tends to slow the nucleus growth. The nucleus energy dissipates at the rate

$$\frac{\partial E}{dt}=\frac{d}{dt}\left(\frac{\partial L}{\partial \dot{R}}\dot{R}-L\right)=-\mu(R)\dot{R}^2,\quad \mu(R)=\mu_\eta(R)+\mu_\varkappa(R)$$

and $L=L(R,\dot{R})$ is the *Rayleigh-Plesset Lagrangian*

$$L(R,\dot{R})=\frac{1}{2}M(R)\dot{R}^2-U(R).$$

Thus, the quasi stationary dynamics of nucleus evolution can be described with introducing an additional drag force $F=-\mu(R)\dot{R}$ which is a linear function of growth rate $\dot{R}$ and hinders the nucleus growth.

As the temperature lowers, the mean free path of elementary excitations $l(T)$ in the medium increases strongly. When the mean free path becomes comparable with the typical critical radius R_c, the hydrodynamic relations prove to be inapplicable for calculating the energy dissipation rate. Here it is necessary to use the transport equation with the boundary conditions at the nucleus surface, which depend on the character of interaction of elementary excitations with the nucleus surface.

In the ballistic $l(T)\gg R_c$ regime the collisions among elementary excitations are sufficiently rare. The excitations collide with the nucleus surface but relax in the region far away from the surface. In the ballistic regime the rate of dissipating the total nucleus energy becomes proportional to the nucleus surface area $4\pi R^2$. The rate can be found with the aid of the general equation of heat transfer

$$-\frac{dE}{dt}=Q=-\int \tau'_{ik}\frac{\partial v_i}{\partial x_k}d^3r-\int \boldsymbol{q}\frac{\nabla T}{T}d^3r,$$

where τ'_{ik} is the dissipative part of momentum flux density tensor and $\boldsymbol{q}$ is the dissipative heat flow. Then we obtain assuming the variations of all quantities to be small due to quasi stationary approximation $\dot{R}\to 0$

$$-\frac{dE}{dt}=Q=4\pi R^2\tau'_{rr}(R)v(R)+4\pi R^2 q(R)\frac{T(R)-T}{T}.$$

Here $\tau'_{rr}(R)$ is the tensor component at $r=R$ in the direction normal to the nucleus surface and $T-T(R)$ is the temperature jump at the surface. The magnitude $\tau'_{rr}(R)$ is determined with the momentum flow transferred by elementary excitations to the nucleus surface. The magnitude $\tau'_{rr}(R)$ is proportional to the fluid velocity $v(R)=-(\Delta\rho/\rho)\dot{R}$ with respect to the nucleus surface

$$\tau'_{rr}(R)=\vartheta v(R).$$

The order-of-magnitude estimate of coefficient ϑ equals the product of excitation density $\rho_{\rm ex}$ by the typical excitation velocity $u_{\rm ex}$, i.e. $\vartheta\sim\rho_{\rm ex}u_{\rm ex}$.

The temperature jump at the boundary between two media is connected with the heat flow via *thermal Kapitza resistance* R_K according to

$$T(R) - T = \frac{q(R)}{R_K}.$$

The magnitude R_K is of the order of the product of specific heat C_p by the typical velocity of excitations u_{ex}. Correspondingly, $R_K \sim C_p u_{\text{ex}}$. We arrive at the result

$$\frac{dE}{dt} = -4\pi R^2 \left[\vartheta v^2(R) + \frac{q^2(R)}{R_K T} \right]$$
$$= 4\pi R^2 \left[\left(\frac{\Delta\rho}{\rho} \right)^2 \vartheta + \frac{(\rho' L)^2}{R_K T} \right] \dot{R}^2 = -\mu(R) \dot{R}^2.$$

Note the dependence of energy dissipation power on the nucleus growth rate $\dot{R}$ remains the same. The behavior of drag coefficient $\mu(R)$ as a function of nucleus radius R varies and ceases to depend on the mean free path $l(T)$.

Emphasize also that the drag coefficient $\mu(R)$ is significantly temperature dependent via the kinetic coefficients of viscosity and heat conduction having different temperature behavior in various condensed media, e.g. normal or superfluid liquids.

The quantum description of nucleus growth reduces to analyzing the effective action $S_{\text{eff}}[R_\tau]$ in imaginary time τ with the periodic paths $R(-\hbar/2T) = R(\hbar/2T)$

$$S_{\text{eff}}[R_\tau] = \int\limits_{-\hbar/2T}^{\hbar/2T} \left(\frac{M(R_\tau)\dot{R}_\tau^2}{2} + U(R_\tau) \right) d\tau$$
$$+ \frac{1}{4\pi} \iint\limits_{-\hbar/2T}^{\hbar/2T} \left[\gamma(R_\tau) - \gamma(R_s) \right]^2 \frac{(\pi T/\hbar)^2}{\sin^2[\pi T(\tau - s)/\hbar]} d\tau\, ds.$$

Here $\gamma(R)$ is determined by the drag coefficient $\mu(R)$ from the relation

$$\mu(R) = \gamma'^2(R) \quad \text{or} \quad \gamma(R) = \int\limits_0^R \sqrt{\mu(R)}\, dR.$$

In the general case the search of extremum paths for the effective action is a very complicated problem. The decay rate $\Gamma(T) \sim \exp[-A(T)]$ of metastable condensed medium within the exponential accuracy can be estimated in accordance with the relation

$$A(T) = \begin{cases} \left[S_0(R_c) + \mu(R_c, T) R_c^2 \right]/\hbar, & T < T_q(R_c), \\ U_0(R_c)/T, & T > T_q(R_c). \end{cases}$$

Here $S_0(R_c)$ is the extremum value of the action in the dissipationless medium found in the previous problem and $U_0(R_c)$ is the potential barrier height. The temperature

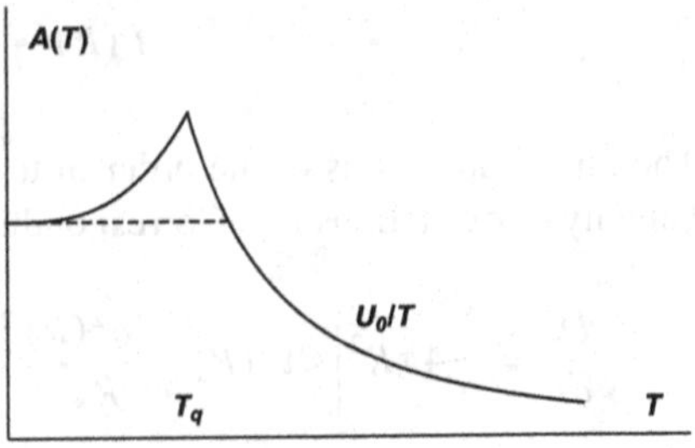

Fig. 3.16 The behavior of exponent $A(T)$ versus the temperature of dissipative medium. The dashed line corresponds to the dissipationless case

T_q determining the thermal-quantum crossover equals approximately

$$T_q \sim \frac{\hbar U_0(R_c)}{S_0(R_c) + \mu(R_c, T_q)R_c^2}.$$

So, the energy dissipation in a condensed medium, firstly, slows down the quantum decay rate and decreases the crossover temperature to the quantum regime. Secondly, the exponent A in the quantum decay region starts significantly to depend on the temperature due to the temperature-dependent kinetic coefficients. The example for the behavior $A(T)$ versus temperature is shown in Fig. 3.16.

3.18 Open Quantum Systems: The Lindblad Equation

52. Let us write the Lindblad equation with operator $\hat{L}_- = \sqrt{\gamma}\sigma_-$

$$\frac{d\hat{\rho}}{dt} = -\frac{i}{\hbar}\big[\hat{H}, \hat{\rho}\big] + \gamma\Big(\sigma_-\hat{\rho}\sigma_+ - \frac{1}{2}\sigma_+\sigma_-\hat{\rho} - \frac{1}{2}\hat{\rho}\sigma_+\sigma_-\Big).$$

Performing a simple calculation, we obtain the matrix equation

$$\begin{pmatrix} \dot{\rho}_{ee}(t) & \dot{\rho}_{eg}(t) \\ \dot{\rho}_{ge}(t) & \dot{\rho}_{gg}(t) \end{pmatrix} = \begin{pmatrix} -\gamma\rho_{ee}(t) & \big(-i\omega - \gamma/2\big)\rho_{eg}(t) \\ \big(i\omega - \gamma/2\big)\rho_{ge}(t) & \gamma\rho_{ee}(t) \end{pmatrix},$$

entailing the following answer:

$$\hat{\rho}(t) = \begin{pmatrix} \rho_{ee} & \rho_{eg} \\ \rho_{ge} & \rho_{gg} \end{pmatrix} = \begin{pmatrix} e^{-\gamma t}\rho_{ee}(0) & e^{(-i\omega-\gamma/2)t}\rho_{eg}(0) \\ e^{(i\omega-\gamma/2)t}\rho_{ge}(0) & 1 - e^{-\gamma t}\rho_{ee}(0), \end{pmatrix}$$

where $\rho_{ge}(0) = \rho^*_{eg}(0)$ due to the Hermitian property of matrix. The matrix elements ρ_{ee} and ρ_{gg} represent the *populations* of excited and ground levels. The nondiagonal elements $\rho_{ge} = \rho^*_{eg}$ can be characterized as *coherences*. In the time limit $t \to \infty$, as we see, the excited level population vanishes but the ground level population tends to unity. As a consequence of such relaxation, the cross terms, responsible for the coherent effects, vanish as well.

53. The solution is similar to the previous problem. The Lindblad equation with operator $\hat{L}_+ = \sqrt{\gamma}\sigma_+$ reads

$$\frac{d\hat{\rho}}{dt} = -\frac{i}{\hbar}[\hat{H},\hat{\rho}] + \gamma\Big(\sigma_+\hat{\rho}\sigma_- - \frac{1}{2}\sigma_-\sigma_+\hat{\rho} - \frac{1}{2}\hat{\rho}\sigma_-\sigma_+\Big).$$

We have the following matrix equation after the straightforward calculation:

$$\begin{pmatrix} \dot{\rho}_{ee}(t) & \dot{\rho}_{eg}(t) \\ \dot{\rho}_{ge}(t) & \dot{\rho}_{gg}(t) \end{pmatrix} = \begin{pmatrix} \gamma\rho_{gg}(t) & (-i\omega - \gamma/2)\rho_{eg}(t) \\ (i\omega - \gamma/2)\rho_{ge}(t) & -\gamma\rho_{gg}(t) \end{pmatrix},$$

resulting in

$$\hat{\rho}(t) = \begin{pmatrix} \rho_{ee} & \rho_{eg} \\ \rho_{ge} & \rho_{gg} \end{pmatrix} = \begin{pmatrix} 1 - e^{-\gamma t}\rho_{gg}(0) & e^{(-i\omega-\gamma/2)t}\rho_{eg}(0) \\ e^{(i\omega-\gamma/2)t}\rho_{ge}(0) & e^{-\gamma t}\rho_{gg}(0), \end{pmatrix}$$

where $\rho_{ge}(0) = \rho^*_{eg}(0)$ due to the Hermitian attribute of matrix. In the time limit $t \to \infty$ the excited level population enhances and reaches the maximum value equal to unity. On the contrary, the ground level population decreases and vanishes completely. Therefore, the Lindblad operator $\hat{L}_+ = \sqrt{\gamma}\sigma_+$ describes the pumping or photon absorption of two-level system. The cross terms, responsible for the coherent effects, disappear alongside this.

54. The solution is performed in the similar way of two previous problems. Let us write the Lindblad equation with operator $\hat{L}_3$

$$\frac{d\hat{\rho}}{dt} = -\frac{i}{\hbar}[\hat{H},\hat{\rho}] + \hat{L}_3\hat{\rho}\hat{L}_3^+ - \frac{1}{2}\hat{L}_3^+\hat{L}_3\hat{\rho} - \frac{1}{2}\hat{\rho}\hat{L}_3^+\hat{L}_3\Big)$$

and then rewrite it in the matrix form

$$\begin{pmatrix} \dot{\rho}_{ee} & \dot{\rho}_{eg} \\ \dot{\rho}_{ge} & \dot{\rho}_{gg} \end{pmatrix} = \begin{pmatrix} 0 & \big(-i\omega - (\gamma_e + \gamma_g)^2/2\big)\rho_{eg} \\ \big(i\omega - (\gamma_e + \gamma_g)^2/2\big)\rho_{ge} & 0 \end{pmatrix}.$$

Solving the equation gives the following behavior in time:

$$\begin{pmatrix} \rho_{ee}(t) & \rho_{eg}(t) \\ \rho_{ge}(t) & \rho_{gg}(t) \end{pmatrix} = \begin{pmatrix} \rho_{ee}(0) & e^{(-i\omega-(\gamma_e+\gamma_g)^2/2)t}\rho_{eg}(0) \\ e^{(i\omega-(\gamma_e+\gamma_g)^2/2)t}\rho_{ge}(0) & \rho_{gg}(0), \end{pmatrix}$$

where $\rho_{ge}(0) = \rho^*_{eg}(0)$ is due to the Hermitian property of matrix. We see here that the populations of the levels remain unchanged. The dissipative effects in the $t \to \infty$

limit, if $\gamma_g \neq -\gamma_e$ (matrix $\hat{L}_3$ is not unit), results in vanishing any coherent effects described by the nondiagonal elements of density matrix.

In conclusion, we note the following. The concurrent implementation of three Lindblad operators $\hat{L}_-$, $\hat{L}_+$ and $\hat{L}_3$ with the three proper time scales allows us to describe phenomenologically the dynamical behavior of two-level systems. For example, employing two operators $\hat{L}_- = \sqrt{\gamma_-}\sigma_-$ and $\hat{L}_+ = \sqrt{\gamma_+}\sigma_+$ responsible for the transitions from the excited level to the ground one and inversely (relaxation and pumping), we can obtain the stationary solution for the population of levels. The formula below gives the stationary or equilibrium value for the population of excited level:

$$(\rho_{ee})_{eq} = \frac{\gamma_+}{\gamma_+ + \gamma_-}.$$

If a two-level atom interacts with the equilibrium electromagnetic field, the probability for absorbing or emitting a photon of frequency ω can be written as $\gamma_+ = \gamma_0 N(\omega)$ and $\gamma_- = \gamma_0(N(\omega) + 1)$, where γ_0 is the probability of photon emission and $N(\omega) = 1/(\exp(\hbar\omega/T) - 1)$ is the equilibrium Planck distribution function. Accordingly, we get the equilibrium population value for the excited level

$$(\rho_{ee})_{eq} = \frac{\gamma_+}{\gamma_+ + \gamma_-} = \frac{N(\omega)}{2N(\omega) + 1} = \frac{1}{e^{\hbar\omega/T} + 1}.$$

For the interaction with the Fermi heat bath, one should write the transition probabilities as $\gamma_+ = \gamma_0 n(\omega)$ and $\gamma_- = \gamma_0(1 - n(\omega))$, where γ_0 is the transition probability with producing one Fermi particle in the heat bath and $n(\omega) = 1/(\exp(\hbar\omega/T) + 1)$ is the Fermi distribution. Eventually, we arrive at the following value for the equilibrium population of the excited level:

$$(\rho_{ee})_{eq} = \frac{\gamma_+}{\gamma_+ + \gamma_-} = \frac{n(\omega)}{n(\omega) + 1 - n(\omega)} = \frac{1}{e^{\hbar\omega/T} + 1}.$$

This simple example shows that the equilibrium population in the open dissipative system is independent of the kind of heat bath which the system interacts.

55. The density matrix $\hat{\rho}$ is calculated from the equation

$$\hat{\rho}(t) = \int_{-\infty}^{\infty} d\theta_e \int_{-\infty}^{\infty} d\theta_g\, P(\theta_e, \theta_g)|\psi(t)\rangle\langle\psi(t)|$$

$$= \int_{-\infty}^{\infty} d\theta_e \int_{-\infty}^{\infty} d\theta_g\, P(\theta_e, \theta_g)\big[aa^*|e\rangle\langle e| + ab^* e^{-i\omega t} e^{i(\theta_e - \theta_g)}|e\rangle\langle g|$$

$$+ a^*b e^{i\omega t} e^{-i(\theta_e - \theta_g)}|g\rangle\langle e| + bb^*|g\rangle\langle e|\big].$$

Calculating the integrals gives the answer

$$\hat{\rho}(t) = aa^*|e\rangle\langle e| + ab^*e^{-i\omega t}e^{-(\lambda_e+\lambda_g)t/2}|e\rangle\langle g| \\ + a^*be^{i\omega t}e^{-(\lambda_e+\lambda_g)t/2}|e\rangle\langle g| + bb^*|g\rangle\langle g|$$

or in the matrix notations

$$\hat{\rho}(t) = \begin{pmatrix} aa^* & ab^*e^{-i\omega t}e^{-(\lambda_e+\lambda_g)t/2} \\ ba^*e^{i\omega t}e^{-(\lambda_e+\lambda_g)t/2} & bb^* \end{pmatrix} = \begin{pmatrix} \rho_{ee} & \rho_{eg} \\ \rho_{ge} & \rho_{gg} \end{pmatrix}.$$

Thus, the level population remains constant and the cross terms, responsible for the coherent effects, decay with the time constant $(\lambda_e + \lambda_g)^{-1}$.

It is possible now to find the derivative of density matrix with respect to time

$$\frac{d\hat{\rho}}{dt} = \begin{pmatrix} 0 & \big(-i\omega - (\lambda_e + \lambda_g)/2\big)\rho_{eg} \\ \big(i\omega - (\lambda_e + \lambda_g)/2\big)\rho_{ge} & 0 \end{pmatrix}.$$

Hence we see that the equation for the time derivative of density matrix coincides with the similar equation obtained on the basis of the Lindblad equation if in the Lindblad operator

$$\hat{L}_3 = \begin{pmatrix} \gamma_e & 0 \\ 0 & -\gamma_g \end{pmatrix}$$

we put $\gamma_e + \gamma_g = \sqrt{\lambda_e + \lambda_g}$. Therefore, the stochastic process of diffusive character, treated above, can be described with the help of the Lindblad equation.

3.19 Elements of Diagrammatic Keldysh Technique for Non-equilibrium Systems

56. (a) Let us consider the solution of equation

$$\begin{cases} \left(i\frac{\partial}{\partial t} - \varepsilon\right)g(t) = f(t), \\ g(t_0) = -g(t_1) \end{cases}$$

with the general right-hand side $f(t)$, time t varying on the Keldysh contour C (Fig. 3.17). According to definition of the Green function $G(t, t')$ the general solution of equation can be expressed in terms of Green function $G(t, t') = G(t - t')$ as follows:

$$g(t) = \int\limits_C G(t - t')f(t')\,dt'.$$

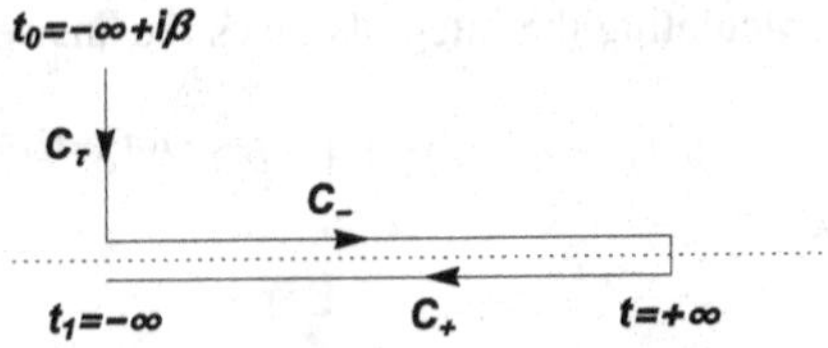

Fig. 3.17 The Keldysh contour

Here the integration is performed along the entire contour C. Let us represent the solution of equation as a sum of solving the homogeneous equation with an arbitrary constant A and the partial solution of equation with the right-hand side $f(t)$

$$g(t) = Ae^{-i\varepsilon t} - ie^{-i\varepsilon t}\int\limits_{t_0}^{t} e^{i\varepsilon t'} f(t')\,dt'.$$

Using the anti-periodic condition, we find an unknown constant A from the relation

$$-Ae^{-i\varepsilon t_0} = Ae^{-i\varepsilon t_1} - ie^{-i\varepsilon t_1}\int\limits_{C} e^{i\varepsilon t'} f(t')\,dt',$$

where the integration is performed along the entire contour C from the starting point to the ending one. Thus,

$$A = i\,n_f(\varepsilon)\int\limits_{C} e^{i\varepsilon t'} f(t')\,dt',$$

where $n_f(\varepsilon) = 1/(e^{\varepsilon/T}+1)$ is the Fermi distribution. Now the solution can be written as

$$g(t) = \int\limits_{C} i\big[n_f(\varepsilon) - \theta_c(t-t')\big]e^{-i\varepsilon(t-t')} f(t')\,dt',$$

where we have introduced the Heaviside function $\theta_c(t-t')$ determined on the Keldysh contour C as

$$\theta_c(t-t') = \begin{cases} 1, & t \geqslant t', \\ 0, & t < t'. \end{cases}$$

The lower index c means the time-contour ordering in the sense of contour C, i.e. inequality $t < t'$ implies that time moment t lies on the contour earlier than the time moment t'. The following properties of the Heaviside contour function are transparent:

$$\int_C \theta_c(t-t')\,dt' = \int_{t_0}^{t} dt', \quad \theta_c(t_0-t')\equiv 0, \quad \theta_c(t_1-t')\equiv 1.$$

So, the electron Green function is given by the simple expression

$$G(t,t') = G(t-t') = i\big[n_f(\varepsilon) - \theta_c(t-t')\big]e^{-i\varepsilon(t-t')}.$$

(b) For the real branch of contour $C_K = C_- + C_+$, there are four possibilities depending on the mutual position of time points t and t' on the contour branches C_- and C_+. Accordingly, it is promising to represent the structure of Green function $G(t,t')$ as a matrix 2×2

$$G(t,t') \to \begin{pmatrix} G^{--} & G^{-+} \\ G^{+-} & G^{++} \end{pmatrix}.$$

(1) Let both the time moments t and t' lie simultaneously on the upper branch C_-, i.e. $t \in C_-$ and $t' \in C_-$. In this case the Heaviside contour function $\theta_c(t-t')$ coincides with the traditional Heaviside function $\theta(t-t')$. Then the corresponding component of Green function equals

$$G(t-t') \equiv G^{--}(t-t') = i\big[n_f - \theta(t-t')\big]e^{-i\varepsilon(t-t')}.$$

Its Fourier transform is determined as

$$G^{--}(\omega) = \int_{-\infty}^{\infty} G^{--}(t)e^{i\omega t}dt = i\int_{-\infty}^{\infty}\big[n_f - \theta(t)\big]e^{i(\omega-\varepsilon)t}dt,$$

$$G^{--}(\omega) = \mathcal{P}\frac{1}{\omega-\varepsilon} - i\pi(1-2n_f)\delta(\omega-\varepsilon),$$

where $\mathcal{P}\frac{1}{x}$ is the *Cauchy principal value* for $1/x$.

(2) Let time moment t lie on the upper branch C_- but the time moment t' lie on the lower branch C_+, i.e. $t \in C_-$ and $t' \in C_+$. In this case the time moment t is always earlier than t' and $\theta_C(t-t') \equiv 0$. The corresponding component of Green function equals

$$G(t-t') \equiv G^{-+}(t-t') = i\,n_f e^{-i\varepsilon(t-t')}.$$

The notation *lesser* $G^{<} = G^{-+}$ is used as well. The Fourier transform is determined as

$$G^{-+}(\omega) = \int_{-\infty}^{\infty} G^{-+}(t)e^{i\omega t}dt = in_f\int_{-\infty}^{\infty} e^{i(\omega-\varepsilon)t}dt = 2\pi i n_f\delta(\omega-\varepsilon).$$

(3) Let time point t lie on the lower contour branch C_+ and time t' be on the upper branch C_-, i.e. $t \in C_+$ and $t' \in C_-$. In this case the time t is always later than t' and $\theta_C(t-t') \equiv 1$. The corresponding component of the Green function equals

$$G(t-t') \equiv G^{+-}(t-t') = -i(1-n_f)e^{-i\varepsilon(t-t')}.$$

The notation *greater* $G^> = G^{+-}$ is also usual. The Fourier transform is determined in the similar way

$$G^{+-}(\omega) = \int_{-\infty}^{\infty} G^{+-}(t)e^{i\omega t}dt = -2\pi i(1-n_f)\delta(\omega-\varepsilon).$$

In the thermodynamic equilibrium one has $G^< = -e^{-\beta\varepsilon}G^>$.

(4) Let both time points t and t' be on the lower contour branch C_+, i.e. $t \in C_+$ and $t' \in C_+$. Since the common practice is to believe that the ordinary time varies from $-\infty$ to $+\infty$ in the direction of the time real axis, the time direction on the contour branch C_+ will be the opposite. Therefore, if the time moment t is earlier than the time moment t' on the ordinary time real axis, the time ordering for the contour branch C_+ becomes inverse, namely, time moment t' will be earlier than the time t. Then, the contour function $\theta_c(t-t')$ will equal the usual Heaviside function but with the arguments reordered, i.e. $\theta_c(t-t') = \theta(t'-t)$. So, we define

$$G(t-t') = G^{++}(t-t') = i\big[n_f(\varepsilon) - \theta(t'-t)\big]e^{-i\varepsilon(t-t')}.$$

The corresponding Fourier transform equals

$$G^{++}(\omega) = \int_{-\infty}^{\infty} G^{++}(t)e^{i\omega t}dt = i\int_{-\infty}^{\infty}\big[n_f - \theta(-t)\big]e^{i(\omega-\varepsilon)t}dt,$$

$$G^{++}(\omega) = -\mathcal{P}\frac{1}{\omega-\varepsilon} - i\pi(1-2n_f)\delta(\omega-\varepsilon),$$

It is easy to see the general property for the Fourier transforms

$$G^{++}(\omega) = -[G^{--}(\omega)]^*.$$

Below we present a number of useful definitions. Let us determine the *retarded Green function* G^R as

$$G^R(t) = G^{--}(t) - G^{-+}(t) = \theta(t)\big[G^>(t) - G^<(t)\big] = -i\theta(t)e^{-i\varepsilon t}$$

and introduce its Fourier transform as usual

$$G^R(\omega) = \int_{-\infty}^{\infty} G^R(t)e^{i\omega t}dt = \frac{1}{\omega - \varepsilon + i\delta}.$$

The Fourier transform pole $\omega = \varepsilon - i\delta$ lies in the lower half-plane ω as it should be for the retarded functions.

The corresponding definition of the *advanced* Green function G^A is introduced according to

$$G^A(t) = G^{-+}(t) - G^{++}(t) = -\theta(-t)\big[G^>(t) - G^<(t)\big] = i\theta(-t)e^{-i\varepsilon t}.$$

Its Fourier transform equals

$$G^A(\omega) = \int_{-\infty}^{\infty} G^A(t)e^{i\omega t}dt = \frac{1}{\omega - \varepsilon - i\delta}.$$

On the contrary, here the Fourier transform pole $\omega = \varepsilon + i\delta$ lies in the upper half-plane ω.

And finally, we determine the *anomalous Keldysh function* G^K

$$G^K(t) = G^{--}(t) + G^{++}(t) = G^>(t) + G^<(t) = -i(1 - 2n_f)e^{-i\varepsilon t}$$

which Fourier transform equals

$$G^K(\omega) = -2\pi i(1 - 2n_f)\delta(\omega - \varepsilon) = -2\pi i \tanh\frac{\omega}{2T}\,\delta(\omega - \varepsilon).$$

Under thermodynamic equilibrium the anomalous function G^K can be expressed via G^R and G^A

$$G^K(\omega) = \big[G^R(\omega) - G^A(\omega)\big]\tanh\frac{\omega}{2T}.$$

This equation is often referred to as a statement of the fluctuation-dissipation theorem implying a rigid relation[9] between the response and correlation functions under thermal equilibrium.

(c) In conclusion, we consider the vertical contour branch when the both time moments t and t' belong to the imaginary time axis. Here it is natural to define the real variable $\tau = it$ and, correspondingly, $\tau - \tau' = i(t - t')$. The time difference $\tau - \tau'$ varies within the region from $-\beta$ to $+\beta$ and the Green function G will be determined in the region $[-\beta, +\beta]$.

Let us define the *Matsubara Green's function* $\mathfrak{G}(\tau)$, putting $\tau = it$

$$\mathfrak{G}(\tau) = -iG(-i\tau).$$

[9] Factor $\tanh(\omega/2T)$ is inherent in the fermionic systems.

Since $\theta_c(t-t') = \theta(\tau-\tau')$ for this branch of contour, the Matsubara Green's function $\mathfrak{G}(\tau)$ will equal

$$\mathfrak{G}(\tau) = \left[n_f(\varepsilon) - \theta(\tau)\right]e^{-\varepsilon\tau}.$$

It is easily to see that

$$\begin{cases} \mathfrak{G}(\tau > 0) = -\left(1 - n_f(\varepsilon)\right)e^{-\varepsilon\tau}, \\ \mathfrak{G}(\tau < 0) = n_f(\varepsilon)e^{-\varepsilon\tau}. \end{cases}$$

Let us check anti-periodic property of the Matsubara Green's function for the Fermi statistics

$$\mathfrak{G}(\tau < 0) = n_f e^{-\varepsilon\tau} = (1 - n_f)e^{-\beta\varepsilon}e^{-\varepsilon\tau} = -\mathfrak{G}(\tau + \beta).$$

Because of the last property the function $\mathfrak{G}(\tau)$ can be expanded into a Fourier series over the *odd* Matsubara frequencies $\omega_n = \pi T(2n+1)$ where $n = 0, \pm 1, \pm 2, \ldots$ are the integers. As a result, we have

$$\mathfrak{G}(\tau) = T\sum_{n=-\infty}^{\infty} \mathfrak{G}(\omega_n)e^{-i\omega_n\tau}, \quad \mathfrak{G}(\omega_n) = \frac{1}{2}\int_{-\beta}^{\beta} \mathfrak{G}(\tau)e^{i\omega_n\tau}d\tau = \frac{1}{i\omega_n - \varepsilon}.$$

57. (a) The solution is analogous to that of the previous problem. The general solution of the equation

$$\begin{cases} \left(i\frac{\partial}{\partial t} - \varepsilon\right)g(t) = f(t), \\ g(t_0) = g(t_1) \end{cases}$$

is given by the relation

$$g(t) = Ae^{-i\varepsilon t} - ie^{-i\varepsilon t}\int_{t_0}^{t} e^{i\varepsilon t'} f(t')\,dt'.$$

The condition of periodicity,

$$Ae^{-i\varepsilon t_0} = Ae^{-i\varepsilon t_1} - ie^{-i\varepsilon t_1}\int_{C} e^{i\varepsilon t'} f(t')\,dt',$$

where the integrating is performed over the entire contour C from the starting point to the ending one, results in

$$A = -in_b(\varepsilon) \int_C e^{i\varepsilon t'} f(t')\, dt'.$$

Here $n_b(\varepsilon) = 1/(e^{\varepsilon/T} - 1)$ is the Bose-Einstein distribution. Eventually, the solution of equation reads

$$g(t) = -i \int_C \big[n_b(\varepsilon) + \theta_c(t - t')\big] e^{-i\varepsilon(t-t')} f(t')\, dt',$$

$\theta_c(t - t')$ being the Heaviside contour function. Then the Green function equals

$$G(t, t') = G(t - t') = -i\big[n_b(\varepsilon) + \theta_c(t - t')\big] e^{-i\varepsilon(t-t')}.$$

(b) On the real contour branch $C_K = C_- + C_+$ there is also four possibilities for the mutual disposition of time moments t and t'. The structure of Green function can also be represented as a matrix 2×2

$$G(t, t') \to \begin{pmatrix} G^{--} & G^{-+} \\ G^{+-} & G^{++} \end{pmatrix}.$$

(1) The both time points t and t' lie on the upper branch C_-. In this case we find

$$G(t - t') \equiv G^{--}(t - t') = -i\big[n_b + \theta(t - t')\big] e^{-i\varepsilon(t-t')},$$

$$G^{--}(\omega) = \int_{-\infty}^{\infty} G^{--}(t) e^{i\omega t} dt = \mathcal{P}\frac{1}{\omega - \varepsilon} - \pi i(1 + 2n_b)\delta(\omega - \varepsilon),$$

where $\mathcal{P}\frac{1}{x}$ is the *Cauchy principal value* for $1/x$.

(2) The time point t lies on the upper branch C_- and t' is on lower branch C_+. Then we have

$$G(t - t') \equiv G^{-+}(t - t') = -in_b e^{-i\varepsilon(t-t')},$$

$$G^{-+}(\omega) = \int_{-\infty}^{\infty} G^{-+}(t) e^{i\omega t} dt = -2\pi i n_b \delta(\omega - \varepsilon).$$

(3) The time point t lies on the lower branch C_+ and t' is on the upper branch C_-. Then we have

$$G(t - t') \equiv G^{+-}(t - t') = -i(n_b + 1) e^{-i\varepsilon(t-t')},$$

$$G^{+-}(\omega) = \int_{-\infty}^{\infty} G^{+-}(t) e^{i\omega t} dt = -2\pi i(n_b + 1)\delta(\omega - \varepsilon).$$

(4) The both time points t and t' lie simultaneously on the lower contour branch C_+. Then we arrive at the following answer:

$$G(t-t') \equiv G^{++}(t-t') = -i\big[n_b + \theta(t'-t)\big]e^{-i\varepsilon(t-t')},$$

$$G^{++}(\omega) = \int_{-\infty}^{\infty} G^{++}(t)e^{i\omega t}dt = -\mathcal{P}\frac{1}{\omega-\varepsilon} - \pi i(1+2n_b)\delta(\omega-\varepsilon).$$

For the bosonic case, the general relation between the Green function components conserves as well

$$G^{++}(\omega) = -[G^{--}(\omega)]^*.$$

The *retarded* $G^R(t)$ and *advanced* $G^A(t)$ Green functions are defined in the way similar to the previous problem

$$G^R(t) = G^{--}(t) - G^{-+}(t) = \theta(t)\big[G^>(t) - G^<(t)\big] = -i\theta(t)e^{-i\varepsilon t},$$
$$G^A(t) = G^{-+}(t) - G^{++}(t) = -\theta(-t)\big[G^>(t) - G^<(t)\big] = i\theta(-t)e^{-i\varepsilon t}.$$

Their Fourier transforms differ by the sign of imaginary part

$$G^{R(A)}(\omega) = \int_{-\infty}^{\infty} G^{R(A)}(t)e^{i\omega t}dt = \frac{1}{\omega-\varepsilon\pm i\delta}.$$

The *anomalous Keldysh function* $G^K(t)$ is introduced as

$$G^K(t) = G^{--}(t) + G^{++}(t) = G^>(t) + G^<(t) = -i(1+2n_b)e^{-i\varepsilon t}.$$

Its Fourier transform equals

$$G^K(\omega) = -2\pi i(1+2n_b)\delta(\omega-\varepsilon) = -2\pi i\coth\frac{\omega}{2T}\,\delta(\omega-\varepsilon).$$

Under thermodynamic equilibrium the anomalous function G^K can be expressed via G^R and G^A

$$G^K(\omega) = \big[G^R(\omega) - G^A(\omega)\big]\coth\frac{\omega}{2T}.$$

This equation is often quoted as a statement of the fluctuation-dissipation theorem implying a rigid relation[10] between the response and correlation functions under thermal equilibrium.

(c) Finally, we analyze the contour branch when the both time points t and t' lie on the imaginary time axis. We introduce the real variable $\tau = it$ as well. The time

[10] Factor $\coth(\omega/2T)$ is inherent in the bosonic systems.

difference $\tau - \tau'$ varies within the interval from $-\beta$ to $+\beta$. The Green function G will be determined in the interval $[-\beta, +\beta]$.

Let us define the *Matsubara Green's function* $\mathfrak{G}(\tau)$, putting $\tau = it$,

$$\mathfrak{G}(\tau) = -iG(-i\tau).$$

Since $\theta_c(t - t') = \theta(\tau - \tau')$ for such contour branch, the Matsubara Green's function $\mathfrak{G}(\tau)$ equals

$$\mathfrak{G}(\tau) = -\big[n_b(\varepsilon) + \theta(\tau)\big]e^{-\varepsilon\tau}.$$

We see readily that

$$\begin{cases} \mathfrak{G}(\tau > 0) = -\big(n_b(\varepsilon) + 1\big)e^{-\varepsilon\tau}, \\ \mathfrak{G}(\tau < 0) = -n_b(\varepsilon)e^{-\varepsilon\tau}. \end{cases}$$

Let us check the periodic property of the Matsubara Green's function for the Bose statistics

$$\mathfrak{G}(\tau < 0) = -n_b e^{-\varepsilon\tau} = -(1 + n_b)e^{-\beta\varepsilon}e^{-\varepsilon\tau} = +\mathfrak{G}(\tau + \beta).$$

Due to the latter property the function $\mathfrak{G}(\tau)$ can be expanded into a Fourier series over the *even* Matsubara frequencies $\omega_n = 2\pi nT$ where $n = 0, \pm 1, \pm 2, \ldots$ are the integers. This results in

$$\mathfrak{G}(\tau) = T\sum_{n=-\infty}^{\infty} \mathfrak{G}(\omega_n)e^{-i\omega_n\tau}, \quad \mathfrak{G}(\omega_n) = \frac{1}{2}\int\limits_{-\beta}^{\beta} \mathfrak{G}(\tau)e^{i\omega_n\tau}d\tau = \frac{1}{i\omega_n - \varepsilon}.$$

58. (a) We are convinced by the direct calculation that matrix R and its inverse matrix R^{-1}

$$R = \frac{1}{\sqrt{2}}\begin{pmatrix} 1 & 1 \\ -1 & 1 \end{pmatrix} \quad \text{and} \quad R^{-1} = \frac{1}{\sqrt{2}}\begin{pmatrix} 1 & -1 \\ 1 & 1 \end{pmatrix}$$

lead to the anti-diagonal representation desired

$$\begin{pmatrix} 1 & -1 \\ 1 & 1 \end{pmatrix}\begin{pmatrix} G^{--} & G^{-+} \\ G^{+-} & G^{++} \end{pmatrix}\begin{pmatrix} 1 & 1 \\ -1 & 1 \end{pmatrix}$$
$$= \begin{pmatrix} (G^{--} - G^{+-} - G^{+-} + G^{++}) & (G^{--} - G^{+-} + G^{-+} - G^{++}) \\ (G^{--} - G^{-+} + G^{+-} - G^{++}) & (G^{--} + G^{++} + G^{-+} + G^{+-}) \end{pmatrix}$$
$$= 2\begin{pmatrix} 0 & G^A \\ G^R & G^K \end{pmatrix}.$$

(b) The required matrices R and P, delivering the diagonal Larkin-Ovchinnikov representation, have the form

$$R=\frac{1}{\sqrt{2}}\begin{pmatrix}1 & 1\\ -1 & 1\end{pmatrix},\quad P=\begin{pmatrix}1 & 0\\ 0 & -1\end{pmatrix}\quad\text{and}\quad R^{-1}P=\frac{1}{\sqrt{2}}\begin{pmatrix}1 & 1\\ 1 & -1\end{pmatrix}.$$

The equality

$$\begin{pmatrix}1 & 1\\ 1 & -1\end{pmatrix}\begin{pmatrix}G^{--} & G^{-+}\\ G^{+-} & G^{++}\end{pmatrix}\begin{pmatrix}1 & 1\\ -1 & 1\end{pmatrix}=2\begin{pmatrix}G^{R} & G^{K}\\ 0 & G^{A}\end{pmatrix}$$

can straightforwardly be verified with the aid of the relations between the components of matrix Green function $\hat{G}$.

59. Let us write the equation

$$\begin{cases}\left(-\frac{\partial^2}{\partial t^2}-\omega_0^2\right)d(t)=\omega_0^2 f(t),\\ d(t_0)=d(t_1)\end{cases}$$

with the general right-hand side $f(t)$ and time t lying on the Keldysh contour C with the initial point t_0 and the ending point t_1. The solution of equation is expressed via the Green function $D_{\omega_0}(t,t')=D_{\omega_0}(t-t')$ according to

$$d(t)=\int\limits_C D_{\omega_0}(t-t')f(t')dt'.$$

Let us represent the general solution as a sum of two solutions, namely, solution of homogeneous equation and partial solution of the starting equation with the right-hand side

$$d(t)=Ae^{-i\omega_0 t}+Be^{i\omega_0 t}-\frac{i\omega_0}{2}e^{-i\omega_0 t}\int\limits_{t_0}^{t}e^{i\omega_0 t'}f(t')\,dt'$$
$$+\frac{i\omega_0}{2}e^{i\omega_0 t}\int\limits_{t_0}^{t}e^{-i\omega_0 t'}f(t')\,dt'.$$

Let us rewrite this expression, using the Heaviside contour function $\theta_c(t-t')$

$$d(t)=Ae^{-i\omega_0 t}+Be^{i\omega_0 t}-\frac{i\omega_0}{2}\int\limits_C\theta_c(t-t')e^{-i\omega_0(t-t')}f(t')\,dt'$$
$$+\frac{i\omega_0}{2}\int\limits_C\theta_c(t-t')e^{i\omega_0(t-t')}f(t')\,dt'.$$

We have the following relation from the condition of periodicity $d(t_0) = d(t_1)$ in the initial $t_0 = -\infty + i\beta$ and final $t_1 = -\infty - i\delta$ points of contour C:

$$Ae^{-i\omega_0 t_0} + Be^{i\omega_0 t_0} = Ae^{-i\omega_0 t_1} + Be^{i\omega_0 t_1}$$
$$-\frac{i\omega_0}{2}e^{-i\omega_0 t_1}\int\limits_C e^{i\omega_0 t'} f(t')\,dt' + \frac{i\omega_0}{2}e^{i\omega_0 t_1}\int\limits_C e^{-i\omega_0 t'} f(t')\,dt'.$$

Since this condition should be valid for an arbitrary frequency $\omega_0 = \omega_0(\boldsymbol{k})$, we arrive at the two equalities for each from the two linearly independent exponential functions

$$Ae^{-i\omega_0 t_0} = Ae^{-i\omega_0 t_1} - \frac{i\omega_0}{2}e^{-i\omega_0 t_1}\int\limits_C e^{i\omega_0 t'} f(t')\,dt',$$
$$Be^{i\omega_0 t_0} = Be^{i\omega_0 t_1} + \frac{i\omega_0}{2}e^{i\omega_0 t_1}\int\limits_C e^{-i\omega_0 t'} f(t')\,dt'.$$

Then we find unknown coefficients A and B

$$A = -\frac{i\omega_0}{2}n_p(\omega_0)\int\limits_C e^{i\omega_0 t'} f(t')\,dt',$$
$$B = -\frac{i\omega_0}{2}\big(1 + n_p(\omega_0)\big)\int\limits_C e^{-i\omega_0 t'} f(t')\,dt',$$

where $n_p(\omega_0) = 1/(e^{\beta\omega_0} - 1)$ is the Planck distribution. So, we obtain the following Green functions for phonons:

$$D_{\omega_0}(t,t') = -\frac{i\omega_0}{2}\Big\{\big[n_p(\omega_0) + \theta_c(t-t')\big]e^{-i\omega_0(t-t')}$$
$$+\big[n_p(\omega_0) + \theta_c(t'-t)\big]e^{i\omega_0(t-t')}\Big\}.$$

Here we have taken $1 - \theta_c(t-t') = \theta_c(t'-t)$ into account. Note the following property of symmetry: $D_{\omega_0}(t,t') = D_{\omega_0}(t',t)$.

As a above, we represent the phonon Green function in the form of matrix 2×2 in order to describe the dependence on the position of time points t and t' at the real branches of Keldysh contour $C_K = C_- + C_+$

$$D_{\omega_0}(t,t') \to \begin{pmatrix} D^{--}_{\omega_0}(t,t') & D^{-+}_{\omega_0}(t,t') \\ D^{+-}_{\omega_0}(t,t') & D^{++}_{\omega_0}(t,t') \end{pmatrix}.$$

(1) If the both time points t and t' lie on the upper contour branch C_-, we have

$$D^{--}_{\omega_0}(t,t') = D^{--}_{\omega_0}(t-t') = -\frac{i\omega_0}{2}\Big\{\big[n_p(\omega_0)+\theta(t-t')\big]e^{-i\omega_0(t-t')} + \big[n_p(\omega_0)+\theta(t'-t)\big]e^{i\omega_0(t-t')}\Big\}.$$

The Fourier transform $D^{--}_{\omega_0}$ equals

$$D^{--}(\omega) = \int\limits_{-\infty}^{\infty} D^{--}_{\omega_0}(t)e^{i\omega t}dt = \mathcal{P}\frac{\omega_0^2}{\omega^2-\omega_0^2} - \frac{i\pi\omega_0}{2}(2n_p(\omega_0)+1)\big[\delta(\omega-\omega_0)+\delta(\omega+\omega_0)\big],$$

where $\mathcal{P}\frac{1}{x}$ is the *Cauchy principal value* for $1/x$.

(2) Let time t be on the upper branch C_- and t' be on the lower branch C_+. Since $\theta_c(t-t')\equiv 0$ and $\theta_c(t'-t)\equiv 1$, we have

$$D^{-+}_{\omega_0}(t,t') = -\frac{i\omega_0}{2}\Big\{n_p(\omega_0)e^{-i\omega_0(t-t')} + \big(n_p(\omega_0)+1\big)e^{i\omega_0(t-t')}\Big\},$$

$$D^{-+}(\omega) = \int\limits_{-\infty}^{\infty} D^{-+}_{\omega_0}(t)e^{i\omega t}dt = -i\pi\omega_0\big[n_p(\omega_0)\delta(\omega-\omega_0) + \big(n_p(\omega_0)+1\big)\delta(\omega+\omega_0)\big].$$

The *retarded* Green function $D^R_{\omega_0}(t)$ is defined as

$$D^R_{\omega_0}(t) = D^{--}_{\omega_0}(t) - D^{+-}_{\omega_0}(t) = -\theta(t)\omega_0\sin\omega_0 t.$$

Its Fourier transform equals

$$D^R(\omega) = \frac{1}{2}\left(\frac{\omega_0}{\omega-\omega_0+i\delta} - \frac{\omega_0}{\omega+\omega_0+i\delta}\right)$$

with the poles $\omega = \pm\omega_0 - i\delta$ lying in the lower half-plane ω.

(3) Let time points t and t' be on the lower C_+ and upper C_- branches of Keldysh contour C_K. Then $\theta_c(t-t')\equiv 1$, $\theta_c(t'-t)\equiv 0$ and in the result we obtain the function $D^{+-}_{\omega_0}(t,t')$ and its Fourier transform

$$D^{+-}_{\omega_0}(t,t') = -\frac{i\omega_0}{2}\left\{\big(n_p(\omega_0)+1\big)e^{-i\omega_0(t-t')} + n_p(\omega_0)e^{i\omega_0(t-t')}\right\},$$

$$D^{+-}(\omega) = \int_{-\infty}^{\infty} D^{+-}_{\omega_0}(t)e^{i\omega t}dt$$
$$= -i\pi\omega_0\big[\big(n_p(\omega_0)+1\big)\delta(\omega-\omega_0) + n_p(\omega_0)\delta(\omega+\omega_0)\big].$$

(4) Let now both time points t and t' be on the lower C_+ branch of Keldysh contour C_K. Then,

$$D^{++}_{\omega_0}(t,t') = D^{++}_{\omega_0}(t-t') = -\frac{i\omega_0}{2}\left\{\big[n_p(\omega_0)+\theta(t'-t)\big]e^{-i\omega_0(t-t')} + \big[n_p(\omega_0)+\theta(t-t')\big]e^{i\omega_0(t-t')}\right\}.$$

The corresponding Fourier transform $D^{++}_{\omega_0}(\omega)$ equals

$$D^{++}(\omega) = \int_{-\infty}^{\infty} D^{++}_{\omega_0}(t)e^{i\omega t}dt = -\mathcal{P}\frac{\omega_0^2}{\omega^2-\omega_0^2}$$
$$-\frac{i\pi\omega_0}{2}\big(2n_p(\omega_0)+1\big)\big[\delta(\omega-\omega_0)+\delta(\omega+\omega_0)\big].$$

The *advanced* Green function $D^A_{\omega_0}(t)$ is defined as

$$D^A_{\omega_0}(t) = D^{-+}_{\omega_0}(t) - D^{++}_{\omega_0}(t) = \theta(-t)\omega_0 \sin\omega_0 t.$$

Its Fourier transform will be equal to

$$D^A(\omega) = \frac{1}{2}\left(\frac{\omega_0}{\omega-\omega_0-i\delta} - \frac{\omega_0}{\omega+\omega_0-i\delta}\right).$$

The both poles $\omega = \pm\omega_0 + i\delta$ are in the upper half-plane ω.

The *anomalous Keldysh function* $D^K_{\omega_0}(t)$ for phonons is introduced as usual

$$D^K_{\omega_0}(t) = D^{--}_{\omega_0}(t) + D^{++}_{\omega_0}(t) = -i\omega_0\big(1+2n_p(\omega_0)\big)\cos\omega_0 t.$$

Its Fourier transform equals

$$D^K(\omega) = -i\pi\omega_0 \coth(\beta\omega_0/2)\big[\delta(\omega-\omega_0)+\delta(\omega+\omega_0)\big]$$
$$= -i\pi\omega_0 \coth(\beta\omega/2)\big[\delta(\omega-\omega_0)-\delta(\omega+\omega_0)\big].$$

Under thermodynamic equilibrium the anomalous function $D^K(\omega)$ is expressed via $D^R(\omega)$ and $D^A(\omega)$ according to

$$D^K(\omega) = \big[D^R(\omega) - D^A(\omega)\big]\coth(\beta\omega/2).$$

(5) The phonon *Matsubara Green's function* $\mathfrak{D}_{\omega_0}(\tau)$ in the imaginary branch of contour C where we put $\tau = it$ is determined by the relation $\mathfrak{D}_{\omega_0}(\tau) = -iD_{\omega_0}(-i\tau)$ and is equal to

$$\mathfrak{D}_{\omega_0}(\tau) = -\frac{\omega_0}{2}\Big\{\big[n_p(\omega_0) + \theta(\tau)\big]e^{-\omega_0\tau} + \big[n_p(\omega_0) + \theta(-\tau)\big]e^{\omega_0\tau}\Big\}.$$

The property of periodicity $\mathfrak{D}_{\omega_0}(\tau < 0) = \mathfrak{D}_{\omega_0}(\tau + \beta)$ for the phonon Matsubara Green's function allows us to expand the function $\mathfrak{D}_{\omega_0}(\tau)$ into a Fourier series over the *even* Matsubara frequencies $\omega_n = 2\pi nT$, where $n = 0, \pm 1, \pm 2, \ldots$ are the integers. The result is the following:

$$\mathfrak{D}_{\omega_0}(\tau) = T\sum_{n=-\infty}^{\infty}\mathfrak{D}(\omega_n)e^{-i\omega_n\tau}, \quad \mathfrak{D}(\omega_n) = \frac{1}{2}\int_{-\beta}^{\beta}\mathfrak{D}(\tau)e^{i\omega_n\tau}d\tau = -\frac{\omega_0^2}{\omega_n^2 + \omega_0^2}.$$

60. (a) Let us rewrite the differential equation for the Green function $G(t, t')$ in the equivalent integral form

$$G(t, t') = G_0(t, t') + \int_{C_K} G_0(t, t'')U(t'')G(t'', t')dt''.$$

The integral is calculated for the entire Keldysh contour C_K. Hence we readily write the first-order correction in the external field potential $U(t)$

$$\delta G(t, t') = \int_{C_K} G_0(t, t'')U(t'')G_0(t'', t')_0 dt''.$$

By changing the integration limits on the lower branch of contour C_K in accordance with the equality

$$\int_{C_K} dt'' = \int_{-\infty}^{\infty} dt'' - \int_{-\infty}^{\infty} dt'',$$

it is profitable to represent the correction δG and potential U in the matrix form as two-component matrices in the Keldysh space

$$\delta\hat{G}(t,t') = \int_{-\infty}^{\infty} \hat{G}_0(t,t'')\hat{U}(t'')\hat{G}_0(t'',t')dt'' \equiv \hat{G}_0 \otimes \hat{U}\hat{G}_0 .$$

Here the corresponding matrices are specified

$$\hat{G} = \begin{pmatrix} G^{--} & G^{-+} \\ G^{+-} & G^{++} \end{pmatrix}, \quad \hat{U} = \begin{pmatrix} U & 0 \\ 0 & -U \end{pmatrix} = U\hat{\tau}_3 \text{ and } \hat{\tau}_3 = \begin{pmatrix} 1 & 0 \\ 0 & -1 \end{pmatrix}.$$

Next, we define the Fourier transforms of Green function as

$$\hat{G}(\omega,\omega') = \iint_{-\infty}^{\infty} \hat{G}(t,t')e^{i\omega t - i\omega' t'}dt\,dt'$$

and, using the homogeneity of free Green function $G_0(t,t') = G_0(t-t')$, we find the frequency representation for first-order correction

$$\delta\hat{G}(\omega,\omega') = \hat{G}_0(\omega)\hat{U}(\omega-\omega')\hat{G}_0(\omega')$$
$$= \begin{pmatrix} G_0^{--}UG_0^{--} - G_0^{-+}UG_0^{+-} & G_0^{--}UG_0^{-+} - G_0^{-+}UG_0^{++} \\ G_0^{+-}UG_0^{--} - G_0^{++}UG_0^{+-} & G_0^{+-}UG_0^{-+} - G_0^{++}UG_0^{++} \end{pmatrix}.$$

Correspondingly, for the coincident time moments t and t', we have

$$\delta\hat{G}(t,t) = \iint \frac{d\omega}{2\pi}\frac{d\omega'}{2\pi}\hat{G}_0(\omega)\hat{U}(\omega-\omega')\hat{G}_0(\omega')e^{-i(\omega-\omega')t}.$$

The Fourier transform reads

$$[\delta\hat{G}(t,t)]_\omega = \int_{-\infty}^{\infty} \frac{d\varepsilon}{2\pi}\hat{G}_0(\varepsilon+\frac{\omega}{2})\hat{U}(\omega)\hat{G}_0(\varepsilon-\frac{\omega}{2}).$$

Hence, the necessary matrix component $[\delta G^{-+}(t,t)]_\omega$ of matrix $[\delta\hat{G}(t,t)]_\omega$, which determines the variation of distribution density $\delta n(\omega)$, will be equal to

$$[\delta G^{-+}(t,t)]_\omega = \int_{-\infty}^{\infty} \frac{d\varepsilon}{2\pi}\Big[G_0^{--}(\varepsilon+\frac{\omega}{2})U(\omega)G_0^{-+}(\varepsilon-\frac{\omega}{2})$$
$$-G_0^{-+}(\varepsilon+\frac{\omega}{2})U(\omega)G_0^{++}(\varepsilon-\frac{\omega}{2})\Big].$$

(b) Let us turn now to determining the distribution density according to $n(t) = -2iG^{-+}(t,t)$. For the equilibrium unperturbed system, one has $G_0(t,t') =$

$in_f(\xi)e^{-i\xi(t-t')}$, where $n_f(\xi)$ is the Fermi distribution, and $n_0 = -2iG_0^{-+}(t,t) = 2n_f$. Thus, the perturbation of distribution density can be written as $\delta n(\omega) = -2i[\delta G^{-+}(t,t)]_\omega$. In order to express this relation in terms of retarded G_0^R, advanced G_0^A and anomalous G_0^K functions, we use the following relations between the components of Green function $\hat{G}$

$$G_0^R = G_0^{--} - G_0^{-+}, \quad G_0^A = G_0^{-+} - G_0^{++}, \quad G_0^K = G_0^{--} - G_0^{++}.$$

Using these relations, we find

$$\delta n(\omega) = \chi(\omega)U(\omega),$$

where susceptibility $\chi(\omega)$ is given by the formula

$$\chi(\omega) = -2i\frac{1}{2}\int\limits_{-\infty}^{\infty}\frac{d\varepsilon}{2\pi}\Big[G_0^A(\varepsilon_+)G_0^A(\varepsilon_-) - G_0^R(\varepsilon_+)G_0^R(\varepsilon_-)$$
$$+G_0^R(\varepsilon_+)G_0^K(\varepsilon_-) + G_0^K(\varepsilon_+)G_0^A(\varepsilon_-)\Big]$$

with denoting $\varepsilon_\pm = \varepsilon \pm \omega/2$. Substituting here the relation $G_0^K(\omega) = [G_0^R(\omega) - G^A(\omega)]\tanh(\omega/2T)$ for the equilibrium fermionic system, we find

$$\chi(\omega) = -i\int\limits_{-\infty}^{\infty}\frac{d\varepsilon}{2\pi}\Big[G_0^A(\varepsilon_+)G_0^A(\varepsilon_-)\Big(1 - \tanh\frac{\varepsilon_+}{2T}\Big)$$
$$- G_0^R(\varepsilon_+)G_0^R(\varepsilon_-)\Big(1 - \tanh\frac{\varepsilon_-}{2T}\Big) + G_0^R(\varepsilon_+)G_0^A(\varepsilon_-)\Big(\tanh\frac{\varepsilon_+}{2T} - \tanh\frac{\varepsilon_-}{2T}\Big)\Big].$$

To give this expression its final form, we will use the following equality:

$$\int\limits_{-\infty}^{\infty} d\varepsilon\, G_0^{R(A)}(\varepsilon_+)G_0^{R(A)}(\varepsilon_-)$$
$$= \int\limits_{-\infty}^{\infty}\frac{d\varepsilon}{(\varepsilon + \omega/2 - \xi \pm i\delta)(\varepsilon - \omega/2 - \xi \pm i\delta)} = 0.$$

This equality is a direct consequence that the both poles lie simultaneously in one of half-planes ε and the integrand function decays at $|\varepsilon| \to \infty$ faster than $1/\varepsilon$. Finally, we have

$$\chi(\omega) = -i\int_{-\infty}^{\infty}\frac{d\varepsilon}{2\pi}\Big[G_0^R(\varepsilon_+)G_0^R(\varepsilon_-)\tanh\frac{\varepsilon_-}{2T} - G_0^A(\varepsilon_+)G_0^A(\varepsilon_-)\tanh\frac{\varepsilon_+}{2T}$$
$$+ G_0^R(\varepsilon_+)G_0^A(\varepsilon_-)\Big(\tanh\frac{\varepsilon_+}{2T} - \tanh\frac{\varepsilon_-}{2T}\Big)\Big].$$

(c) To calculate the Matsubara susceptibility $\chi^M(\omega_n)$, it is useful to shift the integration variable ε by $\pm\omega/2$ and represent the starting expression above as a sum of the following integrals:

$$\chi(\omega) = \chi^{RR}(\omega) + \chi^{RA}(\omega) + \chi^{AR}(\omega) + \chi^{AA}(\omega)$$
$$= -i\int_{-\infty}^{\infty}\frac{d\varepsilon}{2\pi}G_0^R(\varepsilon+\omega)G_0^R(\varepsilon)\tanh\frac{\varepsilon}{2T} + i\int_{-\infty}^{\infty}\frac{d\varepsilon}{2\pi}G_0^R(\varepsilon+\omega)G_0^A(\varepsilon)\tanh\frac{\varepsilon}{2T}$$
$$-i\int_{-\infty}^{\infty}\frac{d\varepsilon}{2\pi}G_0^R(\varepsilon)G_0^A(\varepsilon-\omega)\tanh\frac{\varepsilon}{2T} + i\int_{-\infty}^{\infty}\frac{d\varepsilon}{2\pi}G_0^A(\varepsilon)G_0^A(\varepsilon-\omega)\tanh\frac{\varepsilon}{2T}.$$

Let us start from the simple case $n = 0$ to comprehend the general approach to determining the Matsubara susceptibility $\chi^M(\omega_n)$. For $n = 0$, we have $\omega = i\omega_0 = 0$ and two integrals alone remain

$$\chi^M(0) = \chi_0^{RR} + \chi_0^{AA}$$
$$= -i\int_{-\infty}^{\infty}\frac{d\varepsilon}{2\pi}G_0^R(\varepsilon)G_0^R(\varepsilon)\tanh\frac{\varepsilon}{2T} + i\int_{-\infty}^{\infty}\frac{d\varepsilon}{2\pi}G_0^A(\varepsilon)G_0^A(\varepsilon)\tanh\frac{\varepsilon}{2T}.$$

Treating the integrals χ_0^{RR} and χ_0^{AA}, we will use the analyticity of function $G_0^R(\varepsilon)$ at $\operatorname{Im}\varepsilon > 0$ and analyticity of function $G_0^A(\varepsilon)$ at $\operatorname{Im}\varepsilon < 0$.

While evaluating the integral χ_0^{RR}, we can expand the integration along the real axis ε with augmenting the integration along the infinitely distant semicircle in the upper half-plane Π^+ of complex variable ε (Fig. 3.18). This unchanges the integration result due to sufficiently fast $\sim 1/\varepsilon^2$ decay for the product of $G_0^R(\varepsilon)G_0^R(\varepsilon)$ and $G_0^A(\varepsilon)G_0^A(\varepsilon)$ as $\varepsilon\to\infty$. The calculation of integral for χ_0^{AA} is similar but the integration along the real axis should be augmented with the integration along the infinitely distant semicircle in the lower half-plane of complex variable ε (Fig. 3.18).

Next, we apply the residue theorem for evaluating the integral over the closed curve. Since the product of Green functions $G_0^RG_0^R$ and $G_0^AG_0^A$ is analytical and has no poles in the corresponding half-planes Π^+ and Π^-, the calculation of integrals χ_0^{RR} and χ_0^{AA} reduces to a sum of residues taken at all corresponding poles $\varepsilon = i\varepsilon_k = i\pi T(2k+1)$ of function $\tanh(\varepsilon/2T)$ for integer k.

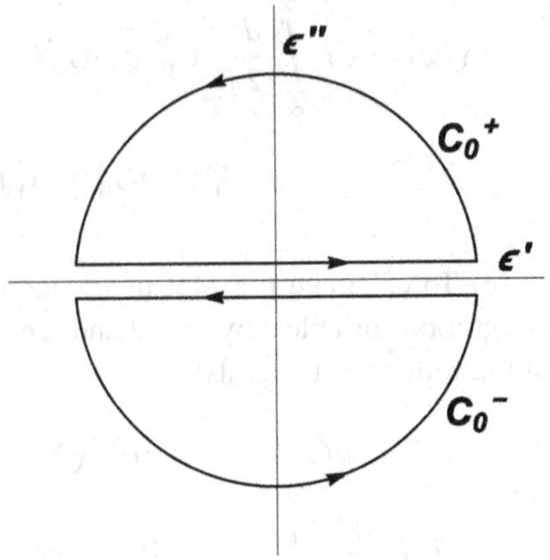

Fig. 3.18 The contours of integration

Eventually we obtain

$$\chi^M(0) = -\frac{i}{2\pi}\left[2\pi i \sum_{k\geqslant 0} 2T G_0^R(i\varepsilon_k)G_0^R(i\varepsilon_k) - (-2\pi i)\sum_{k<0} 2T G_0^A(i\varepsilon_k)G_0^A(i\varepsilon_k)\right].$$

Then we employ the following relation of Matsubara Green's function with the retarded and advanced Green functions:

$$G_0^R(i\varepsilon_k) = \mathfrak{G}_0(\varepsilon_k) \text{ at } \varepsilon_k > 0 \text{ and } G_0^A(i\varepsilon_k) = \mathfrak{G}_0(\varepsilon_k) \text{ at } \varepsilon_k < 0.$$

The answer reads

$$\chi^M(0) = 2T\left(\sum_{k\geqslant 0} + \sum_{k<0}\right)\mathfrak{G}_0(\varepsilon_k)\mathfrak{G}_0(\varepsilon_k) = 2T\sum_{k=-\infty}^{\infty}\mathfrak{G}_0(\varepsilon_k)\mathfrak{G}_0(\varepsilon_k).$$

Let us turn now to the case of an arbitrary integer n and, for definiteness, put the Matsubara frequency to be positive $\omega_n > 0$. We write the first two integrals at $\omega = i\omega_n$ as follows:

$$\chi_n^{RR} + \chi_n^{RA} = -i\int_{-\infty}^{\infty}\frac{d\varepsilon}{2\pi}G_0^R(\varepsilon + i\omega_n)G_0^R(\varepsilon)\tanh\frac{\varepsilon + i\omega_n}{2T} + i\int_{-\infty}^{\infty}\frac{d\varepsilon}{2\pi}G_0^R(\varepsilon + i\omega_n)G_0^A(\varepsilon)\tanh\frac{\varepsilon + i\omega_n}{2T}.$$

Here we have used the fact that

$$\tanh\frac{\varepsilon}{2T} = \tanh\frac{\varepsilon + i\omega_n}{2T} \quad \text{at} \quad \omega_n = 2\pi n T, \quad n = 0, \pm 1, \pm 2 \ldots$$

These integrals can be represented as the following integrals of complex variable ε:

$$-i\int\limits_{C_n}\frac{d\varepsilon}{2\pi}G_0^R(\varepsilon)G_0^R(\varepsilon - i\omega_n)\tanh\frac{\varepsilon}{2T} + i\int\limits_{C_n}\frac{d\varepsilon}{2\pi}G_0^R(\varepsilon)G_0^A(\varepsilon - i\omega_n)\tanh\frac{\varepsilon}{2T}.$$

The integration is performed along the rectilinear contour C_n parallel to the real axis and passing through the point $\mathrm{Im}\,\varepsilon = \omega_n$ on the imaginary axis (Fig. 3.19). Similarly, the remaining two integrals

$$\chi_n^{AR} + \chi_n^{AA}$$
$$= -i\int\limits_{-\infty}^{\infty}\frac{d\varepsilon}{2\pi}G_0^R(\varepsilon)G_0^A(\varepsilon - i\omega_n)\tanh\frac{\varepsilon}{2T} + i\int\limits_{-\infty}^{\infty}\frac{d\varepsilon}{2\pi}G_0^A(\varepsilon)G_0^A(\varepsilon - i\omega_n)\tanh\frac{\varepsilon}{2T}$$

represent the integrals over the contour C_0 coincident with the real axis $\mathrm{Im}\,\varepsilon = 0$ in the complex plane of variable ε (Fig. 3.19).

In order to apply the residue theorem for evaluating the integrals, we divide the entire complex plane ε into three regions as D_{AA} : $\mathrm{Im}\,\varepsilon < 0$, D_{RA} : $0 < \mathrm{Im}\,\varepsilon < \omega_n$, and D_{RR} : $0 < \omega_n < \mathrm{Im}\,\varepsilon$. Then we supplement the four rectilinear integration contours with the corresponding arcs of infinitely distant circle (Fig. 3.19). The integrals over the arcs give zero contribution due to rapid $\sim 1/\varepsilon^2$ decay for the product of two Green functions. So, we obtain three closed integration contours as C_{AA}, C_{RA} and C_{RR} for the multiplications $G_0^A(\varepsilon)G_0^A(\varepsilon - i\omega_n)$, $G_0^R(\varepsilon)G_0^A(\varepsilon - i\omega_n)$, and $G_0^R(\varepsilon)G_0^R(\varepsilon - i\omega_n)$, each of them being analytic in the corresponding region. These three contours in total enclose all the infinite succession of poles $\tanh(\varepsilon/2T)$ existing at points $i\varepsilon_k = \pi i T(2k+1)$ for integer k. The sum of the integrand residues in each pole leads to the answer. So, we have

$$\chi^M(\omega_n) = -\frac{i}{2\pi}2\pi i\Bigg(2T\sum_{k\geqslant n}G_0^R(i\varepsilon_k)C_0^R(i\varepsilon_k - i\omega_n)$$
$$+ 2T\sum_{n>k\geqslant 0}G_0^R(i\varepsilon_k)C_0^A(i\varepsilon_k - i\omega_n) + 2T\sum_{k<0}G_0^A(i\varepsilon_k)C_0^A(i\varepsilon_k - i\omega_n)\Bigg).$$

Next, we use the relation between $G^R(i\varepsilon_k)$, $G^A(i\varepsilon_k)$, and Matsubara Green's function $\mathfrak{G}(\varepsilon_k)$, which has already been applied for evaluating the susceptibility χ_0. The final answer reads

Fig. 3.19 The contours of integration

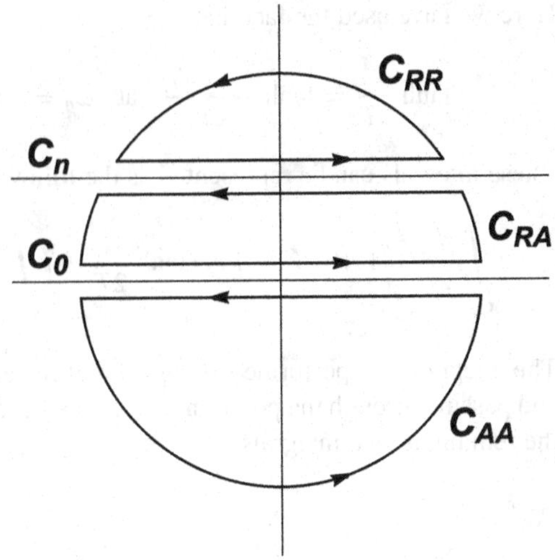

$$\chi^M(\omega_n) = 2T\left(\sum_{k\geqslant n} + \sum_{n>k\geqslant 0} + \sum_{k<0}\right)\mathfrak{G}_0(\varepsilon_k)\mathfrak{G}_0(\varepsilon_k - \omega_n)$$
$$= 2T\sum_{k=-\infty}^{\infty}\mathfrak{G}_0(\varepsilon_k)\mathfrak{G}_0(\varepsilon_k - \omega_n) = 2T\sum_{k=-\infty}^{\infty}\mathfrak{G}_0(\varepsilon_k + \omega_n)\mathfrak{G}_0(\varepsilon_k).$$

The case $\omega_n < 0$ can be treated in the similar way or, using the general properties of Green functions, one can show that the Matsubara susceptibility $\chi^M(\omega_n)$ is an even function of Matsubara frequencies

$$\chi^M(\omega_n) = \chi^M(|\omega_n|).$$

Thus, we are convinced of the general relationship between the Matsubara susceptibility $\chi^M(\omega_n)$ and the usual retarded susceptibility $\chi(\omega)$, namely:

$$\chi^M(\omega_n) = \chi(\omega = i|\omega_n|).$$

The similar relationship between the Matsubara response function and the retarded response function is traceable, for example, in problem 48 about macroscopic quantum tunneling. The property $\mathfrak{G}_0^*(\varepsilon_k) = \mathfrak{G}_0(-\varepsilon_k)$ means that the Matsubara susceptibility χ^M is a real quantity.

The inverse problem of reconstructing the susceptibility or linear response function $\chi(\omega)$ at real frequencies ω, if one knows the Matsubara response function $\chi^M(\omega_n)$ given on the countable set of Matsubara frequencies ω_n, is referred to as *analytic continuation from the imaginary frequencies to the real ones.*

Bibliography

L.D. Landau, E.M. Lifshitz, *Statistical Physics, Part I*, 3rd edn. (Pergamon, Oxford, 1980)
E.M. Lifshitz, L.P. Pitaevskii, *Statistical Physics, Part II* (Pergamon, Oxford, 1980)
E.M. Lifshitz, L.P. Pitaevskii, *Physical Kinetics* (Pergamon, Oxford, 1981)
L.D. Landau, E.M. Lifshitz, *Fluid Mechanics*, 2nd edn. (Pergamon, Oxford, 1987)
Yu.M. Belousov, S.N. Burmistrov, A.I. Ternov, *Problem Solving in Theoretical Physics* (Wiley-VCH, Berlin, 2020)
A.A. Abrikosov, *Fundamentals of the Theory of Metals* (Groningen, Elsevier, North-Holland, 1980)
J.M. Ziman, *Electrons and Phonons: The Theory of Transport Phenomena in Solids* (Oxford University Press, Oxford, 2003)
F.J. Blatt, *Theory of Mobility of Electrons in Solids* (Academic Press Inc., New York, 1957)
V.F. Gantmakher, *Electrons and Disorder in Solids* (Oxford University Press, Oxford, 2005)
B.M. Askerov, *Electron Transport Phenomena in Semiconductors* (World Scientific, Singapore, 1994)
Y. Imry, *Introduction to Mesoscopic Physics* (Oxford University Press, Oxford, 1997)
I.M. Khalatnikov, *An Introduction to the Theory of Superfluidity* (Westview Press, Colorado, 2000)
S.J. Putterman, *Superfluid Hydrodynamics* (North-Holland, Amsterdam, 1974)
U. Weiss, *Quantum Dissipative Systems*, 2nd edn. (World Scientific, Singapore, 1999)
H.P. Breuer, F. Petruccione, *The Theory of Open Quantum Systems* (Oxford University Press, Oxford, 2002)

S. N. Burmistrov, *Physical Kinetics*,
https://doi.org/10.1007/978-981-19-1649-6

Index

S. N. Burmistrov, *Physical Kinetics*,
https://doi.org/10.1007/978-981-19-1649-6

Printed in the United States
by Baker & Taylor Publisher Services